Formation of the First Black Holes

Formation of the First Black Holes

Editors

Muhammad Latif
United Arab Emirates University, Al-Ain, UAE

Dominik Schleicher
Universidad de Concepciòn, Chile

World Scientific

NEW JERSEY • LONDON • SINGAPORE • BEIJING • SHANGHAI • HONG KONG • TAIPEI • CHENNAI • TOKYO

Published by

World Scientific Publishing Co. Pte. Ltd.

5 Toh Tuck Link, Singapore 596224

USA office: 27 Warren Street, Suite 401-402, Hackensack, NJ 07601

UK office: 57 Shelton Street, Covent Garden, London WC2H 9HE

Library of Congress Cataloging-in-Publication Data

Names: Latif, Muhammad, 1980– | Schleicher, Dominik (Dominik R. G.)

Title: Formation of the first black holes / [edited by] Muhammad Latif (United Arab Emirates University, Al-Ain, UAE), Dominik Schleicher (Universidad de Concepciòn, Chile).

Description: New Jersey : World Scientific, 2019. | Includes bibliographical references and index.

Identifiers: LCCN 2018030665| ISBN 9789813227941 (hardcover : alk. paper) | ISBN 981322794X (hardcover : alk. paper)

Subjects: LCSH: Black holes (Astronomy) | Red shift. | Astrophysics.

Classification: LCC QB843.B55 F6755 2019 | DDC 523.8/875--dc23

LC record available at https://lccn.loc.gov/2018030665

British Library Cataloguing-in-Publication Data

A catalogue record for this book is available from the British Library.

For any available supplementary material, please visit
https://www.worldscientific.com/worldscibooks/10.1142/10652#t=suppl

Typeset by Stallion Press
Email: enquiries@stallionpress.com

Preface

The origin of the first black holes is a very active area of research, and editing such a book has been quite a challenge. At least to our knowledge, there has been no comparable textbook before. A book, "The First Stars", has been edited by Volker Bromm, and a book, "The First Galaxies: Theoretical Predictions and Observational Clues" has been compiled by Tommy Wiklind and Bahram Mobasher. Another book on the subject has been compiled by Immo Appenzeller "High-Redshift Galaxies", focusing more on the observational side. In the context of black holes, the available textbooks tend to focus on them as compact objects, such as in the excellent book by Shapiro & Teukolsky, or on their treatment as astrophysical objects within a variety of different contexts (see e.g. the remarkable book by Romero & Vila).

A book on the first black holes so far was missing, and possibly for a variety of reasons. First, the origin of black holes is a very actively developing field, involving a broad range of physics, and a large number of different processes. Different scenarios exist to explain the formation of the first black holes, which may either all be correct or at least provide viable alternatives, in case some of them fail. An important problem certainly is that direct observations of the very first black holes are so far not available, even though observations have reached out until $z \sim 7$ to find the most massive ones. The prospects for the future however are bright, both due to the new facilities that will be available during the next decade, as well as the new branch of gravitational wave astronomy, that may significantly contribute to probe the origin of black holes in the stellar mass range, and potentially also at higher masses in the future.

For the purpose of this book, when talking about the first black holes, we wish to clarify that here we restrict ourselves to black holes forming in the context of cosmological structure formation, within the first dark matter halos at redshifts between 10 and 30, where for the first time cosmological fluctuations became non-linear and turned into gravitationally bound objects. What we specifically do not consider are early Universe scenarios related to black hole formation in cosmic phase transitions or in the epoch of Inflation, which strongly depend on physics beyond our standard model and are therefore not well constrained. The focus in this book is rather on black hole formation in the context of known physics, considering processes such as hydrodynamics, stellar dynamics, chemistry and cooling, and to infer under which conditions they lead to the formation of a massive black hole.

We also note that here we do not strongly focus on the relativistic aspects of the collapse. The reason for that is that this aspect has been treated extensively in many other textbooks, and historically was the main focus of black hole formation research, to understand how the horizon is forming and the transition between a baryonic object and a black hole, i.e. a vacuum solution in spacetime, can occur. The latter is treated extensively in more advanced textbooks on General Relativity, as well as in dedicated books, for instance by Demetrios Christodoulou. Once entering the strongly relativistic regime, the formation of a black hole becomes almost unavoidable, and rather than describing this regime in detail, the question we ask here is how to enter this regime, i.e. how to accumulate enough mass on a sufficiently small scale, so that General Relativistic effects become relevant.

As we will see, the formation of massive black holes involves a broad range of processes; it requires the understanding of self-gravitating fluids, the chemistry and thermodynamics of these fluids, the ambient conditions affecting them, but also the dynamics in star clusters and merger processes in clusters. It is this rich variety of physical processes that forms the core of this book, and which we aim to describe to foster the understanding regarding the origin of the first black holes. The first half of the book is therefore dedicated to the physics and the models, starting from the general cosmological framework, basic physical and chemical processes, up to the formation and evolution of supermassive stars, the possible progenitors of massive black holes. The second half of the book is dedicated to possibilities for probing the models, including dedicated statistical predictions, a description of the current observational status of high redshift black holes, prospects for the detection of gravitational waves, and future observational prospects in the next decade.

While working on such a book, it has quickly become clear that it will never be complete, and the rapid progress in the field also means that specific aspects may become outdated even while writing. This is both due to new and ongoing discoveries, but also theoretical developments and the testing of models via numerical simulations, the development in specific subareas for instance in the context of chemistry or stellar radiation spectra, etc. As it is a matter of ongoing research, the aim of the book also cannot be to tell the final story, but rather to provide the most relevant ingredients that will allow students and future researchers to work in the field and to understand modern articles on the topic.

Due to the vast range of topics covered here, both due to the range of physics, models and the different areas of expertise, it was indispensable for the creation of the book to have dedicated chapters from experts on the different topics. We have carefully selected our contributors to make sure that each of them has valuable contributions in the field, and can provide a didactic introduction that is accessible to advanced students in astronomy. We would like to take this opportunity to thank all of them for the time invested in this effort, which was the most central and necessary ingredient for providing the first book on this topic. We also thank the Project PIA-CONICYT Anillo code ACT172033 entitled 'Formation and growth of supermassive black holes' for the support in the development of this book.

Dominik Schleicher & Muhammad Latif

Contents

Chapter 1

Astrophysical black holes

Pedro R. Capelo
Center for Theoretical Astrophysics and Cosmology
Institute for Computational Science
University of Zurich
Winterthurerstrasse 190, CH-8057 Zürich, Switzerland
pcapelo@physik.uzh.ch

In this chapter, we introduce the concept of a black hole (BH) and recount the initial theoretical predictions. We then review the possible types of BHs in nature, from primordial, to stellar-mass, to supermassive BHs. Finally, we focus on the latter category and on their intricate relation with their host galaxies.

1. The concept of a black hole and the first predictions

All objects in the Universe are gravitational "holes". Given the mass and size of a system, there is always a finite speed that needs to be overcome in order to escape its gravitational well. This is the well-known Newtonian concept of the escape velocity, which can be derived by equating the sum of the kinetic and gravitational potential energy of a particle at the surface of the object to zero, and applies to all matter — from apples to planets, from stars to galaxies. Such a simple calculation was performed by Michell (1784) — see also Laplace (1796) — when considering light particles with a finite mass and speed, to speculate on the existence of objects (so-called "dark stars") from which light could not escape. These objects must therefore have a radius smaller than

$$r_{\rm Schw} = \frac{2GM_{\bullet}}{c^2}, \tag{1}$$

where $M_\bullet$ is the mass of the object, G is the gravitational constant, c is the speed of light in vacuum, and the subscript "Schw" refers to Schwarzschild, for reasons that will become apparent below.

The key feature of these objects is not their mass nor their size, but their compactness C, defined as the ratio between the value given by Equation (1) and the actual size of the system. Planets, stars, and galaxies have values of C of the order of 10^{-9}–10^{-6}, whereas the known stellar compact objects — white dwarfs and neutron stars — have $C \sim 10^{-4}$–10^{-1}. Only when C reaches 1 do we have a system in which gravity is strong enough that nothing can escape (assuming that c cannot be overcome).

Michell's remarkable intellectual achievement, made using simple Newtonian physics (and assuming the corpuscular theory of light), was confirmed a little more than a century later, shortly after the publication of the theory of general relativity (GR; Einstein, 1915, 1916), when the first exact vacuum solution to Einstein's field equations was found by Schwarzschild (1916) — see also Droste (1917); Weyl (1917) — for the case of a spherical, non-electrically charged, non-rotating system. The Schwarzschild metric can be described by (Hilbert, 1917)

$$\mathrm{d}s^2 = -\left(1 - \frac{r_{\mathrm{Schw}}}{r}\right) c^2 \mathrm{d}t^2 + \left(1 - \frac{r_{\mathrm{Schw}}}{r}\right)^{-1} \mathrm{d}r^2 + r^2(\mathrm{d}\theta^2 + \mathrm{d}\phi^2 \sin^2\phi), \quad (2)$$

where $\mathrm{d}s^2$ is the space-time line element and the solution is written in spherical coordinates (t, r, θ, and ϕ), using the Landau–Lifshitz spacelike convention (Landau and Lifshitz, 1962; Misner *et al.*, 1973).

Even though the right-hand side of Equation (2) diverges at both $r = 0$ and r_{Schw}, only the former is a true physical singularity (i.e. the Riemann curvature tensor is infinite only at $r = 0$), with the space-time being non-singular at the so-called Schwarzschild radius.[a] However, the Schwarzschild radius is of fundamental importance, as the radial coordinate of a particle travelling towards the centre changes from spacelike to timelike when crossing r_{Schw}, meaning that the only possible future of that particle is the singularity.[b] Meanwhile, an external static observer will never observe such a boundary (or event horizon) crossing, as the observed time will be infinite (even though the proper time of the particle is finite). Moreover, any

[a]This can be easily proven by transforming the coordinates to, e.g. the Kruskal–Szekeres (Kruskal, 1960; Szekeres, 1960) coordinates.

[b]This is the main difference between the GR solution and Michell's Newtonian dark stars: dark stars can in principle be stable systems, whereas the GR solution *must* collapse towards $r = 0$.

radiation sent from such particle and reaching any external observer will be infinitely redshifted. In other words, a photon sent from r_{Schw} would need infinite energy to reach the observer, effectively making the space-time region within the event horizon causally disconnected from the rest of the Universe. For these reasons, objects with an event horizon (i.e. with $C = 1$) are called BHs.

After the publication of the Schwarzschild solution, other exact solutions to Einstein's field equations were found, in the case of electrically charged (Reissner, 1916; Nordström, 1918), rotating (Kerr, 1963), and rotating, electrically charged BHs (Newman *et al.*, 1965). One peculiarity of BHs is that they are extremely simple. In fact, they can be described at most by three parameters: mass, spin (i.e. angular momentum), and electric charge — this is referred to as the 'no hair' theorem (see, e.g. Israel, 1967). Moreover, in typical astrophysical environments, it is believed that electrically charged BHs cannot exist, as any existing electric charge would be quickly cancelled by the charges in the surrounding plasma (or by spontaneous production of pairs of oppositely charged particles; see, e.g. Gibbons, 1975; Blandford and Znajek, 1977). For this reason, the most complete description for an astrophysical BH is the Kerr metric, which depends only on mass and spin.

Using the Boyer–Lindquist (Boyer and Lindquist, 1967) coordinates, one can write down the Kerr metric such that the radial coordinate of the event horizon[c] is

$$r_{\mathrm{Kerr}} = \frac{r_{\mathrm{Schw}}}{2}\left(1 + \sqrt{1 - a^2}\right), \qquad (3)$$

where $a \equiv c\mathcal{L}/(GM_\bullet^2)$ is the BH spin and $\mathcal{L}$ its angular momentum. The value of $|a|$ can vary between 0 (recovering the Schwarzschild BH) and 1 (for a maximally spinning BH[d]), and its sign depends on the particle orbit we consider: 1 and -1 for corotating and counterrotating orbits (with respect to the BH angular momentum), respectively. If we take a BH with $|a| = 1$,

[c]In the Kerr metric, there are actually two event horizons (internal and external). However, the external event horizon described here is, for all intents and purposes, the event horizon of the BH, as it is the first one-way membrane an external particle crosses.

[d]The actual maximum value depends on the nature and geometry of the accreting matter (e.g. Bardeen, 1970; Thorne, 1974; Popham and Gammie, 1998). If $|a| > 1$, the BH would have no event horizon [see Equation (3)] and the singularity would be visible by the rest of the Universe, i.e. 'naked'. The 'cosmic censorship' conjecture (e.g. Penrose, 1969) prohibits such cases (similar constraints exist also for electrically charged BHs).

the radial coordinate of the event horizon is half of that of a Schwarzschild BH (i.e. $r_{\rm Kerr} = r_{\rm Schw}/2$).

More importantly, the orbits of massive particles around Kerr BHs vary depending on the value of a. When $a = -1$, 0, and 1, the innermost stable circular orbit (ISCO) a massive particle can have is of a radius $r_{\rm ISCO} = 4.5$, 3, and $0.5 \times r_{\rm Schw}$, respectively. For $r < r_{\rm ISCO}$, a particle can only spiral inwards (or outwards, if it has enough velocity to do so) and cannot maintain a stable circular orbit.[e]

The position of the ISCO has significant consequences on how much gravitational energy can be extracted from the gas in the vicinity of the BH (via accretion processes; e.g. Shakura and Sunyaev, 1973; Blandford and Payne, 1982), as the energy lost by particles increases as the distance from the BH decreases.[f]

In other words, the radiative efficiency $\epsilon_{\rm r}$ (i.e. how much of the rest energy of the accreting particle is released), given by

$$L = \epsilon_{\rm r} \dot{M}_{\bullet\,\rm accr} c^2, \tag{4}$$

where L and $\dot{M}_{\bullet\,\rm accr}$ are the accretion power (or luminosity) and the mass accretion rate, respectively, depends on the BH spin (e.g. in a Novikov–Thorne disc, $0.06 \lesssim \epsilon_{\rm r} \lesssim 0.42$ for $0 \leq a \leq 1$; Novikov and Thorne, 1973).

Indeed, astrophysical accreting BHs are believed to be normally surrounded by accretion discs (e.g. Shakura and Sunyaev, 1973), in which matter differentially rotates around the BH, a fraction of it loses angular momentum and moves towards the centre, and, through viscous forces, becomes hot and radiates (e.g. X-rays). The ISCO roughly coincides with the inner edge of the accretion disc. For more on the concept of accretion in astrophysics, see, e.g. Frank *et al.* (2002).

Accretion on to BHs is of fundamental importance because (a) it is through the results of accretion that we can detect most BHs (e.g. in X-ray binaries and active galactic nuclei — AGN); (b) it affects the BH nature,

[e] We note that, even when $a = 1$, $r_{\rm ISCO}$ is always greater than $r_{\rm Kerr}$ (Bardeen *et al.*, 1972).

[f] Additionally, for high values of a, the ISCO finds itself within the ergosphere, a region of space-time in which, due to the frame-dragging effect (also known as the Lense–Thirring effect; Lense and Thirring, 1918), rotation is inevitable, and this BH rotational energy can be also extracted (e.g. Penrose, 1969; Blandford and Znajek, 1977).

by changing its mass, with the mass growth rate

$$\dot{M}_{\bullet\,\mathrm{growth}} = (1 - \epsilon_{\mathrm{r}})\dot{M}_{\bullet\,\mathrm{accr}}, \tag{5}$$

and spin (e.g. Perego *et al.*, 2009; Dotti *et al.*, 2013; Dubois *et al.*, 2014b; Fiacconi *et al.*, 2018); and (c) its power can potentially affect the surrounding material (see Section 3). Accretion depends on the availability of low-angular momentum gas[g] in the vicinity of the BH. There is, however, a theoretical limit to how much a BH can accrete, regardless of the availability of "fuel", which is given by the radiation pressure — associated to the accretion power L — counteracting gravity. When approximating accretion to be spherical (e.g. Hoyle and Lyttleton, 1939; Bondi and Hoyle, 1944; Bondi, 1952), one can equate the inward force on an electron–proton pair — $GM_\bullet m_{\mathrm{p}}/r^2$ — to the outward force — $L\sigma_{\mathrm{T}}/(4\pi c r^2)$ — and obtain (Eddington, 1916)

$$L_{\mathrm{Edd}} = \frac{4\pi G M_\bullet m_{\mathrm{p}} c}{\sigma_{\mathrm{T}}}, \tag{6}$$

where m_{p} is the proton mass and σ_{T} is the electron Thomson cross-section. The corresponding mass accretion rate is simply given by $\dot{M}_{\bullet\,\mathrm{Edd}} = L_{\mathrm{Edd}}/(\epsilon_{\mathrm{r}} c^2)$, where ϵ_{r} is usually chosen equal to 0.1. A BH always accreting at the Eddington level would grow exponentially on a time-scale $\tau_{\mathrm{Salp}} = 4.5 \times 10^8\, \epsilon_{\mathrm{r}}/(1 - \epsilon_{\mathrm{r}})$ yr, known as the Salpeter time (Salpeter, 1964). Assuming, e.g. $\epsilon_{\mathrm{r}} = 0.1$, this BH would need 7×10^8 yr to increase its own mass by six orders of magnitude (e.g. from 10^3 to 10^9 $\mathrm{M}_\odot$). We note that, under certain circumstances (mostly depending on the geometry of the accreting flow), it is still possible to achieve $\dot{M}_{\bullet\,\mathrm{accr}} > \dot{M}_{\bullet\,\mathrm{Edd}}$ (see Chapter 11).

For a long time after the first GR solution for a BH was found, this was considered only a mathematical curiosity (see, e.g. Einstein, 1939). After all, the size of the Schwarzschild radius is extremely small, ranging from the width of the Manhattan island, when the mass is 1 $\mathrm{M}_\odot$, to the average distance between the Sun and Uranus, when the mass is 10^9 times larger. How can so much mass be entrapped in such a small region?

[g]BHs can also accrete other kinds of matter. For example, BHs with a mass above $\sim 10^8$ $\mathrm{M}_\odot$ can "eat" whole (solar-mass) stars, without disrupting them. However, in that case the entire mass of the star gets lost inside the BH, and there is no possible extraction of energy (i.e. $\epsilon_{\mathrm{r}} = 0$).

The first answers to this question were given only in the 1930s, with advances in the theory of stellar evolution, when it was postulated that stellar compact objects (e.g. white dwarfs and neutron stars) could indeed exist in nature as the final outcome of the evolution of stars (e.g. Chandrasekhar, 1931; Landau, 1932; Baade and Zwicky, 1934; Tolman, 1939; Oppenheimer and Volkoff, 1939), when gravity is not counter-balanced any longer by thermo-nuclear processes. For relatively low-mass stars, gravity is still balanced by some force (electron and neutron degeneracy pressure for white dwarfs and neutron stars, respectively). For higher-mass stars, no force is able to counteract gravity, material flows across the Schwarzschild radius,[h] and a BH forms (Oppenheimer and Snyder, 1939). For a review on the relation between the progenitor's stellar mass and its final fate, see, e.g. Heger *et al.* (2003a,b).

For the first time, it was theorised how BHs could indeed form in nature.

2. Black holes in nature

A BH can in principle have any mass, as long as there is a mechanism able to compress such mass inside its own event horizon. In this section, we divide the BHs believed to exist in nature in the order of their lower-mass limit, starting from the highly speculative primordial BHs (PBHs; Section 2.1), whose masses could range from $\sim 10^{-5}$ g to $\sim 10^5$ $M_\odot$ (or more), continuing with stellar-mass BHs ($\sim$3–100 $M_\odot$; Section 2.2), and ending with intermediate-mass BHs (IMBHs; $\sim$100–10^5 $M_\odot$; Section 2.3) and supermassive BHs (SMBHs; $\gtrsim 10^5$ $M_\odot$; Section 2.4).

There are several methods for measuring the mass of a BH, ranging from dynamical measurements of individual stars (e.g. X-ray binaries and Sagittarius A*; see Sections 2.2 and 2.4) and gas clouds (e.g. through water masers; NGC 4258; see Section 2.4), ensembles of stars (from stellar absorption lines), and gas (from gas emission lines in, e.g. circumnuclear discs), to less direct methods such as reverberation (or echo) mapping (e.g. Blandford and McKee, 1982; Peterson, 1993), simple Eddington-luminosity arguments, or even using virial estimates from known scaling relations (see Section 3). The technique used mostly depends on whether the BH is accreting and/or if we can resolve its sphere of influence, defined by the

[h] As mentioned above, because of the diverging relation between proper and observer time at $r_{\rm Schw}$, matter infalling through the event horizon seems frozen in time to an observer at infinity, hence the alternative definition of 'frozen star' (especially in the Soviet literature).

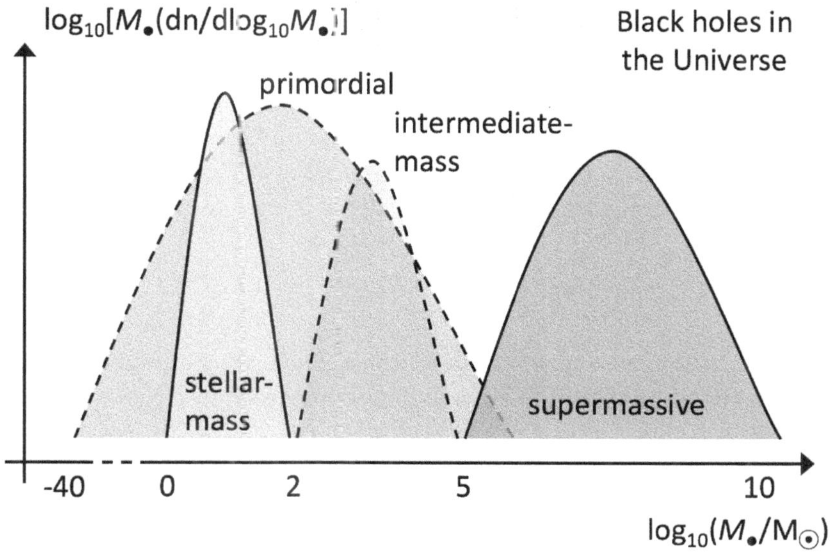

Fig. 1. Sketched number density — $M_\bullet(dn/d\log_{10}M_\bullet)$ – of BHs in the Universe, in the order of lower-mass limit: (from left to right) PBHs, stellar-mass BHs, IMBHs, and SMBHs. The shapes of the distributions of PBHs and IMBHs (dashed lines) are arbitrary, as a reliable census does not exist. In particular, PBHs are highly speculative (see Section 2.1) and IMBHs have not been detected yet (although there are a few strong candidates; see Section 2.3). The small number of detected stellar-mass BHs does not allow for an accurate measurement of their mass function, but see, e.g. Merloni (2008); Kovetz *et al.* (2017). The mass function of SMBHs is inspired by Shankar *et al.* (2009). Figure inspired by García-Bellido (2017).

radius[i]

$$r_{\rm infl} = \frac{GM_\bullet}{\sigma^2}, \tag{7}$$

where σ is the stellar velocity dispersion of the central part of the galaxy (assumed to be constant with radius), within which the BH gravitational potential dominates the motion of the surrounding matter. E.g. a BH of mass 10^8 M$_\odot$ in a galaxy with $\sigma = 200$ km s^{-1} has a sphere of influence of radius ∼10 pc.

In Figure 1, we show a rough sketch of the mass distribution of BHs in the Universe.

[i]An alternative definition is given by $M_\star(r < r_{\rm infl}) = 2M_\bullet$, where $M_\star$ is the stellar mass. The two definitions coincide if the galaxy can be described by a singular isothermal sphere (e.g. Merritt, 2013). A related (and of the same order of magnitude) quantity is the Bondi radius (Bondi, 1952), given by $GM_\bullet/c_{\rm s}^2$, where $c_{\rm s}$ is the sound speed of the surrounding gas, within which the gas is gravitationally bound to the BH.

2.1. *Primordial black holes*

According to the most accepted theory,[j] PBHs are believed to be the by-product of the extremely high densities and inhomogeneities present in the first instants after the Big Bang (Zel'dovich and Novikov, 1966; Hawking, 1971), when the Universe was radiation-dominated. The PBH mass is presumed to be of the order of the particle horizon mass (e.g. Carr, 2005):

$$M_\bullet \sim \frac{c^3 t_{\rm form}}{G} \simeq 4 \times 10^{38} \frac{t_{\rm form}}{\rm s}\ {\rm g} \simeq 2 \times 10^5 \frac{t_{\rm form}}{\rm s}\ {\rm M_\odot}, \tag{8}$$

where $t_{\rm form}$ is the moment when they formed. Depending on when the PBH forms, its mass can then span an enormous range of values, from the Planck mass (when $t_{\rm form} = t_{\rm Planck}$) to 10^5 $\rm M_\odot$ (when $t_{\rm form} \sim 1$ s) or larger.

A particularly important value is 10^{15} g (when $t_{\rm form} \sim 10^{-23}$ s), because it is approximately the mass of PBHs which should be evaporating today due to the Hawking (1974) effect,[k] since BHs should evaporate with a time-scale $\sim 10^{71}(M_\bullet/{\rm M_\odot})^3$ s. Larger BHs will evaporate in the future, whereas smaller BHs may have already evaporated.[l] So far, however, no firm detection of Hawking radiation has been reported in the extragalactic or Galactic γ-ray backgrounds (e.g. Carr *et al.*, 2016b).

PBHs with a mass larger than 10^{15} g should in principle be still present today, and they have also been touted as partly or fully responsible for the dark matter in the Universe. The recent detections of gravitational waves (GWs) from the advanced Laser Interferometer Gravitational-wave Observatory (LIGO) of the inspiral and coalescence of BHs of masses somewhat larger than usually expected from stellar evolution (e.g. Abbott *et al.*, 2016b; see also Section 2.2 and Figure 2) have given new life to this scenario, which dates back to the MACHO — massive, compact, halo object — era (e.g. Paczynski, 1986; Alcock *et al.*, 2000). To date, however, there are several gravitational-lensing and dynamical constraints that likely make such a scenario implausible (for a recent review, see, e.g. Carr *et al.*, 2016a).

---

[j] Amongst the alternative theories, we list, e.g. cosmic string loops (e.g. Hawking, 1989) and bubble collisions (e.g. Hawking *et al.*, 1982).

[k] According to the Hawking effect, a particle–antiparticle pair can be produced from the BH's gravitational energy in the vicinity of the event horizon. If, instead of immediately recombining, one of the particles falls into the BH and the other escapes, the net effect is that the BH has lost some of its energy (i.e. its mass), corresponding to the energy of the escaping particle.

[l] We note, in passing, that any BH that may be created in the laboratory (e.g. in the Large Hadron Collider), would be so small that it would evaporate almost instantly (e.g. Dimopoulos and Landsberg, 2001).

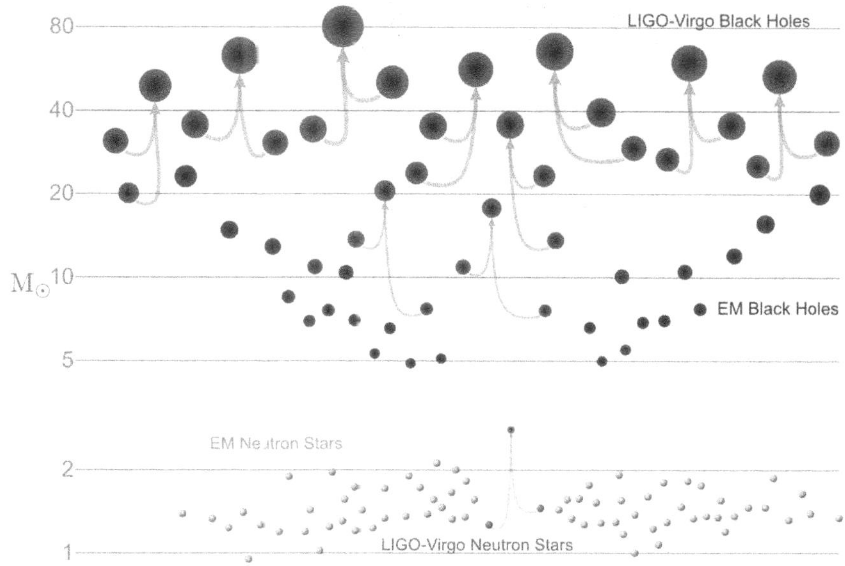

Fig. 2. Masses of stellar-mass BHs (mass $\gtrsim 3\ M_\odot$) and neutron stars (mass $\lesssim 3\ M_\odot$) detected through electromagnetic (EM) and GW observations. The heaviest stellar-mass BHs in the plot have masses in the range $\sim$60–80 $M_\odot$ (GW150914, GW170823, and GW170729; Abbott *et al.*, 2016b; The LIGO Scientific Collaboration & the Virgo Collaboration, 2018). The nature of the remnant of the LIGO-Virgo neutron-star system (GW170817; Abbott *et al.*, 2017e), marked by a question mark (of mass $\sim$3 $M_\odot$), is still unknown. Image credit: LIGO-Virgo/Frank Elavsky/Northwestern University.

Finally, PBHs could actually be the seeds of the SMBHs found at the centre of massive galaxies (see, e.g. Bernal *et al.*, 2018).

2.2. *Stellar-mass black holes*

Stellar-mass BHs have masses of the order of $\sim$3–100 $M_\odot$ and are believed to be the final outcome of the evolution of massive stars (see Section 1). Given their size and distance, it is currently (and for the foreseeable future) extremely difficult to directly detect such systems in isolation.

Luckily, most massive stars are in binaries and, as they coevolve, the end state of one may be a compact object, which accretes mass from the other that is in its late evolutionary stages and is blowing off its outer envelope. Hence the presence of a BH (or neutron star) is inferred from evidence for episodic accretion, inferred from the light curves and spectral energy distributions of variable sources.

Giacconi *et al.* (1962), in the first X-ray exploration of space, discovered the source Cygnus X-1, which was later confirmed to be a binary system composed of a stellar-mass BH (of mass ~15 $M_\odot$; Orosz *et al.*, 2011) and a companion star. Nowadays, several X-ray binaries have been detected (see Figure 2), in which the X-ray radiation is believed to be originated from the companion-star gas being accreted on to the compact object.

In many cases, it is assumed that the compact object is a neutron star. In some cases, however, as in the case of Cygnus X-1, it is highly probable that the compact object is a BH, since the calculated mass is very likely above the upper limit for neutron stars (Tolman, 1939; Oppenheimer and Volkoff, 1939). The uncertainty arises from the fact that we cannot measure directly the mass, but only the mass function, defined as

$$f(M_\bullet) \equiv \frac{M_\bullet \sin^3 i}{(1+q)^2} = \frac{P v_{\rm r}^3}{2\pi G}, \tag{9}$$

where P is the orbital period, q is the mass ratio, $v_{\rm r}$ is the radial velocity of the companion star, and i is the orbital inclination. Radial-velocity measurements provide only a lower limit on the mass, mostly due to the difficulty of determining the angle i (unless the binary is eclipsing).

When a stellar-mass compact object is not accreting gas, it is virtually impossible to detect such an object.[m] Only very recently, thanks to the first detections of GWs by the advanced LIGO, it has been possible to detect pairs of stellar-mass compact systems (both BHs and neutron stars) in the final stages of a merger.[n] By comparing the detection from interferometers to numerical-GR models, both the mass of the merging systems and that of the remnant were inferred, yielding a LIGO-BH mass range that was slightly on the upper end of what usually thought possible from stellar-evolution modelling (Heger *et al.*, 2003a,b; see also Section 2.1). GW detections have provided a new way to study mergers of compact objects. For example, stellar-mass compact-object mergers (BH–neutron star or neutron star–neutron star) are considered an important pathway for the production of the elements beyond iron group nuclei (r-process nucleosynthesis; see, e.g. Lattimer and Schramm, 1974; Rosswog, 2015; Bonetti *et al.*, 2018). Very recently, the first neutron star–neutron star merger was

[m] However, microlensing surveys such as MACHO (Bennett *et al.*, 2002) have provided a few candidates.

[n] With the planned Laser Interferometer Space Antenna (LISA), it will also be possible to detect extreme mass ratio inspiral (EMRI) events between a stellar-mass BH and a SMBH (e.g. Babak *et al.*, 2017).

detected (GW170817; Abbott *et al.*, 2017e). Hopefully, we will soon be able to detect also BH–neutron star coalescences.

2.3. *Intermediate-mass black holes*

IMBHs have a mass range of ~100–10^5 $M_\odot$ and are, amongst the non-primordial BHs, the most elusive of all.

At high redshift z, they are believed to be one of the types of BH seeds that will eventually become SMBHs. Given their relatively low mass, it is currently extremely difficult to detect them and measure their mass both by gas or stellar dynamics methods (as their sphere of influence can be smaller than 1 pc, depending on the host) or by emission from accreting gas [as the Eddington luminosity scales linearly with $M_\bullet$; see Equation (6)].

In this section, we bring our attention to those seed BHs which, for several reasons, did not grow to become SMBHs and should therefore have masses in the IMBH range today.

One obvious place to look for IMBHs are local dwarf galaxies: these galaxies are thought to have had a relatively quiet life, with no major mergers nor significant gas accretion, hence they are not too dissimilar from their high-redshift counterparts. It is plausible, therefore, that an existing central BH has not accreted much during its life and has retained its initial relatively low mass. Detections of such BHs in local dwarf galaxies would be very important also to understand how they formed in the first place, since different formation scenarios would produce different BH occupation fractions and scaling relations in the low-mass range (e.g. Volonteri, 2010). Several IMBH candidates have been identified in dwarf galaxies (e.g. RGG 118; ~5×10^4 $M_\odot$; Baldassare *et al.*, 2015) and populations of IMBHs in dwarf galaxies have been recently detected up to $z \sim 2$ (e.g. Mezcua *et al.*, 2018).

Another site where IMBHs could hide are globular clusters, since BHs could form from runaway collapse of stellar clusters (see Chapter 7). However, so far, all BH-mass estimates from kinematic measurements were shown to be also explained without a BH (e.g. NGC 7078; van den Bosch *et al.*, 2006). Moreover, strong evidence for accretion has not been found, as globular clusters are likely devoid of gas. Recently, Kızıltan *et al.* (2017) used a novel method, by measuring the locations and accelerations of pulsars in the globular cluster 47 Tucanae (NGC 104), comparing them to what was modelled assuming no BH, and computing the mass of the central BH necessary to match the observations, obtaining a mass of ~2.3×10^3 $M_\odot$.

IMBHs have also been linked to ultra-luminous X-ray sources (ULXs), which are non-nuclear, point-like extragalactic sources with an apparent isotropic luminosity $>10^{39}$ erg s^{-1} in the 0.3–10 keV band (e.g. Feng and Soria, 2011). Such luminosities require either a relatively massive BH accreting at Eddington levels [see Equation (6)] or stellar-mass compact objects accreting at super-Eddington levels. The growing consensus is that most ULXs are indeed super-Eddington accreting stellar-mass compact objects (either BHs or neutron stars; e.g. Bachetti *et al.*, 2014; Fiacconi *et al.*, 2017b) with only the highest-luminosity ULXs (hyper-luminous X-ray sources – HLXs; $\gtrsim 5 \times 10^{40}$ erg s^{-1}) possibly related to IMBHs (e.g. HLX-1, with a mass range $\sim 6 \times 10^3$–2×10^5 $M_\odot$, depending on inclination and BH spin; Farrell *et al.*, 2009; Straub *et al.*, 2014).

IMBHs could be found also in other ways and environments: within accretion discs of SMBHs (e.g. Bellovary *et al.*, 2016); hidden inside high-velocity clouds near SMBHs (e.g. Oka *et al.*, 2016); or detected by future gravitational-wave interferometers (e.g. LISA; Miller, 2009; Tamfal *et al.*, 2018) when two of them coalesce. Given their relatively small mass, they are also able to tidally disrupt (solar-mass) stars (e.g. Hills, 1975; Rees, 1988) — as opposed to eating them whole, as is the case for SMBHs with mass $\gtrsim 10^8$ $M_\odot$ — possibly producing an ultraviolet or X-ray signal when the debris gets accreted (e.g. Evans and Kochanek, 1989; Renzini *et al.*, 1995; Bade *et al.*, 1996; Bloom *et al.*, 2011; Mainetti *et al.*, 2016).

For a recent review on IMBHs and their observational evidence, see e.g. Mezcua (2017).

2.4. *Supermassive black holes*

SMBHs have masses greater than $\sim 10^5$ $M_\odot$ and can reach values of a few 10^{10} $M_\odot$. In principle, there should be no upper limit to the mass, but both observations and theoretical reasoning (see, e.g. Natarajan and Treister, 2009; King, 2016; Inayoshi and Haiman, 2016; Pacucci *et al.*, 2017a) seem to imply that SMBHs heavier than a few 10^{10} $M_\odot$ do not exist (or, if they do, they grow their mass above the upper limit by non-luminous means such as SMBH mergers; King, 2016).

The first hint that extremely large BHs could exist in nature came with the first observations of quasars in the early 1960s.[o] These systems were extremely distant (e.g. Schmidt, 1963), implying that they had luminosities

[o]Quasars, or quasi-stellar radio sources, are systems in which the host galaxy is not resolved and only the central point source's emission is detected. They are a subset of AGN (see reviews by Peterson, 1997; Krolik, 1998; Ho, 2008; Padovani *et al.*, 2017). In this chapter, we use the two terms interchangeably.

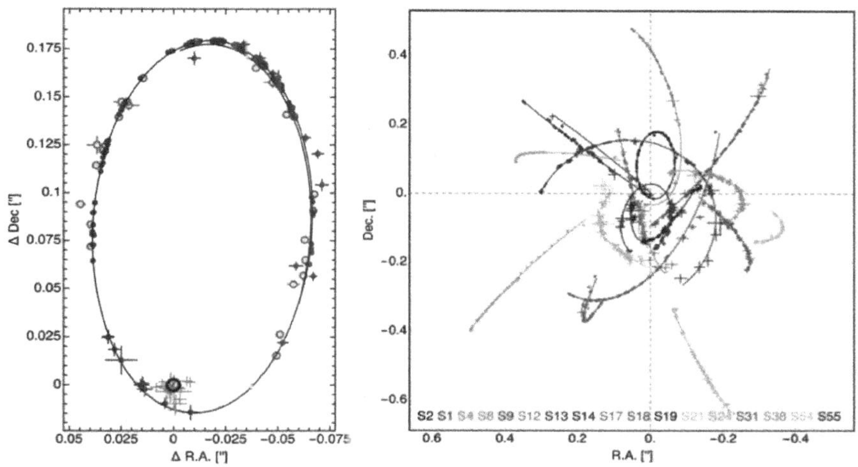

Fig. 3. Orbit of the star S2 (left-hand panel) and of 17 S-stars (right-hand panel) around Sagittarius A*. Figures from Gillessen *et al.* (2017). ©AAS. Reproduced with permission.

exceeding even 10^3 times the typical luminosities of galaxies. In many cases, they had jets whose direction was the same over distances ranging from sub-pc to several tens of kpc. In some of these cases, the jet showed (apparent) superluminal motion, which can be explained by a jet pointing towards us and with velocities very close to c (e.g. Rees, 1966). Also, many sources showed time variability of the order of even weeks.

All these pieces of indirect evidence point to the possibility that quasars are indeed powered by SMBHs (Salpeter, 1964; Zel'dovich, 1964; Lynden-Bell, 1969). Amongst all possible astrophysical objects that we know of, only BHs have such a deep potential well that can produce so much accretion energy. Also, only a relativistic potential can create jets with velocities close to the speed of light, and the spin of a BH is a plausible cause for the stability of a jet over large time- and length-scales. Finally, time variability of the order of weeks implies sizes of the order of 'c × week' and, consequently, high values of $\mathcal{C}$.

Since the first observations of quasars, a plethora of systems, all under the umbrella-name of AGN, have been observed at all wavelengths. Additionally, systems observed earlier than the first quasars (e.g. Seyfert galaxies; Seyfert, 1943) were also catalogued as AGN. AGN are by definition accreting SMBHs (and IMBHs). Only when there is enough gas fuelling (i.e. enough gas losing angular momentum), a SMBH can shine through its accretion disc as an AGN and be easily detected.

When SMBHs are not accreting, their existence can still be inferred (and their masses can still be measured) through dynamical methods. Here we list only the two strongest cases.

The best example is the SMBH at the centre of our own Milky Way, coincident with the radio source Sagittarius A*. Repeated observations of individual stellar proper motions have shown almost perfect Keplerian orbits (see Figure 3), implying the existence of a very compact object. In particular, the complete orbit of the S2 (or S0-2) star shows that there is a mass of $\sim 4.3 \times 10^6$ $M_\odot$ within a region of ~ 100 AU (e.g. Ghez *et al.*, 2008; Genzel *et al.*, 2010).

Another strong case is in the galaxy NGC 4258, in which water masers (water molecules in gas clouds, excited by the central AGN) were found to trace a well-organised (Keplerian) rotating thin disc, viewed edge-on, whose kinematics could be explained only by a central compact object of mass $\sim 4 \times 10^7$ $M_\odot$ (Miyoshi *et al.*, 1995).

In both these cases, it is still in principle possible to imagine alternatives to SMBHs:[p] e.g. a cluster of non-luminous objects (e.g. stellar-mass BHs, neutron stars, white dwarfs, brown dwarfs) could in fact be the dark system at the centre of both the Milky Way and NGC 4258. However, it was shown by Maoz (1998) that such systems would collapse on a time-scale much shorter than the galaxy's age. Moreover, recent detailed observations of the orbit of the S2 star around the Galactic centre provided the detection of gravitational redshift (GRAVITY Collaboration, 2018). Very recently, the Event Horizon Telescope, a global very long baseline interferometry array, resolved the shadow of the SMBH in the giant elliptical galaxy M87 (NGC 4486), providing for the very first time a "picture" of a BH (The Event Horizon Telescope Collaboration, 2019).

One fundamental and still unanswered question is: how did SMBHs form? Several alternatives have been put forward for the origin of the seeds (e.g. Rees, 1978a; Volonteri, 2012), including, e.g. direct collapse of pristine gas (Haehnelt and Rees, 1993, see Chapters 5 and 6), population III BHs (Madau and Rees, 2001, see Chapter 4), dense stellar clusters (Begelman and Rees, 1978, see Chapter 7), or galaxy mergers (Mayer *et al.*, 2010, see Chapter 5). SMBH formation is the main topic of this book, on which therefore we will not dwell further in this section. In the remainder of this chapter, we will discuss the importance of SMBHs in relation to their host galaxies.

[p]More exotic alternatives include, e.g. fermion stars (currently ruled out; Schödel *et al.*, 2002), boson stars, and gravastars.

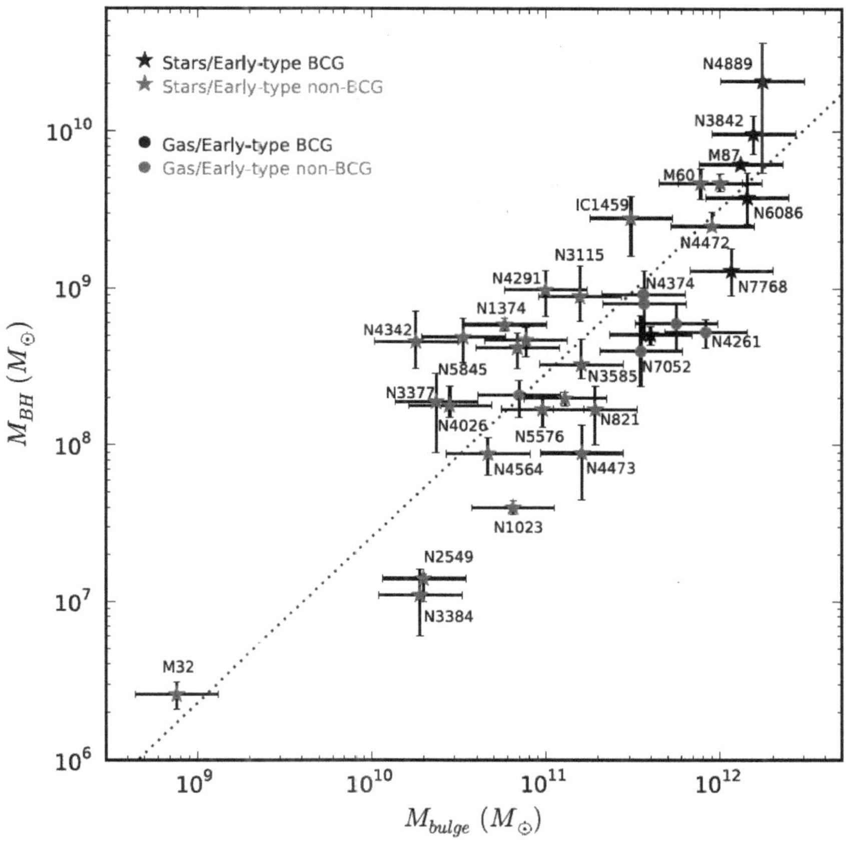

Fig. 4. Relation between the mass of the central SMBH $M_\bullet$ and the bulge mass $M_{\rm bulge}$ of the host galaxy, for the 35 early-type galaxies with dynamical measurements of the bulge stellar mass in the sample of McConnell and Ma (2013), divided in brightest cluster galaxies (BCG; grey) and non-BCG (black). The SMBH masses are measured using the dynamics of stars (stars) or gas (circles). The error bars indicate 68 per cent confidence intervals. The black, dotted line represents the best-fitting power-law relation, which is in good agreement with Equation (10). Figure from McConnell and Ma (2013). ©AAS. Reproduced with permission.

3. Supermassive black holes and their host galaxies

It is now widely accepted that SMBHs exist at the centre of several, if not most, nearby massive galaxies (e.g. Kormendy and Ho, 2013). Additionally, SMBHs as the engines of AGN are believed to exist in many high-redshift galaxies.

In many instances, the mass of the central SMBH has been estimated via a wide variety of methods (see Sections 2 and 2.4). Accurate mass measurements in the local Universe led to the finding of scaling relations between the mass of the central object and several galactic quantities: stellar velocity dispersion, bulge stellar mass (see Figure 4), bulge optical and near-infrared luminosity, total luminosity, bulge concentration, dark matter halo mass, number and total mass of globular clusters in the host galaxy, light deficit, spiral-arm pitch angle, etc. (see, e.g. McConnell and Ma, 2013, and references therein).

In particular, tight scaling relations have been found between the SMBH mass and a few properties of the host, including the bulge luminosity (e.g. Kormendy and Richstone, 1995), bulge mass $M_{\rm bulge}$ (the so-called Magorrian relation; e.g. Magorrian *et al.*, 1998),

$$\frac{M_\bullet}{10^9{\rm M}_\odot} = 0.49^{+0.06}_{-0.05}\left(\frac{M_{\rm bulge}}{10^{11}{\rm M}_\odot}\right)^{1.17\pm0.08}, \qquad (10)$$

and effective velocity dispersion σ (the so-called $M_\bullet$–σ relation; e.g. Ferrarese and Merritt, 2000; Gebhardt *et al.*, 2000),

$$\frac{M_\bullet}{10^9{\rm M}_\odot} = 0.310^{+0.037}_{-0.033}\left(\frac{\sigma}{200\ {\rm km\ s}^{-1}}\right)^{4.38\pm0.29}, \qquad (11)$$

where the coefficients of the fits of Equations (10) and (11) are valid for the classical bulges and elliptical galaxies listed in Kormendy and Ho (2013).

Moreover, high-redshift observations have shown that the evolution of the cosmic star formation rate (SFR) and the SMBH accretion rate densities over cosmic time follow very similar trends (e.g. Madau and Dickinson 2014; Aird *et al.* 2015; see Figure 5): both rise up to $z \sim 2$, where they reach their peak, before steeply declining down to local values. Remarkably, the ratio between the two remains more or less the same across all cosmic time, with the SFR $\sim 10^3$ times larger than the SMBH accretion rate.

All the above considerations are consistent with a scenario in which SMBHs and their host galaxies may co-evolve[q] (e.g. Silk and Rees, 1998; Haehnelt *et al.*, 1998): SMBHs grow through gas accretion and mergers and feed back part of their accretion energy to the host galaxy.

[q] We caution that there are alternative explanations for the origin of scaling relations, which do not require any co-evolution. E.g. the tightness of the Magorrian relation could be partly or fully caused simply by the hierarchical assembly of SMBH and stellar mass through galaxy merging, as a consequence of the central-limit theorem (e.g. Peng, 2007; Hirschmann *et al.*, 2010; Jahnke and Macciò, 2011).

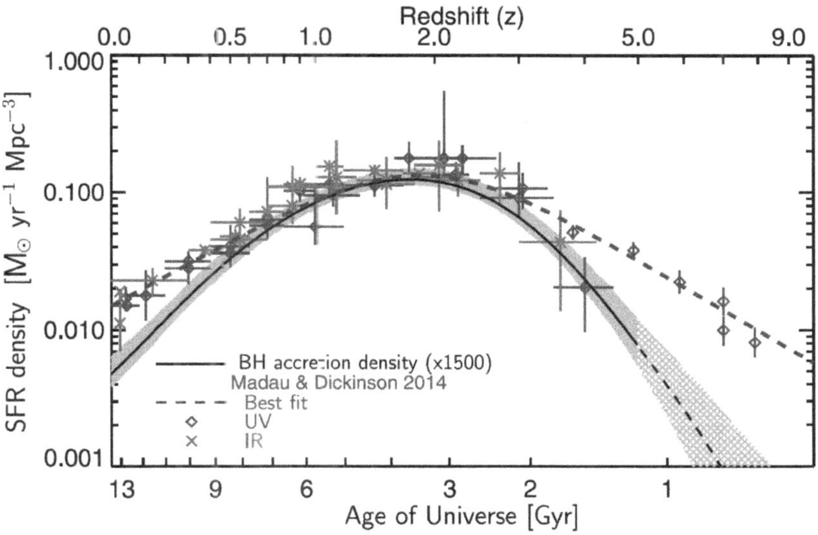

Fig. 5. SFR density [measurements based on rest-frame ultraviolet (diamonds) and infrared (crosses) observations, together with their best fit (dashed line), from Madau and Dickinson (2014)] and SMBH accretion rate density [black line and 99 per cent confidence shaded region, from the model of Aird *et al.* (2015), extrapolated for $z > 5$], scaled up by a factor of 1.5×10^3, as a function of redshift. See also the second and third panel of Figure 6. Figure from Aird *et al.* (2015), reproduced by permission of Oxford University Press, on behalf of the Royal Astronomical Society.

Assuming that SMBHs grow mostly due to radiative accretion (a reasonable assumption, as we will see below), the energy released by a SMBH to grow its mass to $M_\bullet$ is $E_\bullet \simeq \epsilon_r M_\bullet c^2/(1-\epsilon_r)$ [see Equations (4) and (5)]. The binding energy of the host galaxy bulge is instead $E_{\rm bulge} \simeq M_{\rm bulge}\sigma^2$. Combining Equations (10) and (11) and taking $\epsilon_r = 0.1$, we obtain $E_\bullet/E_{\rm bulge} \simeq 8 \times 10^2 (M_\bullet/10^9 M_\odot)^{-0.3}$. The energy released by the growth of a SMBH of mass 10^7 $M_\odot$ (10^{10} $M_\odot$) is $\sim 3 \times 10^3$ ($\sim 4 \times 10^2$) times the binding energy of the host bulge. If even a small part of $E_\bullet$ couples to the gas of the galaxy, all this gas would be blown away (e.g. Fabian, 2012; King and Pounds, 2015). Clearly, there is enough energy in the accretion process to have a significant effect on the host galaxy. This so-called AGN feedback can be radiative and/or mechanical (jets/winds) and has been invoked to explain, e.g. the above-mentioned scaling relations and the mismatch between the dark matter halo mass function and the observed luminosity function (at large masses). Feedback is usually negative (i.e. SF is quenched) and is either ejective (with the gas being blown away from the galaxy) or preventive (with the gas being kept hot), but can also be positive (i.e. SF is enhanced), mostly in the case of jets/winds. The topic of AGN feedback is at the same time still hotly debated and so vast that it is beyond the scope of this introductory chapter. We refer the reader to recent reviews by, e.g. Alexander and Hickox (2012); Fabian (2012); Kormendy and Ho (2013); Heckman and Best (2014); King and Pounds (2015).

One can check the assumption that SMBHs grow due to radiative accretion, by comparing the AGN activity at high redshift to the mass of local SMBHs. This is the basis of the so-called Sołtan argument (Sołtan, 1982). By integrating the AGN (bolometric) luminosity over redshift (thus obtaining the comoving energy density from AGN), one can then obtain the SMBH mass density at $z = 0$ and compare it to local observations (which rely on the above-mentioned scaling relations). The overall agreement between the two numbers (see, e.g. Yu and Tremaine, 2002; Marconi *et al.*, 2004), when assuming a radiative efficiency $\epsilon_{\rm r} \sim 0.1$, strengthens the hypothesis that SMBHs grow mostly by radiative accretion.[r]

If almost all the mass of a SMBH is indeed provided by radiative accretion, this brings us to the question: how can gas get accreted efficiently?

The existence of high-redshift quasars with SMBH masses of the order of 10^9 $\rm M_\odot$ implies that a comparable amount of gas was funnelled into the very inner galactic regions in a very short time: the age of the Universe at $z \sim 7.54$, the redshift of the current record-holder quasar (Bañados *et al.*, 2018b), was ~0.7 Gyr.

The main issue to overcome is not the quantity of available gas, but the amount of angular momentum that such gas needs to lose in order to travel from kpc scales (galactic or even extra-galactic regions) to sub-pc scales (the accretion disc). If we take a typical galaxy with a flat rotation velocity $v = 200$ km s^{-1}, gas at $r = 10$ kpc from the centre has a specific angular momentum magnitude $l = rv \simeq 2 \times 10^3$ kpc^2 Gyr^{-1}. If we now compute the same quantity at the ISCO radius of a maximally spinning Kerr BH ($r = r_{\rm Schw}/2$; see Section 1), we obtain $l = (GM_\bullet r)^{1/2} \simeq 1.5 \times 10^{-11}(M_\bullet/\rm M_\odot)$ kpc^2 Gyr^{-1}. Taking, e.g. a BH of mass 10^9 $\rm M_\odot$, this means that the gas needs to lose 99.999 per cent of its specific angular momentum.

[r]An individual SMBH can also grow via SMBH mergers, radiatively inefficient accretion (e.g. in advection-dominated accretion flows; Narayan and Yi, 1994), or by swallowing whole stars (see also Section 2.3), although the local SMBH mass density is barely affected by these mechanisms. We note, in passing, that SMBHs can affect individual stars also in opposite fashion, by disrupting a tight binary stellar system and causing one of the stars to be ejected at velocities of the order of 10^3 km s^{-1}, hence taking the name of hyper-velocity star (e.g. Hills, 1988).

Gas can lose its angular momentum through many different mechanisms, depending on the state of the system (e.g. if it is isolated or interacting) and on the distance from the centre (for a review, see, e.g. Jogee, 2006). Most of these mechanisms are linked to deviations from axial symmetry in the gravitational potential, so that a torque can act on the gas (but can also rely on, e.g. dynamical friction, as individual gas clouds travel inwards).

Galactic-scale stellar bars (e.g. Sanders and Huntley, 1976; Roberts *et al.*, 1979; Athanassoula, 1992) are the most studied example of such mechanisms in the case of isolated systems.[s] However, they can only manage to funnel the gas down to distances of ~0.1–1 kpc, where the bar ceases to be efficient because the gas has reached the inner Lindblad resonance radius (e.g. Binney and Tremaine, 2008; Kormendy, 2013). In order to reach the nuclear regions of the galaxy, additional mechanisms are needed, such as systems of nested non-axisymmetric structures (e.g. Hopkins and Quataert, 2010), including nested bars (e.g. Shlosman *et al.*, 1989), or bar-driven nuclear spirals (e.g. Maciejewski, 2004; Fanali *et al.*, 2015).

Bars are not the only mechanism of note in isolated systems. E.g. dense, massive gas clouds in gas-rich clumpy discs can migrate inwards due to mutual angular momentum exchange with other density perturbations and stochastically collide with the central SMBH (Gabor and Bournaud, 2013), simply travel inwards due to dynamical friction, or move ballistically inwards due to turbulence driven by supernova explosions (Hobbs *et al.*, 2011).

Another well-studied mechanism invoves galaxy mergers, whose dynamics have been investigated since the 1940s (Holmberg, 1941). During the interaction, the tidal field of one galaxy causes the other galaxy to form non-axisymmetric structures which then torque the stars (e.g. Toomre and Toomre, 1972) and gas (e.g. Barnes and Hernquist, 1996). Additionally, gas can also lose angular momentum as a consequence of hydrodynamically driven (ram-pressure) shocks, during the closest pericentric passages (Barnes, 2002; Capelo and Dotti, 2017): by doing so, gas can effectively become decoupled from the stars and can populate phase-space regions of extremely low angular momentum, possibly inaccessible to the tidal-torque process if, e.g. there exist orbital resonances. The merger-driven gas inflows result in intense bursts of central star formation (e.g. Cox *et al.*, 2008)

[s]It is still debated if the origin of bars is internal/secular (e.g. Zana *et al.*, 2018) or external/tidal (e.g. Romano-Díaz *et al.*, 2008). However, the role of bars is more pronounced in isolated galaxies because they are likely the only mechanism that can produce large-scale torques.

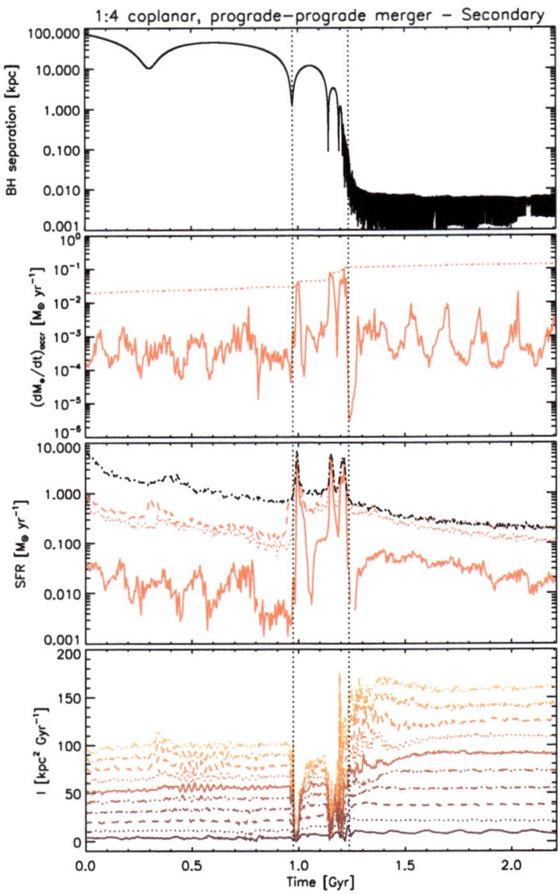

Fig. 6. Temporal evolution of the main quantities of and around the secondary SMBH of a minor merger from Capelo *et al.* (2015). In all panels, the two vertical dotted lines show (from left to right) the end of the stochastic stage and the beginning of the remnant stage (see text for more details). First panel: separation between the two SMBHs. Second panel: SMBH accretion rate (solid) and SMBH Eddington accretion rate (dotted). Third panel: global SFR across both galaxies (dash-dotted, black), and SFR in concentric spheres around the SMBH: 0–0.1 (solid, red), 0–1 (dotted, red), and 0–10 (dashed, red) kpc. Fourth panel: gas specific angular momentum magnitude in ten concentric 100-pc shells (0–1 kpc) around the local centre of mass near the SMBH. The colour varies from dark to light orange as the radius of the shell increases. The SMBH accretion rate peaks are correlated with the troughs of gas specific angular momentum. Moreover, nuclear SFR and SMBH accretion rate are proportional to each other (see Volonteri *et al.*, 2015a,b). Figure adapted from Capelo *et al.* (2015), reproduced by permission of Oxford University Press, on behalf of the Royal Astronomical Society.

and (assuming that the inflow proceeds down to sub-pc scales) SMBH accretion.

The study of SMBH accretion in galaxy mergers is inextricably linked to that of SMBH dynamics (specifically SMBH pairing), since the location of a SMBH can greatly influence the amount of gas available for accretion, especially in the case of the secondary SMBH.

In the past two decades, the dynamics and/or growth of SMBHs have been included in several merger simulations (e.g. Di Matteo *et al.*, 2005; Hopkins *et al.*, 2006; Younger *et al.*, 2008; Johansson *et al.*, 2009; Callegari *et al.*, 2009, 2011; Hayward *et al.*, 2014; Van Wassenhove *et al.*, 2014; Capelo *et al.*, 2015; Gabor *et al.*, 2016; Pfister *et al.*, 2017; Khan *et al.*, 2018), yielding a scenario exemplified in Figure 6, where the evolution of a minor merger from Capelo *et al.* (2015) is shown. The accretion rate on to both the primary and secondary SMBH (capped at a few times the Eddington rate), as well as the SFR, are initially both stochastic and relatively low. When the two galaxies reach the second pericentre (left-hand vertical dotted line), the gas loses much of its angular momentum and both the SMBH accretion rate (especially of the secondary) and SFR reach very high peaks, concurrent to a few pericentric passages. It is during this stage that single and dual (e.g. Van Wassenhove *et al.*, 2012; Blecha *et al.*, 2013; Capelo *et al.*, 2017; Blecha *et al.*, 2018) AGN activity are at their highest. Eventually, a remnant galaxy forms (right-hand vertical dotted line), the gas regains its angular momentum, and SMBH accretion rate and SFR go back to their initial levels.

The SMBHs continue their pairing process in the remnant galaxy, first on galactic scales (e.g. Tamburello *et al.*, 2017), then on circumnuclear scales (Dotti *et al.*, 2007; Fiacconi *et al.*, 2013; Souza Lima *et al.*, 2017) and, finally, on sub-pc scales, where they eventually coalesce (e.g. Begelman *et al.*, 1980), radiating GWs[t] which will be hopefully detected by LISA (e.g. eLISA Consortium *et al.* 2013; Klein *et al.* 2016; Barack *et al.* 2018; see Chapter 13).

[t]The anisotropic emission of GWs leads to a net emission of linear momentum and to a recoil of the SMBH remnant which, depending on the magnitude and orientation of the spins of the two merging SMBHs, can reach velocities of the order of 10^3 km s^{-1} (e.g. Campanelli *et al.*, 2007), comparable to the escape velocity of most hosts, thus leading to the possible ejection of the SMBH from its host.

4. Conclusions

A lot has happened since Michell first speculated on the existence of dark stars in 1784. For the past century, we have had a solid theoretical ground (Einstein's GR) to understand BH physics, i.e. the nature and structure of the BHs themselves. We have also had decades of observations across the EM spectrum of sources from all environments and up to $z > 7$, contributing to our understanding of BH astrophysics, i.e. the interaction between these dark objects and the surrounding matter. In 2015, we had the first detection of GWs, letting us listen to the inspiral and coalescence of stellar-mass BHs. Soon, with the current Pulsar Timing Arrays (PTAs; Hobbs *et al.*, 2010) and planned LISA, we should be able to listen to GW events related to IMBHs and SMBHs.

Despite the enormous advances in BH physics and astrophysics, we still do not know how these objects have formed in the first place. The remainder of this book will be mostly devoted to addressing this fundamental question. We begin laying down the framework in which the first SMBHs formed: how the first dark structures formed (Chapter 2), the chemistry of the primordial gas (Chapter 3), and how the first stars formed (Chapter 4). We then review several modes of SMBH formation — from the direct collapse of gas (Chapters 5 and 6), stellar clusters (Chapter 7), and supermassive stars (Chapter 8) — and discuss the statistical predictions from all these methods in Chapter 9. In Chapters 10 and 11, we review both the accretion and feedback of the first SMBHs. Finally, in Chapters 12–14, we describe the current and future observational status, both in the EM and GW domains.

Chapter 2

Formation and evolution of the cosmic large-scale structure

Björn Malte Schäfer

Zentrum für Astronomie der Universität Heidelberg,
Astronomisches Recheninstitut
Philosophenweg 12, 69120 Heidelberg, Germany
bjoern.malte.schaefer@uni-heidelberg.de

The formation of first objects in the Universe cannot be understood without the required context on cosmological structure formation. This chapter discusses the expansions dynamics of homogeneous and isotropic Friedmann–Lemaître–Robertson–Walker universes and the evolution of cosmic structures in the regime of weak gravity on small scales. The discussion includes the homogenous solution of the structure formation equations in the linear regime, which conserves the statistical properties of the initial conditions with their near-Gaussianity given by inflation, as well as processes in the nonlinear regime of structure formation with the accompanying generation of non-Gaussianities. Eulerian and Lagrangian perturbation theory are discussed as mathematical tools for describing structure formation in the weakly nonlinear regime, as well as the statistical tools for describing correlated random fields, both with Gaussian and weakly non-Gaussian statistics. Finally, halo-formation by spherical gravitational collapse and the Press–Schechter formalism for predicting the number density of dark matter haloes is introduced, as well as models for tidal interaction of haloes with the surrounding cosmic large-scale structure. The chapter thereby provides the tools to explore the formation of the dark matter halos in which the first black holes are forming.

1. Introduction

A discussion of the formation of the first black holes must begin with a discussion of cosmic structure formation, which sets the cosmic boundary conditions for astrophysical objects to form. The formation of cosmic structures is a fluid-mechanical phenomenon of self-gravity: Perturbations in the distribution of (dark) matter are amplified by accretion, and this accretion process reflects properties of the cosmological model.

This chapter gives an overview over cosmological models in Section 2 and statistical concepts for describing the distribution of matter in Section 3. Fluid mechanics in its application is discussed in Sections 4, 5 and 6. Open questions are reviewed in Section 7 before a summary in Section 8 concludes the chapter.

2. Cosmology

Modern cosmology (Peacock, 1999; Liddle and Lyth, 2000; Boerner, 2003; Cheng, 2005; Hobson *et al.*, 2006; Longair, 2008) is based on General Relativity as the gravitational theory which links the metric on the largest scales to the gravitating fluids. On scales larger than a few hundred Mpc/h the distribution of matter appears uniform to a freely falling observer who perceives the metric to be homogeneous and isotropic (the Copernican principle), and to vary only with time. The line element $\mathrm{d}s^2 = g_{\mu\nu}\mathrm{d}x^\mu \mathrm{d}x^\nu$ which incorporates the Copernican principle is the Robertson–Walker line element, is most conveniently expressed in spherical coordinates (t, χ, θ, ϕ) with the comoving distance χ as the radial coordinate,

$$\mathrm{d}s^2 = c^2\mathrm{d}t^2 - a^2(t)\left(\frac{1}{1-K\chi^2}\mathrm{d}\chi^2 + r^2\mathrm{d}\theta^2 + r^2\sin^2\theta\mathrm{d}\varphi^2\right), \tag{1}$$

with the scale factor $a(t)$, which describes the scaling of all physical distances with cosmic time t that every freely falling observer reads off from her or his clock. By convention, the scale factor assumes a value of $a = 1$ today and the relation to redshift is given by $a = 1/(1+z)$.

The gravitational field equation links the local curvature encoded in the Ricci-tensor $R_{\mu\nu}$ and the Ricci-scalar $R = g^{\mu\nu}R_{\mu\nu}$ to the energy-momentum tensor $T_{\mu\nu}$ of the cosmological fluids,

$$G_{\mu\nu} = \frac{8\pi G}{c^4}T_{\mu\nu} \quad \text{with} \quad G_{\mu\nu} = R_{\mu\nu} - \frac{R}{2}g_{\mu\nu}, \tag{2}$$

while obeying local energy-momentum conservation $\nabla^\mu T_{\mu\nu} = 0$ expressed through a covariant divergence. The source $T_{\mu\nu}$ of the gravitational field is assumed to be an ideal, relativistic fluid which is parameterised by density and pressure,

$$T_{\mu\nu} = \left(\rho c^2 + p\right)\upsilon_\mu\upsilon_\nu - g_{\mu\nu}p, \tag{3}$$

and depends on the 4-velocity v_μ which is simply $(c, \vec{0})$ for a comoving observer. An ideal fluid follows naturally from a Lagrange density $\mathcal{L} = \mathcal{L}(\varphi, \nabla^\mu\varphi)$ that depends only on the fields φ and their derivatives $\nabla^\mu\varphi$, but not explicitly on the coordinates x^μ.

Substitution of the Robertson–Walker-metric into the field equation yields the Friedmann equations. Firstly,

$$\left(\frac{\dot{a}}{a}\right)^2 = \frac{8\pi G}{3}\rho - \frac{K}{a^2}, \tag{4}$$

as an expression for the time-derivative of the logarithmic scale factor, i.e. the Hubble function $H(a) = \dot{a}/a$. The second Friedmann-equation expresses the deceleration parameter $q(a) = -\ddot{a}a/\dot{a}^2$ in terms of the trace of the energy-momentum tensor,

$$\frac{\ddot{a}}{a} = -\frac{4\pi G}{3}(\rho + 3p). \tag{5}$$

These two relations constitute the Friedmann–Lemaître cosmologies that are based on the spherically symmetric, spatially homogeneous metric as a solution to the field equation of General Relativity, with homogeneous ideal fluids, whose density ρ and pressure p contribute to the energy-momentum tensor $T_{\mu\nu}$. It is worth noting that they are among the simplest solutions of General Relativity and constitute a system with pure Ricci-curvature.

Relativistic local energy-momentum conservation $\nabla^\mu T_{\mu\nu} = 0$ can be recovered from the Friedmann-equations, because the gravitational field equation is constructed to respect it, either in the form of the adiabatic equation,

$$\frac{\mathrm{d}}{\mathrm{d}a}\left(a^3\rho(a)\right) + p\frac{\mathrm{d}}{\mathrm{d}a}\left(a^3\right) = 0 \tag{6}$$

or, equivalently, in the shape of a continuity equation,

$$\dot{\rho} + 3H(a)\,(p + \rho) = 0, \tag{7}$$

under the assumption of ideal, relativistic fluids and the Friedmann–Lemaître–Robertson–Walker (FLRW)-symmetries.

Empirically, the relation between pressure and density is parametrised by the equation of state parameter w:

$$p = w\,\rho c^2, \tag{8}$$

because this relation is the only free choice under the FLRW-symmetry assumptions. Pressureless dark matter is characterised by $w = 0$ and the value $w \equiv -1$ corresponds to the cosmological constant Λ. For

Fluid	$\rho(a)$	$H(a)$	w	q
radiation	$\propto a^{-4}$	$\propto a^{-2}$	$+1/3$	1
matter	$\propto a^{-3}$	$\propto a^{-3/2}$	0	$1/2$
curvature	$\propto a^{-2}$	$\propto a^{-1}$	$-1/3$	0
dark energy	$\propto a^{-2\ldots 0}$	$\propto a^{-1\ldots 0}$	$-1/3\ldots-1$	$0\ldots-1$
Λ	$=$ const	$=$ const	-1	-1
phantom energy	$\propto a^{q}, q>0$	grows	<-1	<-1

the curvature K to vanish, the density ρ has to be equal to the critical density $\rho_{\rm crit}(a) = 3H_0^2/(8\pi G) \simeq 1.8784 \times 10^{-26}$ kg/m^3 $\simeq 2.7745 \times 10^{11}h^2\ M_\odot/\mathrm{Mpc}^3$, which implies the scaling of the density parameter $\Omega_w(a) = \rho(a)/\rho_{\rm crit}(a)$ of each fluid,

$$\frac{\Omega_w(a)}{\Omega_w} = \frac{H_0^2}{a^{3(1+w)}H^2(a)}. \tag{9}$$

The relation between equation of state w and deceleration parameter q is given by

$$2(1+q) = 3(1+w), \tag{10}$$

if the density of the fluid corresponds to the critical density. Fluids with $w < -1/3$ are interesting as an explanation for the accelerated expansion of the universe observed today, as $\ddot{a} > 0$ and $q < 0$, and the cosmological constant in particular is described by a fluid $w = -1$, although it is formally already included as a term $\Lambda g_{\mu\nu}$ in the field equation.

Curvature can be formally described by including a fluid with density $\Omega_K = 1 - \Omega_w$ and assigning it an equation of state of $w = -1/3$, although curvature is not a physical substance and although Ω_K can be negative, in contrast to all other density parameters. In addition, Nature does not use this particular degree of freedom, as cosmic inflation has driven the parameter Ω_K to very small values, as current limits on Ω_K are well below the percent level.

Table 1 summarises the most important cosmological fluids, their equation of state, their dependence on scale factor in the Hubble function as well as their deceleration parameter.

2.1. *Dark energy and its parametrisation*

Using the results from the last section, the Hubble function $H(a) = \mathrm{d}\ln a/dt$ for a universe filled with radiation, pressureless dark matter and dark energy with a time-evolving equation of state $w(a)$ can be constructed to be

$$\frac{H^2(a)}{H_0^2} = \frac{\Omega_\gamma}{a^4} + \frac{\Omega_m}{a^3} + \Omega_\varphi \exp\left(3\int_a^1 \mathrm{d}\ln a\ [1+w(a)]\right)$$

$$\overset{w(a)=w_0}{=} \frac{\Omega_\gamma}{a^4} + \frac{\Omega_m}{a^3} + \frac{\Omega_\varphi}{a^{3(1+w_0)}}, \tag{11}$$

in a spatially flat dark energy cosmology with the the matter density Ω_m and the dark energy density $\Omega_\varphi = 1 - \Omega_\gamma - \Omega_m$. Today's estimates $\Omega_m = 0.25$ and $\Omega_\varphi = 0.75$ allow to quantify the time of matter-dark energy equality $a_{m\varphi} \simeq 0.7$ (for $w = -1$), defined by $\Omega_m(a_{m\varphi}) = \Omega_\varphi(a_{m\varphi})$, while the radiation density is small, $\Omega_\gamma \simeq 10^{-5}$.

The relation between comoving distance χ (given in terms of the Hubble distance $\chi_H = c/H_0 = 2.9969$ Gpc$/h$) and scale factor a is then given by

$$\chi = c\int_a^1 \frac{\mathrm{d}a}{a^2 H(a)}, \tag{12}$$

with the speed of light c. The dark energy equation of state $w(a)$ is conveniently approximated by its first order Taylor expansion with respect to the scale-factor a,

$$w(a) = w_0 + (1-a)w_a, \tag{13}$$

introduced by Chevallier and Polarski (2001) and Linder and Jenkins (2003), for non-interacting dark energy models with a slow time evolution $w(a)$. Incidentially, this parameterisation allows a complete integration of the Hubble-function $H(a)$. The conformal time η in units of the Hubble time $t_H = 1/H_0$ follows in analogy to the comoving distance χ from the differential $\mathrm{d}\eta = \mathrm{d}t/a = \mathrm{d}a/(a^2H(a))$,

$$\eta = \int_a^1 \frac{\mathrm{d}a}{a^2 H(a)}, \tag{14}$$

such that $\chi = c\eta$ and one recovers Minkowskian light propagation in these coordinates.

Figure 1 shows the logarithmic derivative $\mathrm{d}\ln\tilde{H}/\mathrm{d}\ln a = 3/2 + \mathrm{d}\ln H/\mathrm{d}\ln a$ of the Hubble function, where the added term 3/2 removes the rapid scaling $H \propto a^{-3/2}$ in the matter dominated epoch, as well as the matter density parameter $\Omega_m(a)$ as a function of time and for a range of dark energy equation of state parameters.

3. Statistical description of structures

The fluctuations of the cosmic density field $\delta(\mathbf{x})$, for which one chooses the relative deviation of the density field $\rho(\mathbf{x})$ from the mean background

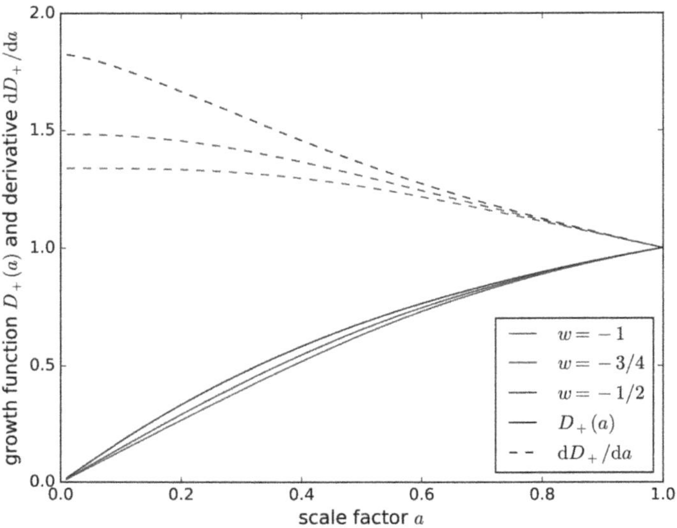

Fig. 1. The logarithmic derivative $3/2{+}\mathrm{d}\ln H/\mathrm{d}\ln a$ (solid lines) and the matter density parameter $\Omega_m(a)$ (dashed lines), for a cosmology with a cosmological constant Λ with (from bottom to top): $w_0 = -1$, dark energy with $w_0 = -3/4$ and with $w_0 = -1/2$.

density $\langle\rho\rangle = \Omega_m \rho_{\mathrm{crit}}$,

$$\delta(\mathbf{x}) = \frac{\rho(\mathbf{x}) - \langle\rho\rangle}{\langle\rho\rangle}, \tag{15}$$

are assumed to be Gaussian with a certain correlation length, meaning that the probability of finding the amplitudes $\delta(\mathbf{x}_1)$ and $\delta(\mathbf{x}_2)$ and positions $\mathbf{x}_1$ and $\mathbf{x}_2$ in a hypothetical ensemble of universes is given by a bivariate Gaussian probability density,

$$p(\delta(\mathbf{x_1}), \delta(\mathbf{x_2})) = \frac{1}{\sqrt{(2\pi)^2 \det(Q)}} \exp\left[-\frac{1}{2}\begin{pmatrix}\delta(\mathbf{x}_1)\\ \delta(\mathbf{x}_2)\end{pmatrix}^t C^{-1}\begin{pmatrix}\delta(\mathbf{x}_1)\\ \delta(\mathbf{x}_2)\end{pmatrix}\right] \tag{16}$$

with the covariance matrix C:

$$C = \begin{pmatrix}\langle\delta(\mathbf{x}_1)^2\rangle & \langle\delta(\mathbf{x}_1)\delta(\mathbf{x}_2)\rangle\\ \langle\delta(\mathbf{x}_2)\delta(\mathbf{x}_1)\rangle & \langle\delta(\mathbf{x}_2)^2\rangle\end{pmatrix}. \tag{17}$$

The off-diagonal variance in C is the correlation function $\xi(\mathbf{x}_1, \mathbf{x}_2) \equiv \langle\delta(\mathbf{x}_1)\delta(\mathbf{x}_2)\rangle$ of the random field, which describes how fast with increasing distance $|\mathbf{x}_2 - \mathbf{x}_1|$ the field loses its memory on the amplitude at $\mathbf{x}_1$. A length scale in $\xi(\mathbf{x}_1, \mathbf{x}_2)$ can be interpreted as a correlation length. Due to the Cauchy–Schwarz inequality,

$$\langle\delta(\mathbf{x}_1)\delta(\mathbf{x}_2)\rangle^2 \leq \langle\delta(\mathbf{x}_1)^2\rangle\langle\delta(\mathbf{x}_2)^2\rangle \quad \rightarrow \quad r = \frac{\langle\delta(\mathbf{x}_1)\delta(\mathbf{x}_2)\rangle}{\sqrt{\langle\delta(\mathbf{x}_1)^2\rangle\langle\delta(\mathbf{x}_2)^2\rangle}} \tag{18}$$

the correlation function is always smaller than the geometrical mean of the variances at a single point, i.e. the covariance C is positive definite, and the Pearson-correlation coefficient $|r|$ is smaller than unity.

Clearly, if the correlation function $\xi(\mathbf{x}_1, \mathbf{x}_2)$ vanishes, the Gaussian probability density separates $p(\delta(\mathbf{x_1}), \delta(\mathbf{x_2})) = p(\delta(\mathbf{x_1}))p(\delta(\mathbf{x_2}))$ and the amplitudes are mutually uncorrelated and the field amplitudes follow independently from univariate Gaussian distributions.

In homogeneous Gaussian random fields the correlation function $\xi(\mathbf{x}_1, \mathbf{x}_2)$ only depends on the separation $\mathbf{r} = \mathbf{x}_2 - \mathbf{x}_2$ between the two points and not the location. Then, the two variances $\langle\delta(\mathbf{x}_1)^2\rangle$ and $\langle\delta(\mathbf{x}_2)^2\rangle$ are identical. If in addition the random field has isotropic fluctuation properties, the correlation function only depends on the magnitude of $\mathbf{r}$ and not its direction. In all of the above, the averaging brackets $\langle\ldots\rangle$ denote the average over realisations of the density field.

The knowledge of the variance is sufficient because all moments of a Gaussian distributed random variable with zero mean are proportional to the variance, $\langle\delta^{2n}\rangle \propto \langle\delta^2\rangle^n$. Hence the characteristic function $\varphi(t) = \int \mathrm{d}\delta p(\delta)\exp(\mathrm{i}t\delta) = \sum_n \langle\delta^n\rangle(\mathrm{i}t)/n!$ only requires the estimation of the variance $\langle\delta^2\rangle$ for reconstructing $p(\delta)\mathrm{d}\delta$ from the moments $\langle\delta^{2n}\rangle$ by inverse Fourier transform.

Homogeneous random fields possess the interesting property of mutually uncorrelated Fourier modes. Applying a Fourier-transform to the density contrast $\delta(\mathbf{x})$,

$$\delta(\mathbf{k}) = \int \mathrm{d}^3x\, \delta(\mathbf{x})\exp(-\mathrm{i}\mathbf{kx}) \quad \leftrightarrow \quad \delta(\mathbf{x}) = \int \frac{\mathrm{d}^3k}{(2\pi)^3}\, \delta(\mathbf{k})\exp(+\mathrm{i}\mathbf{kx}), \tag{19}$$

and computing the variance between two Fourier modes $\delta(\mathbf{k}_1)$ and $\delta(\mathbf{k}_2)$

$$\langle\delta(\mathbf{k}_1)\delta^*(\mathbf{k}_2)\rangle = (2\pi)^3\delta_D(\mathbf{k}_1 - \mathbf{k}_2)P(\mathbf{k}_1), \tag{20}$$

shows that it is proportional to the Dirac δ-function, indicating their statistical independence. The proportionality is given by the spectrum,

$$P(\mathbf{k}) = \int \mathrm{d}^3r\, \xi(\mathbf{r})\exp(-\mathrm{i}\mathbf{kr}), \tag{21}$$

which follows as the Fourier-transform of the correlation function $\xi(\mathbf{r})$. If in addition the random field is isotropic, $P(k)$ only depends on the wave number k instead of the wave vector $\mathbf{k}$. In this case, the angular integrations in Equation (20) can be carried out by introducing spherical coordinates in Fourier-space, yielding:

$$P(k) = 2\pi \int_0^\infty r^2 \mathrm{d}r\, \xi(r) j_0(kr), \tag{22}$$

with the spherical Bessel function of the first kind $j_0(kr)$ of order $\ell = 0$, being equal to $\mathrm{sinc}(kr) = \sin(kr)/(kr)$ (Abramowitz and Stegun, 1972).

Cosmological inflation provides a mechanism for generating Gaussian fluctuation fields with the spectrum $P(k)$,

$$P(k) \propto k^{n_s} T^2(k) \tag{23}$$

(Baumann and Peiris, 2008; Langlois, 2010), with the cold dark matter (CDM) transfer function $T(k)$. $T(k)$ describes the scale-dependent suppression of the growth of small-scale modes between horizon-entry and matter-radiation equality by the Meszaros mechanism. It is well approximated with the polynomial fit proposed by Bardeen *et al.* (1986),

$$T(q) = \frac{\ln(1 + 2.34q)}{2.34q} \left(1 + 3.89q + (16.1q)^2 + (5.46q)^3 + (6.71q)^4\right)^{-\frac{1}{4}},$$

or the more accurate fit described by Eisenstein and Hu (1998, 1999) incorporating baryonic wiggles for flat cosmological models with low matter density Ω_m. The asymptotic behaviour of the transfer function is such that $T(k) \propto$ const for $k \ll 1$ and $T(k) \propto k^{-2}$ at $k \gg 1$, such that $P(k) \propto k^{n_s}$ on large scales and $P(k) \propto k^{n_s - 4}$ on small scales. In particular the scale-free behaviour of $P(k)$ on small scales is a feature of cold dark matter models, because any thermal motion would wipe out structures on small scales and would lead to an exponential cut-off in the spectrum.

The wave vector is rescaled with the shape parameter $\Gamma \simeq \Omega_m h$, which corresponds to the horizon size at the time of matter-radiation equality $a_{\gamma m}$, and describes the peak shape of the CDM power spectrum $P(k)$. The corrections of a non-negligible baryon density Ω_b on the peak shape are described by Sugiyama (1995),

$$\Gamma = \Omega_m h \exp\left[-\Omega_b \left(1 + \frac{\sqrt{2h}}{\Omega_m}\right)\right], \tag{24}$$

where Γ is measured in units of $(\mathrm{Mpc}/h)^{-1}$, such that $q = k/\Gamma$ is a dimensionless wave vector. The spectrum is usually normalised to the variance of the linearly evolved density field at zero redshift on a scale of $R = 8$ Mpc/h,

$$\sigma_R^2 = \frac{1}{2\pi^2} \int_0^\infty \mathrm{d}k\, k^2 P(k) W^2(kR), \tag{25}$$

with a Fourier-transformed spherical top hat filter function, $W(x) = 3j_1(x)/x$. $j_1(x)$ is the spherical Bessel function of the first kind of order $\ell = 1$ (Abramowitz and Stegun, 1972; Arfken and Weber, 2005). This particular definition of $\sigma(R)$, along with the fact that the power spectrum has the dimension of a volume, motivates the definition of the dimensionless power spectrum $\Delta^2(k) \propto k^3 P(k)$,

$$\Delta^2(k) = \frac{k^3}{2\pi^2} P(k) \quad \rightarrow \quad \sigma_R^2 = \int_0^\infty \mathrm{d}\ln k\, \Delta^2(k) W^2(kR), \tag{26}$$

such that $\Delta^2(k)$ reflects the fluctuation variance per logarithmic band in k, $\mathrm{d}\sigma_R^2/\mathrm{d}\ln k \propto \Delta^2$. As the spectrum $P(k)$ has the unit of a volume and is conventionally given in units of $(\mathrm{Mpc}/h)^3$, the variance $\Delta^2(k)$ is dimensionless.

4. Relativistic fluid mechanics and the non-relativistic limit

The equations of relativistic fluid mechanics results from relativistic energy momentum conservation $\nabla^\mu T_{\mu\nu} = 0$ by projecting this relation onto the 4-velocity υ_μ of an observer (which in our case is a FLRW-observer) and perpendicular to it. Specifically, $\upsilon^\nu \nabla^\mu T_{\mu\nu} = 0$ yields the continuity equation and $(\upsilon^\alpha \upsilon^\nu / c^2 - g^{\alpha\nu}) \nabla^\mu T_{\mu\nu} = 0$ the Euler-equation.

If the perturbations of the gravitational field are weak, $\Phi \ll c^2$, the metric will only contain terms involving the Newtonian gravitational potential Φ which enters the Euler-equation through a gradient $\nabla\Phi$ as the remnant of the covariant derivative ∇^μ, and the error by using an FLRW-observer's 4-velocity is small, implying in particular that one can use cosmic time instead of the proper time if in addition the fluid velocities are small compared to c. The projections will only use in this limit the covariant derivative ∇^μ defined by the FLRW-metric and ignore perturbations in the metric, reducing the equations of fluid mechanics by Newtonian relations on an expanding background, with the cosmic time as the global time variable.

In the limit of small metric perturbations the gravitational field equation will be replaced by $\Box\Phi = 4\pi G\rho$, with only the matter density being active as the source of the gravitational field, while pressure is suppressed by a factor of $1/c^2$ relative to the matter density. Furthermore, since most of the gravitating matter is non-relativistic, pressure does not play a role as a source of gravitational perturbations. If in addition one assumes that retardation effects in the propagation of the gravitational field is not important on small scales, one recovers the classical Poisson-equation $\Delta\Phi = 4\pi G\rho$ as the gravitational field equation. The retardation scale in this is set by the Hubble scale c/H_0.

Specifically, the Poisson-equation reformulated in terms of comoving distances reads for the dimensionless potential $\varphi = \Phi/c^2$

$$\Delta\varphi = \frac{3}{2}\Omega_m(\eta)\,\frac{(aH)^2}{c^2}\delta = \frac{3\Omega_m}{2\chi_H^2}\frac{\delta}{a}, \tag{27}$$

where the strength of the gravitational potential is influenced by the cosmology through the evolution of the ambient matter density and the scale is set by the Hubble distance $\chi_H = c/H_0$. Newton's constant G has been replaced with the definition of the critical density, and absorbed into the ambient density into $\Omega_m(\eta)$ with the conformal time η as the time variable. The replacement of $\Omega_m(\eta)$ by Ω_m uses the adiabatic evolution of the matter density in Equation (9).

5. Linear structure formation

5.1. *Linearised structure formation equations*

The formation of cosmic structures is based on gravitational amplification of seed perturbations in the cosmic density field. Following Bernardeau *et al.* (2002), the most convenient choice of variables is to use the physical density field, the velocity fields and the potential on one side, the conformal time as the time variable and the comoving distance for the differential operators on the other. Being a hydrodynamical self-gravitating phenomenon, structure formation is described in the comoving frame by the system of differential equations composed of (*i*) the continuity equation

$$\frac{\partial}{\partial\eta}\delta + \mathrm{div}\,[(1+\delta)\upsilon] = 0, \tag{28}$$

which relates the time-evolution of the density field to the divergence of the matter fluxes $(1+\delta)\upsilon$, (*ii*) the Euler-equation

$$\frac{\partial}{\partial\eta}\upsilon + aH\upsilon + (\upsilon\nabla)\upsilon = -\nabla\Phi, \tag{29}$$

which describes the evolution of the peculiar velocity field υ from the gradient $\nabla\Phi$ of the peculiar gravitational potential Φ, acting on a fluid element, and finally (*iii*) the comoving Poisson-equation

$$\Delta\Phi = \frac{3H_0^2\Omega_m}{2a}\delta, \tag{30}$$

which gives the gravitational potential Φ induced by the matter distribution δ. The three equations are sufficient to describe the dynamics of the three relevant fields δ, υ and Φ, because there are no dissipative and pressure forces due to the collisionlessness of dark matter, and it is not necessary to track the energy balance or to introduce an equation of state parametrising the pressure-density relation. The expanding background is described by the Hubble-function $H(a)$ and the usage of Newtonian dynamics and Newtonian gravity is well justified due to the small velocities $\upsilon \ll c$ and the smallness of gravitational potentials $\Phi \ll c^2$ on sub-horizon scales $k \geq 2\pi/\chi_H$, where retardation effects do not play a role.

Linearisation of the structure formation equations by substituting a perturbative expansion of the density- and velocity field yields the linearised continuity equation,

$$\frac{\partial}{\partial \eta}\delta + \mathrm{div}\upsilon = 0, \tag{31}$$

and the linearised Euler-equation,

$$\frac{\partial}{\partial \eta}\upsilon + aH\upsilon = -\nabla\Phi, \tag{32}$$

which are valid if the overdensity δ is small, $\delta \ll 1$. The Newtonian Poisson-equation is always linear, or the superposition principle would not hold.

5.2. *Growth equation and its solution*

The divergence of the linearised Euler-equation can be combined with the linearised continuity equation and its time derivative as well as the Poisson equation to yield the growth equation as a second-order ordinary differential equation. As spatial derivatives cancel, the solution describes the homogeneous growth $\delta(\mathbf{x}, a) = D_+(a)\delta(\mathbf{x}, a = 1)$ of the density field in the linear regime of structure formation, $|\delta| \ll 1$ (Peebles, 1980a; Turner and White, 1997; Linder and Jenkins, 2003),

$$\frac{\mathrm{d}^2}{\mathrm{d}a^2}D_+(a) + \frac{1}{a}\left(3 + \frac{\mathrm{d}\ln H}{\mathrm{d}\ln a}\right)\frac{\mathrm{d}}{\mathrm{d}a}D_+(a) = \frac{3}{2a^2}\Omega_m(a)D_+(a). \tag{33}$$

The growing solution $D_+(a)$ of the growth equation's two solutions is referred to as the growth function and assumes the simple solution $D_+(a) = a$ in a Standard Cold Dark Matter (SCDM) cosmology ($\Omega_m = 1$, $\Omega_\varphi = 0$) , where $\Omega_m \equiv 1$ and $H(a)/H_0 = a^{-3/2}$. This scaling motivates the usage of the scale factor a as the preferred time variable, and suggests to transform the differentials with $\mathrm{d}/\mathrm{d}\eta = a^2H(a)\mathrm{d}/\mathrm{d}a$. For that reason, the initial conditions for solving the growth equation are given by $D_+(0) = 0$ and $\mathrm{d}D_+(0)/\mathrm{d}a = 1$, due to the domination of the Ω_m-term in $H(a)$ at early times. It is a peculiar result that, due to the additional factor $1/a$ in the Poisson-equation, gravitational potentials are constant during linear structure formation in cosmologies with $\Omega_m = 1$. By convention, the growth function is normalised to the value $D_+ = 1$ today, $a = 1$.

$\Omega_m(a)$ on the right-hand side acts as a driving term, because gravitational potentials are strong if $\Omega_m(a)$ is large, making it easier for the cosmic large-scale structure to form. The term $3 + \mathrm{d}\ln H/\mathrm{d}\ln a$ is sometimes referred to as Hubble-drag, which suppresses the formation of structures if their dynamical time-scale is larger than the time-scale of the Hubble expansion. It is worth emphasising, however, that only a change in the expansion rate of the Universe affects structure formation: Expansion at a constant Hubble-rate would not affect the contrast δ of structures at all.

The second solution $D_-(a)$ scales $\propto 1/a$ in the matter dominated phase of the Universe and does not play a role at late times due to its fast decay.

It is worth noting that the Hubble function $H(a)$ and the growth function $D_+(a)$ convey the same information, as one can compute one from the other. In fact, Peebles (1980b) shows that $D_+(a)$ is given by

$$D_+(a) \propto H(a) \int_a^1 \frac{\mathrm{d}a}{(aH(a))^3}. \tag{34}$$

This is due to the fact that in standard cosmologies the dependence $\Omega_m(a)$ on time is given entirely by the Hubble function $H(a)$, and that in Newtonian gravity the same term determines the term on the right-hand side driving structure formation. This ceases to be applicable, however, in modified gravity theories, or in cosmologies with interactions between the cosmological fluids. Naturally, a signature of these cosmologies would be a mismatch between the information conveyed by $H(a)$ and $D_+(a)$.

Homogeneous structure formation corresponds to independently growing Fourier modes,

$$\delta(\mathbf{x}, a) = D_+(a)\delta(\mathbf{x}, a = 1) \longrightarrow \delta(\mathbf{k}, a) = D_+(a)\delta(\mathbf{k}, a = 1), \tag{35}$$

conserving every statistical property of the initial conditions, in particular Gaussianity. The Gaussianity of the initial density perturbations is a consequence of inflation, where a large number of uncorrelated quantum fluctuations are superimposed, yielding a Gaussian amplitude distribution due to the central limit theorem. In fact, homogeneous growth in the linear regime is the reason why investigation of inflationary processes in structure is possible by observing the large-scale structure today, even after the cosmic time $1/H_0$ has passed.

Figure 2 shows the growth function $D_+(a)$ in comparison to the derivative D_+/a for cosmologies with $\Omega_m = 0.25$ and different values of the equation of state w of the dark energy fluid.

A convenient way for approximating the growth function is the γ-parameter, introduced by Peebles (1980b) in the study of peculiar velocities:

$$\frac{\mathrm{d}\ln D_+}{\mathrm{d}\ln a} \simeq \Omega_m(a)^\gamma, \tag{36}$$

with $\gamma \simeq 0.6$ in ΛCDM. Solving this equation for the growth function yields

$$D_+(a) = \exp\left(\int_0^a \mathrm{d}\ln a\,\Omega_m(a)^\gamma\right). \tag{37}$$

In dynamic dark energy models, γ can be approximated by $\gamma = 0.55 + 0.05(1 + w(z = 1))$ with the dark energy equation of state parameter taken at unit redshift (Linder, 2005).

The effect of adding a fluid with a negative equation of state is a slower growth in the recent cosmic past and a faster growth in the remote past (if the growth function is normalised to unity today). Figure 2 shows the growth function $D_+(a)$ and the evolution of the potentials D_+/a. This dependence on the cosmological model is generated by the scaling of Ω_m with time and the magnitude of the term $3+\mathrm{d}\ln H/\mathrm{d}\ln a$, which can be shown to be equal to $2 - q$ with the deceleration parameter q. Physically, they originate from the comoving Poisson equation and the cosmology-dependence of the relativistic Euler-equation: A higher matter density generates stronger gravitational fields for a given perturbation in δ, thus enhancing structure formation. An accelerating universe makes it difficult for structures to form, and ultimately cosmic structure formation is truncated by the dark energy-domination of the cosmic expansion. Conversely, the early decelerating phase of the Universe during matter domination made it easy for structures to form.

5.3. *Peculiar velocities*

Matter fluxes in the large-scale structure drive structure formation: If they converge, they transport matter into a volume and increase the local density, according to the continuity equation. In order to investigate the properties of matter fluxes one can carry out a Helmholtz-decomposition of the velocity field into its curl and gradient components $\theta = \mathrm{div}\upsilon$ and $\omega = \mathrm{rot}\upsilon$. From the Euler-equation one obtains the evolution equation for the divergence of the matter fluxes (Bernardeau *et al.*, 2002),

$$\frac{\partial}{\partial\eta}\theta + aH\theta + \frac{3H_0^2\Omega_m}{2a}\delta = 0 \tag{38}$$

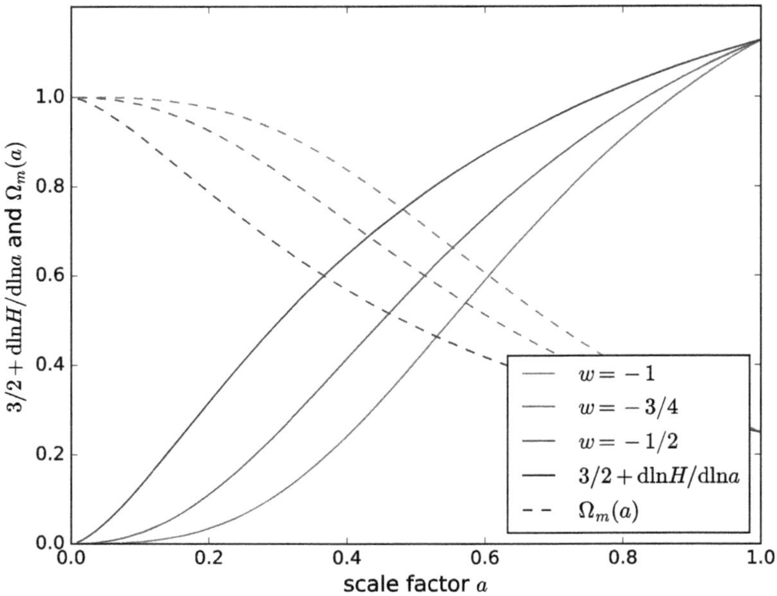

Fig. 2. Linear growth function $D_+(a)$ (solid lines) and the evolution D_+/a of potentials (dashed lines), for varying equation of state parameter w of the dark energy fluid: cosmological constant Λ with $w_0 = -1$, dark energy with $w_0 = -3/4$ and with $w_0 = -1/2$. The linear growth function is highest for lower values of ω_0; vice versa for the potentials.

and the corresponding equation for the vorticity ω,

$$\frac{\partial}{\partial \eta}\omega + aH\omega = 0. \tag{39}$$

With the definition of the differential of the conformal time, $\mathrm{d}a = a^2 H \mathrm{d}\eta$, one immediately notices that $\mathrm{d}\ln\omega = -\mathrm{d}\ln\ a$, and hence $\omega \propto 1/a$ during matter domination: Vorticity cannot be generated in linear structure formation in collisionless fluids, and the flows are necessarily laminar. The divergence θ can be linked to the evolution of the density field using the continuity equation

$$\theta = -aH\frac{\mathrm{d}\ln D_+}{\mathrm{d}\ln a}\delta, \tag{40}$$

which underlines the fact that in the linear regime of structure formation, the velocity field is the gradient of a potential, and the velocity divergence is proportional to the density contrast.

6. Nonlinear structure formation

6.1. *Phenomenology of nonlinear structure formations*

As long the structure formation is linear, the growth is homogeneous and conserves the Gaussianity of the initial conditions. Nonlinear structure formation implies inhomogeneous growth and the emergence of non-Gaussian features, which is illustrated by a number of arguments: Nonlinearity implies inhomogeneity, because if e.g. a void reaches underdensities close to $\delta \simeq -1$ (corresponding to $\rho \simeq 0$), the linearisation fails and the growth has to slow down locally. Inhomogeneity implies non-Gaussianity because the initially Gaussian distribution $p(\delta)\mathrm{d}\delta$ becomes wider with increasing amplitudes δ, but the density contrast δ cannot be more negative than -1, requiring the amplitude distribution $p(\delta)\mathrm{d}\delta$ to become asymmetric and to acquire a nonzero skewness. For completing the argument one immediately notices that in inhomogeneous growth, i.e. a position dependence of the growth rate $D_+(\mathbf{x}, a)$, the Fourier-modes $\delta(\mathbf{k}, a)$ become coupled, violating the central limit theorem such that the superposition of Fourier-modes yields a non-Gaussian amplitude distribution.

6.2. *Eulerian perturbation theory*

The nonlinearities in the continuity and Euler-equation make a closed analytical solution impossible. It is possible, however, to obtain a perturbative solution (Buchert and Götz, 1987; Bouchet *et al.*, 1992; Sahni and Coles, 1995; Catelan *et al.*, 1995; Ehlers and Buchert, 1997; Bernardeau *et al.*, 2002) to the structure formation equations, which contains the mode coupling mechanism and describes the generation of non-Gaussianities in nonlinear structure formation. The starting point is a perturbative expansion of the type

$$\delta(\mathbf{x}, a) = \sum_{n=1}^{\infty} \delta^{(n)}(\mathbf{x}, a) \simeq \sum_{n=1}^{\infty} D_+^n(a)\delta^{(n)}(\mathbf{x}) \tag{41}$$

where the last step holds exactly in universes with $\Omega_m = 1$ and is approximately valid in dark energy cosmologies. $\delta^{(n)}(\mathbf{x})$ is proportional to the nth power $\delta(\mathbf{x})^n$ of the initial conditions. The nonlinearities in the continuity- and the Euler-equation translate to convolutions of the density and the velocity fields in Fourier space which couple the individual Fourier modes, violating the central limit theorem and therefore violating Gaussianity. It is worth noting that in the perturbative expansion each field $\delta^{(n)}$ grows homogeneously at the rate $D_+^n(a)$, but the sum does not.

Substituting a perturbation series of the type of Equation (41) for the density and velocity fields into the fully nonlinear, Fourier-transformed structure formation equations, and sorting the factors according to the exponent n yields:

$$\delta^{(n)}(\mathbf{k}) = \int \mathrm{d}^3 q_1 \ldots \int \mathrm{d}^3 q_n \, \delta_D \left(\mathbf{k} - \sum_{i=1}^{n} \mathbf{q}_i\right) F_n(\mathbf{q}_1, \ldots, \mathbf{q}_n) \prod_{i=1}^{n} \delta(\mathbf{q}_i) \tag{42}$$

with the mode coupling function $F_n(\mathbf{q}_1, \ldots, \mathbf{q}_n)$, for which a recursion relation can be obtained. The lowest order symmetrised solutions for F_n are $F_1 = 1$ and

$$F_2(\mathbf{q}_1, \mathbf{q}_2) = \frac{5}{7} + \frac{x}{2}\left(\frac{q_1}{q_2} + \frac{q_2}{q_1}\right) + \frac{2}{7}x^2 \quad \text{with} \quad x = \frac{\mathbf{q}_1 \cdot \mathbf{q}_2}{q_1 q_2} \tag{43}$$

being the cosine of the angle between $\mathbf{q}_1$ and $\mathbf{q}_2$. Assuming $q_1 = q_2$ for simplicity, the mode coupling function F_2 attains the largest value of $F_2 = 2$ if the wave vectors are parallel ($x = +1$), an intermediate value of $F_2 = 5/7$ if $\mathbf{q}_1 \perp \mathbf{q}_2$ ($x = 0$) and the smallest value of $F_2 = 0$ if the wave vectors are antiparallel ($x = -1$). Varying the wave numbers at fixed separation angle x shows that F_2 is smallest if $q_1 = q_2$, and that the mode coupling increases if the wave numbers are chosen differently. From this point of view, mode-coupling bears resemblance to a resonance phenomenon, where modes with identical direction of propagation experience the strongest coupling. The perturbative solution to the system of equations (28) and (29) in terms of a perturbation series in δ and υ is possible due to their renormalisation properties, which hold exactly in the case of SCDM ($\Omega_m = 1$, $\Omega_\varphi = 0$) and approximately for dark energy cosmologies which are the topic of a number of papers (Crocce and Scoccimarro, 2006a,b; Bernardeau *et al.*, 2008). In these cosmologies, the mode-coupling kernels acquire a slow time dependence themselves, which is measured in numerical simulations.

6.3. *Lagrangian perturbation theory*

An alternative way of formulating the perturbative, translinear dynamics of the cosmic density field is Lagrangian perturbation theory, where the central objects are the particle trajectories which link the initial positions $\mathbf{q}$ to their positions $\mathbf{x}$ at time η rather than the density- and velocity fields (Buchert and Götz, 1987; Ehlers and Buchert, 1997; Buchert, 1989, 1994). The mapping between the initial position $\mathbf{q}$ of a particle and the position $\mathbf{x}$ ad time η is at lowest order given by

$$\mathbf{x}(\mathbf{q}, \eta) = \mathbf{q} - \nabla\Psi(\mathbf{q}, \eta), \tag{44}$$

with the displacement potential Ψ which describes the potential flows developing in the large-scale structure in linear structure formation. The lowest order perturbative mapping is referred to as the Zel'dovich-approximation (Zel'dovich, 1970; Buchert, 1992). The linear solution to Ψ can be derived as $\Delta\Psi(\mathbf{q},\eta) = D_+(\eta)\delta(\mathbf{q})$, using the solution $D_+(\eta)$ to the homogeneous growth equation (Catelan, 1995).

Since the equation of motion of a particle in comoving coordinates is given by

$$\frac{\mathrm{d}^2}{\mathrm{d}\eta^2}\mathbf{x} + aH\frac{\mathrm{d}}{\mathrm{d}\eta}\mathbf{x} = -\nabla\Phi, \tag{45}$$

it is possible to establish a relationship between the displacement potential Ψ and the physical gravitational potential Φ by taking the divergence of this equation, allowing the replacement of $\Delta\Phi$ with the Poisson equation. Using mass conservation $1 + \delta(\mathbf{x},\eta) = 1/J(\mathbf{q},\eta)$ and the Jacobian

$$J(\mathbf{q},\eta) \equiv \frac{\mathrm{d}\mathbf{x}}{\mathrm{d}\mathbf{q}} = \left(\det\left[\delta_{ij} + \partial_i\partial_j\Psi\right]\right)^{-1}, \tag{46}$$

of the mapping between $\mathbf{q}$ and $\mathbf{x}$ yields the relationship

$$J(\mathbf{q},\eta)\,\mathrm{div}\left[\frac{\mathrm{d}^2}{\mathrm{d}\eta^2} + aH\frac{\mathrm{d}}{\mathrm{d}\eta}\right]\nabla\Psi = \frac{3}{2}\Omega_m(\eta)(aH)^2(J(\mathbf{q},\eta)-1), \tag{47}$$

which illustrates that the trace of the tidal forces $\partial_i\partial_j\Psi$ is responsible for a compression of the cosmic density field. The limit of applicability of Lagrangian perturbation theory is reached when $J = 0$, because of the divergence of the density field. This conditions corresponds to the non-invertibility of the mapping $\mathbf{q} \to \mathbf{x}$ which occurs when two trajectories cross.

6.4. *Non-Gaussian statistics*

In application to statistics, any correlation function of nonlinear fields can be reduced to a higher-order correlation function of the linearly evolving fields which obey Gaussian statistics, integrated over momentum space with the mode coupling function as a weighting function. While odd n-point correlation functions of Gaussian random fields are equal to zero, even n-point functions can be decomposed into products of two-point functions by virtue of the Wick-theorem,

$$\langle\delta(\mathbf{k}_1)\ldots\delta(\mathbf{k}_n)\rangle = \sum_{\mathrm{pairs}}\prod_{i,j\in\mathrm{pairs}}\langle\delta(\mathbf{k}_i)\delta(\mathbf{k}_j)\rangle \tag{48}$$

for which a proof can be found in e.g. Durrer (2008) and which constitutes an extension of the well-known relation $\langle\delta^{2n}\rangle = (2n-1)!!\langle\delta^2\rangle^n$ for the higher moments of a Gaussian random variable δ with $\langle\delta\rangle = 0$. In this way it is possible to describe the generation of non-Gaussian statistical properties by substituting the perturbative expansion equation (41), which links n-point correlation functions or their corresponding polyspectra to higher-order expressions, which render a vanishing odd moment nonzero and add correction terms to the even moments, effectively destroying the Gaussian recursion relation thus making the statistics non-Gaussian.

From a physical point of view the nonlinearities in the continuity- and Euler-equations are responsible for mode-coupling and ultimately, for the generation of non-Gaussian statistics, similarly to the Karman–Hovarth–hierarchy in turbulence theory. The multiplication of the fields ρ and υ translates to a convolution in Fourier-space, which links all terms in the perturbative expansion. This effectively causes non-Gaussian statistical property, even for initially Gaussian fields, by the generation of corrections to the moments.

6.5. *Halo formation*

Haloes of dark matter form from isolated high-density peaks of the fluctuating matter field, by gravitational collapse under their own gravity. In the limiting case of spherical protohaloes it is possible to solve the collapse equation for an $\Omega_m = 1$-cosmology exactly in terms of a parametric solution, and the numerical solution of the collapse equation for dark energy cosmologies shows that there is only a small difference in the collapse overdensity δ_c. Due to the Birkhoff-theorem, only the matter interior of a spherically symmetric perturbation drives the collapse according to an equation of motion of the type

$$\ddot{R} = -\frac{GM}{R^2}. \tag{49}$$

Solving this equation in a dimensionless form yields a minimal overdensity for spherical collapse of $\delta_c \simeq 1.69$ for $\Omega_m = 1$-cosmologies. Consequently, one would consider all regions in the initial field that, under linear growth would reach amplitudes $\delta \lesssim \delta_c$ today as collapsed into haloes. The influence of tidal shear fields on the collapse dynamics is minor, but they are responsible for the generation of angular momentum in haloes through the process of tidal torquing.

The number density of haloes of a certain mass can be derived from the fluctuation statistics of a random field (Bardeen *et al.*, 1986): The cumulative distribution function $P_\delta(M)$ of the amplitudes δ gives the probability that at a given point is larger than δ,

$$P_\delta(M) = \int_{\delta_c}^{\infty} \mathrm{d}\delta\, p(\delta) \tag{50}$$

where the variance of the field is obtained through an integration over the spectrum of the field, smoothed on a spatial scale which reflects the mass-scale of the halo,

$$\sigma_R^2 = \int \frac{k^2 \mathrm{d}k}{2\pi^2} P(k) W(kR)^2 \tag{51}$$

with a smoothing function $W(kR)$ that can be chosen to be a top-hat or a Gaussian smoothing kernel. The relationship between R and M is given by $M = 4\pi/3\,\Omega_m \rho_{\mathrm{crit}} R^3$. The distribution of the halo masses is then given by the derivative $\partial P/\partial M$. For a Gaussian distribution of amplitudes of the density field one obtains

$$n(M) = \frac{\partial P}{\partial M} = \sqrt{\frac{2}{\pi}} \frac{\delta_c}{\sigma_R D_+} \frac{\mathrm{d}\ln \sigma_R}{\mathrm{d}M} \exp\left(-\frac{\delta_c^2}{\sigma_R^2 D_+^2}\right) \tag{52}$$

with the corrected normalisation. The mass function $n(M)$ can be approximated by a power-law with an exponential cutoff in mass.

6.6. *Tidal torquing and the generation of angular momentum*

The angular momentum of a halo follows from the integration of the momentum density over the physical volume of the halo weighted by the distance of each mass element from the centre of gravity

$$\mathbf{L}(\eta) = \int_V \mathrm{d}^3 r\ (\mathbf{r} - \bar{\mathbf{r}}) \times \upsilon(\mathbf{r}, \eta) \rho(\mathbf{r}, \eta), \tag{53}$$

where $\upsilon(\mathbf{r}, \eta)$ is velocity of the mass element with density $\rho(\mathbf{r}, \eta) = \langle\rho\rangle(1 + \delta(\mathbf{r}, \eta))$ at position $\mathbf{r}$ around the centre of gravity $\bar{\mathbf{r}}$.

Simplifying this relationship by assuming a constant density over the volume of the protohalo $\langle\rho\rangle = \Omega_m \rho_{\mathrm{crit}}$ inside the protogalactic region, White (1984), Catelan and Theuns (1996), Theuns and Catelan (1997) and Crittenden *et al.* (2001) arrive at a concise expression by describing the particle motion in the Zel'dovich-approximation:

$$\mathbf{x}(\mathbf{q}, t) = \mathbf{q} - D_+(\eta)\nabla\Psi(\mathbf{q}) \rightarrow \dot{\mathbf{x}} = -\dot{D}_+ \nabla\Psi, \tag{54}$$

where the dot denotes a derivative with respect to cosmic time t. In the centre of mass frame the expression for the angular momentum becomes

$$\mathbf{L} = \rho_0 a^5 \int_{V_L} \mathrm{d}^3 q \; (\mathbf{x} - \bar{\mathbf{x}}) \times \dot{\mathbf{x}} \simeq \rho_0 a^5 \int_{V_L} \mathrm{d}^3 q \, (\mathbf{q} - \bar{\mathbf{q}}) \times \dot{\mathbf{x}}, \tag{55}$$

when switching from physical to comoving coordinates.

The displacement field $\nabla\Psi(\mathbf{q})$ can be Taylor-expanded in the vicinity of the centre of mass $\bar{\mathbf{q}}$

$$\partial_\alpha \Psi(\mathbf{q}) \simeq \partial_\alpha \Psi(\bar{\mathbf{q}}) + \sum_\beta (\mathbf{q} - \bar{\mathbf{q}})_\beta \Psi_{\alpha\beta}, \tag{56}$$

such that the tidal shear

$$\Psi_{\sigma\gamma}(\bar{\mathbf{q}}) = \partial_\sigma \partial_\gamma \Psi(\bar{\mathbf{q}}), \tag{57}$$

becomes one of the central quantities responsible for angular momentum generation. Identifying the tensor of second moments of the mass distribution of the protogalactic object as the moment of inertia $I_{\beta\sigma}$,

$$I_{\beta\sigma} = \rho_0 a^3 \int_{V_L} \mathrm{d}^3 q \, (\mathbf{q} - \bar{\mathbf{q}})_\beta (\mathbf{q} - \bar{\mathbf{q}})_\sigma \tag{58}$$

one obtains the final expression of the angular momentum L_α:

$$L_\alpha = a^2 \dot{D}_+ \epsilon_{\alpha\beta\gamma} \sum_\sigma I_{\beta\sigma} \Psi_{\sigma\gamma}. \tag{59}$$

Physically this corresponds to the case when inertia and tidal shear do not share a common eigensystem: The product X of inertia $I_{\beta\sigma}$ and tidal shear $\Psi_{\sigma\gamma}$ can be decomposed into its symmetric part $X^+_{\beta\gamma} = (I_{\beta\sigma}\Psi_{\sigma\gamma} + \Psi_{\beta\sigma}I_{\sigma\gamma})/2 = \{I, \Psi\}_{\beta\gamma}/2$ and into the corresponding antisymmetric part $X^-_{\beta\gamma} = (I_{\beta\sigma}\Psi_{\sigma\gamma} - \Psi_{\beta\sigma}I_{\sigma\gamma})/2 = [I, \Psi]_{\beta\gamma}/2$. The contraction with $\epsilon_{\alpha\beta\gamma}$ will be proportional to X^-, i.e. the commutator between the shear and inertia tensors. Hence, angular momentum is only generated if those two tensors do not share a common eigensystem. Clearly, no object would start rotating in a spherically symmetric potential, and one needs a misalignment between the two eigensystems for tidal torquing to be effective.

7. Open questions

The description of cosmic structure formation worked under the assumption of fluid mechanics, which is a questionable concept in the case of dark matter: Dark matter particles have a very small cross section for elastic collisions and therefore, systems of dark matter will not exibit properties like pressure and viscosity whose microscopic origin are particle collisions. A significant progress in this direction has been made by employing statistical field theory (Bartelmann *et al.*, 2014).

At the stage when the cosmic density field reaches large amplitudes all perturbative approaches fail: This is the regime of numerical simulations, which have reached an incredible level of sophistication and go in their description much beyond dark matter dynamics, by including a multiphase baryonic component and by tracking its evolution in terms of temperature and chemical composition. How exactly the baryonic component reacts back onto the dark matter component through gravitational forces is still unclear, likewise the exact thermal evolution of the baryonic component is still a matter of research, and different feedback mechanisms are investigated that keep star formation at an acceptably low level. There are, in addition, a number of observations which are being debated, for instance the abundance of subhaloes in dark matter structures and large bulk flows. Whether they are indicative of a failure of the assumption of cold dark matter or of a different gravitational theory, is yet unclear. And it should be emphasised that algorithmic advances in the discretisation have solved problems which were thought to be of physical origin, for instance the very thin discs of spiral galaxies which were previously not reconcilable with numerical simulations (Springel, 2010).

On larger scales one hopes to investigate deviations of the gravitational law from General Relativity by measuring the structure formation rate on large scales, which would provide a test of gravity in a weak-field, slow-motion and large-scale limit. And lastly, it is generally accepted without a detailed proof that the structured metric on small scales transitions to the smooth metric with the FLRW-symmetries on larger scales, and that the effects of the averaging process are small and not compatible with the phenomenology of dark energy.

8. Summary

The gravitational dynamics of the Universe is governed by General Relativity, which is the most general metric theory in gravity in four dimensions with a second-order local and energy-momentum conserving field equation. Assuming spatial homogeneity and isotropy leads to the Friedmann–Lemaître-cosmologies, which link the expansion dynamics of the Universe to the density of the gravitating substances and the cosmological constant. The expansion dynamics is well described by a FLRW-model with no spatial curvature, with a matter density $\Omega_m \simeq 0.3$ and a cosmological constant whose value corresponds to $\Omega_\Lambda \simeq 1 - \Omega_m$.

Cosmic inflation is a physical mechanism for pushing spatial curvature to very small values, and for generating fluctuations in the distribution of (predominantly dark) matter. These structures seeded in the early universe grew by self-gravity. Cosmic structure formation is influenced by the time evolution of the background cosmological model and shows for small amplitudes a straightforward solution in terms of the growth function D_+. In this limit, the growth is homogeneous and conserves all statistical properties of the density field. In addition, the velocity field is a gradient field whose divergence is proportional to the density field.

When the amplitudes in the density field become large, the growth turns nonlinear and generates new, in particular non-Gaussian statistical properties, through the mode-coupling mechanism.

Isolated peaks in the density field collapse under their own gravity and form haloes, which ultimately host galaxies. Their number density can be computed from the fluctuation statistics of a random field, which in cosmology is in its statistical properties close to Gaussian. Tidal fields can introduce angular momentum into aspherical haloes up to the moment of gravitational collapse.

While the formation of the large-scale structure is determined by gravity and the dynamics of dark matter, additional processes are important on small scales. In particular, the role of chemistry and cooling will be discussed in Chapter 3, the role of the baryons during gravitational collapse in Chapters 4 and 5 and the potential role of radiation backgrounds in Chapter 6. Statistical predictions on the first black holes, based on our cosmological framework, are presented in Chapter 9.

Chapter 3

Thermodynamics and chemistry of the early universe

Stefano Bovino[1] and Daniele Galli[2]

[1] *Astronomy Department,*
Universidad de Concepción,
Barrio Universitario,
Concepción, Chile,
stefanobovino@udec.cl

[2] *INAF – Osservatorio Astrofisico di Arcetri,*
Largo E. Fermi 5, I-50125 Firenze, Italy
galli@arcetri.astro.it

The interplay between chemistry and thermodynamics determines the final outcome of the process of gravitational collapse and sets the conditions for the formation of the first cosmological objects, including primordial supermassive black holes. In this chapter we will review the main chemical reactions and the most important heating/cooling taking place in a gas of primordial composition, including the effects of local and cosmological radiation backgrounds.

1. Introduction

The primordial environment out of which the first objects were formed is characterised by well established chemical initial conditions set by the recombination era, the absence of heavy elements (metals/dust), no turbulence before the formation of the first structures and no or very weak magnetic fields. For these reasons the early Universe can be considered a simplified model of the current interstellar medium (ISM). Molecular hydrogen is the main cooling agent in this scenario and its presence (or absence) can disentangle different dynamical paths. In this chapter, we will present the most relevant processes affecting the chemistry of H_2 with particular focus on the main ions (H^- and H_2^+) which determine its abundance. We will introduce the basic cooling/heating contributions which impact the thermodynamics of the gas and discuss high-energy processes (X-ray and cosmic-ray) that under certain conditions can influence the chemistry of the early Universe. Particular emphasis will be given to the radiation-induced processes which affect the dynamics of a primordial halo and its outcome.

1.1. *Gas kinetics*

The motion and the behaviour of an ideal gas (fluid) is regulated by the momentum, the mass and the energy conservation laws. These equations can be written both in Eulerian as well as in Lagrangian form (integral equations) where in the latter case a gas element is followed along its path. In this section we will focus on the energy equation as we are interested on the thermal and chemical evolution of the gas.

The energy equation in Lagrangian notation reads as

$$\rho\frac{D\mathcal{E}}{Dt} = -P\nabla\cdot\mathbf{u} + \Gamma(\rho,T) - \Lambda(\rho,T), \tag{1}$$

where $\mathcal{E}$ is the specific internal energy (in erg g^{-1}), $P = (\gamma-1)\rho\mathcal{E}$ is the gas pressure (with the adiabatic index $\gamma = 5/3$ for a monoatomic gas and $\gamma = 7/5$ for a fully molecular gas), $\mathbf{u}$ is the gas velocity, Γ and Λ are the volumetric heating and cooling rates, respectively.[a] In Eq. (1) the first term on the right-hand side represents the rate of doing $P\,dV$ work in adiabatic compressions of the gas. The heating and cooling rates are functions of the number densities n_i of the chemical species i (represented here by the mass density ρ) and the gas temperature T. Both terms are strongly coupled to the chemical evolution of the gas, i.e. its kinetics.

Reactions between different species (ions, neutral, or photons) occur through collisions and their velocity/efficiency are regulated by the reaction rate coefficient k that in general is a function of the gas temperature.[b] The kinetics of a given species i, i.e. the *time* variation of its abundance dn_i/dt, is expressed via a *rate equation* that specifies the formation and destruction channels for that species. For example, let us consider a simple chemical system composed by two reactions, the photoionisation and the recombination of atomic hydrogen

$$\mathrm{H} + \gamma \xrightarrow{k_1} \mathrm{H}^+ + \mathrm{e}^-, \tag{2}$$

$$\mathrm{H}^+ + \mathrm{e}^- \xrightarrow{k_2} \mathrm{H} + \gamma, \tag{3}$$

[a]Notice that the symbol Γ is also often employed to represent photorates.

[b]In some specific cases the rate coefficient can also depend on other parameters (e.g., the density).

where k_1 and k_2 are the rate coefficients. Note that for a photochemical reaction the rate is expressed in s^{-1} while standard reactions have units of $cm^{3(m-1)}$ s^{-1}, with m representing the number of reactants; so, for example, for 2-body reactions ($m = 2$) the units are cm^3 s^{-1} while for 3-body reactions $m = 3$ and the rate is in cm^6 s^{-1}. In addition, the photoionisation rate k_1 is a function of the *radiation temperature* $T_{\rm rad}$ characterising the radiation field. The rate equation for the evolution of hydrogen in this example is

$$\frac{dn_{\rm H}}{dt} = k_2(T)n_{\rm H^+}n_{\rm e} - k_1(T_{\rm rad})n_{\rm H}, \tag{4}$$

showing that, given an initial $n_{\rm H}$ abundance, it will increase as long as the *reaction flux* (in cm^{-3} s^{-1}) for the recombination is larger than the photoionisation flux. The example above can be generalised to a multi-species case which includes any type of reactions,

$$\frac{dn_i}{dt} = \sum_{j\in F_i}\left(k_j \prod_{r\in R_j} n_r(j)\right) - \sum_{j\in D_i}\left(k_j \prod_{r\in R_j} n_r(j)\right), \tag{5}$$

where the first sum includes all the terms which form the i-th species (belonging to the set F_i), while the second part is the analogous for the reactions that destroy the i-th species (set D_i). The j-th reaction has a set of reactants (R_j), and depends on the number density $n_r(j)$ of each reactant at time t.

The energy equation (1) is usually solved by a standard operator-splitting technique, where the first term on the right-hand side is integrated alongside the mass and momentum conservation equations, while the energy change due to cooling and heating processes is coupled to the chemistry (Eq. (5)) and added to the hydrodynamics as a source term. The thermal and chemical evolution together represent a stiff system of first-order ordinary differential equations (ODEs) that requires the use of accurate and adaptive numerical solvers (Bovino *et al.*, 2013a). Besides their stiffness, astrophysical networks are also sparse (i.e. most of the Jacobian elements are zero) and this adds further challenges to an already expensive numerical problem. In addition, the system of ODEs can easily diverge when moving from zero to solar metallicity systems where metals, dust, and hundreds of species have to be taken into account to properly describe the kinetics of the gas. A reduced network of CO can already include ~ 40 species and ~ 400 reactions (Glover *et al.*, 2010; Grassi *et al.*, 2017).

Fortunately, primordial environments are relatively simple and can be considered as well established problems, with few uncertainties and a clear network of reactions. An example is given in the graph reported in Fig. 1, showing a network of 9 species and roughly 20 reactions that can be employed for instance to accurately follow the formation of the first cosmological objects (see Latif *et al.*, (2015) and references therein).

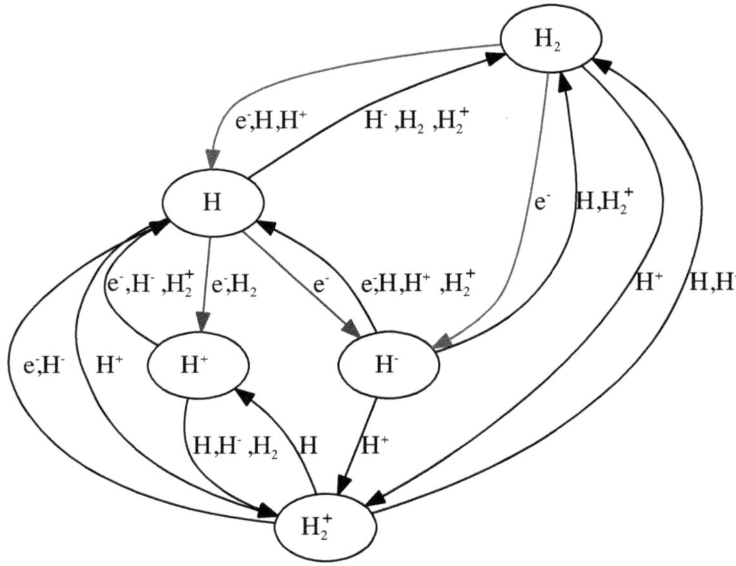

Fig. 1. Graph of a simple H_2-oriented primordial network. Each arrow represents a reaction path where the reactants involved are sketched as hub or over the arrows. For example $H_2 + e^-$ produces $H + H^-$ as indicated by the arrows where H_2 is reported as a hub and electron as a reactant, or $H + e^-$ gives $H^+ + H^-$ as indicated by the arrows. This graph has been obtained with the *docmake* tool released with the astrochemical package KROME which makes use of the Graphviz software (`http://www.graphviz.org`).

Following the chemical evolution of the primordial gas is very important because it allows to establish the connection between the microphysics (chemistry and cooling) and the macrophysics represented by the hydrodynamical equations (energy, mass, and momentum), and has a strong impact on relevant physical processes, as for instance the star-formation process. Chemistry is also needed when comparing theoretical work with observational data. However, chemical solvers can represent a bottleneck in modern numerical astrophysics. For this reason, fast and efficient astrochemical packages aimed at solving the chemical and thermal evolution of the gas have been developed in the last years, as for instance the KROME package (Grassi *et al.*, 2014).

1.2. *Ultraviolet radiation sources*

Chemistry is also strongly connected to the radiation field, through photochemistry and photoheating, as discussed in this section, and with additional emphasis on the radiation spectrum in Chapter 6. Photochemical processes in the early Universe are mainly induced by ultra-violet (UV) radiation produced by the first (Pop. III) and second (Pop. II) generation of stars, and quasar-like ionising sources. In particular the Lyman-Werner (LW) background (at frequencies of 11.2–13.6 eV, corresponding to 912–1100 Å) produced by Pop III stars determines the onset of the soft UV background in the Universe and has a strong effect on the subsequent formation of stars (Haiman *et al.*, 1997; Omukai, 2001).

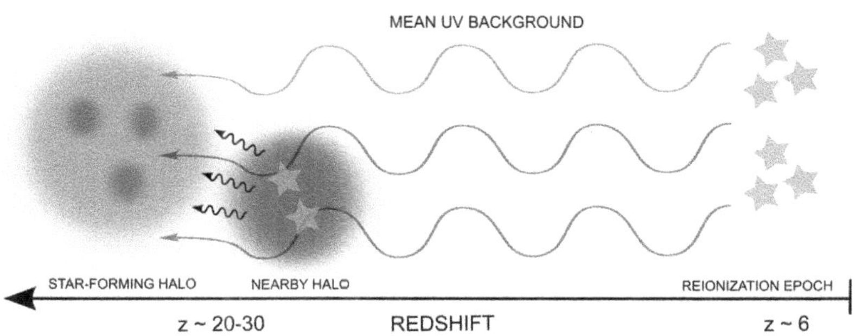

Fig. 2. (Color online) Pictorial representation of the different radiation sources which can irradiate a primordial star-forming halo (in pink) at high-redshift, namely the mean UV background from Pop. III stars emitted at the epoch of reionisation, and the strong UV radiation produced by Pop. III/Pop. II stars in a nearby halo. Redshifts and relative distances are indicative as well as the direction of the photons.

The total UV radiation that could be experienced by the gas in a dark-matter halo is the summation of the mean cosmic UV background produced just before the epoch of cosmic reionisation by massive stars (Bromm and Loeb, 2003) and a local background coming from individual stellar sources hosted by nearby halos (see sketch in Fig. 2). Population III stars in the host halo (galaxy) could also contribute to the amount of radiation together with the metal enrichment. As we are here interested in a metal-free environment the latter is not taken into consideration.

In the following we assume that most of the radiation above the Lyman limit (13.6 eV) is absorbed by HI, while LW photons can travel longer due to the low optical depth of the gas at those energies. The total LW background is then expressed as

$$J_{\rm LW} = J_{\rm bg}^{\rm III} + J_{\rm loc}^{\rm III} \tag{6}$$

where $J_{\rm loc}^{\rm III}$ is the local UV radiation coming from a nearby star-forming galaxy (Visbal *et al.*, 2014; Dijkstra *et al.*, 2008) produced by individual Pop. III stars, and $J_{\rm bg}^{\rm III}$ is the mean cosmological background, i.e. the minimum level of LW radiation that a halo is exposed to at any given redshift z. The latter is usually defined as (Greif and Bromm, 2006; Omukai, 2001; Omukai *et al.*, 2008)

$$J_{\rm bg}^{\rm III} \approx \frac{1}{f_{\rm esc}} \left(\frac{hc}{4\pi m_{\rm H}} \right) N_{\rm LW} \rho_*^{\rm III}(z)(1+z)^3 \tag{7}$$

where $f_{\rm esc}$ is the escape fraction of LW photons from the halo, $N_{\rm LW}$ is the number of photons emitted in the LW bands per stellar baryon ($N_{\rm LW} \approx 2 \times 10^4$, see (Greif and Bromm, 2006)), h the Planck constant, c the speed of light, $m_{\rm H}$ the proton mass, and $\rho_*^{\rm III}(z)$ the comoving density of Pop III stars at a given redshift z. Similar expressions can be written for the radiation produced by Pop II stars.[c] Different approximations to the above spectra have been elaborated over the years. Quite often Pop III and Pop II spectra are approximated by uniform constant sources in terms of black-body radiation expressed as $J_{\rm LW}(E) = J_{21} \times 10^{-21} B(E, T_\star)/B(E_0, T_\star)$, where $E_0 = 13.6$ eV, and $T_\star = 10^5$ K and 10^4 K for Pop III and Pop II star, respectively. Quasar-like sources are modelled as a simple power-law, $J_{\rm LW}(E) = J_{21} \times 10^{-21}(E/E_0)^{-1}$. Typical levels of LW background range from $J_{21} \approx 10^{-3}$ to 10^{-1}, and depend on the stellar density at a given redshift (Greif and Bromm, 2006), while local sources can have larger values $J_{21} > 10^3$.

Beside these simplified models, more realistic spectra have been proposed by different groups. In Wolcott–Green *et al.* (2017); Sugimura *et al.* (2014); Agarwal and Khochfar (2015); Agarwal *et al.* (2016b), for example, more realistic spectra produced by Pop. III and Pop. II stars have been computed by employing stellar codes like STARBURST99. However these spectra should be taken with caution as they depend on many parameters/uncertainties, the most important being the initial mass function (IMF) of Pop. III stars which is difficult to constrain. The spatial and temporal dependency of the LW background has been considered by e.g. Visbal *et al.* (2014); Dijkstra *et al.* (2008); Agarwal *et al.* (2012). Time-dependent

[c]Throughout this book we express the radiation background in terms of the flux J_{21} at the Lyman limit (912 Å, i.e. 13.6 eV): $J_{\rm LW} = J_{21} \times 10^{-21}$ erg s^{-1} cm^{-2} sr^{-1} Hz^{-1}.

average LW backgrounds have been proposed in simple parametrised form as a decreasing function of the redshift (Visbal *et al.*, 2014) and from more accurate calculations. A summary on this topic including the current uncertainties is therefore given in Chapter 6. The LW background is expected to increase with time (Visbal *et al.*, 2014) but it was established that neglecting time-dependency does not affect significantly the calculations of the final photorates, in particular as gravitational collapse will occur within a cosmologically short time span.

As we will see in the subsequent sections a UV radiation background in a primordial gas has the capability to photodissociate H_2 and H_2^+, and photodetach H^-. The shape of the assumed spectrum is strongly connected to the amplitude of the photorates for the radiatively-induced reactions, which are in turn relevant for the destruction/survival of molecules and ions in the primordial Universe. In the following section we will examine in detail the processes that affect the formation of the most important species in the early Universe, molecular hydrogen and its associated ions.

2. Key species: H_2, H^-, and H_2^+ chemistry

Basically, the chemistry of the early Universe reduces to three main species, H_2, H^-, and H_2^+, whose fates are strongly connected. Chemistry of deuterium is not discussed in this chapter but it is worth mentioning that it might be relevant for the formation of primordial low-mass stars through for example halo mergers (Prieto *et al.*, 2014; Bovino *et al.*, 2014a; Shchekinov and Vasiliev, 2006) and the cooling of post-shock gas (Greif and Bromm, 2006; Nakauchi *et al.*, 2014).

2.1. H_2 *chemistry*

H_2 is the simplest and the dominant molecule in the Universe and plays a crucial role in many astrophysical environments. It was the first species formed in the Universe and had great relevance in the formation of the first cosmological objects (see also discussion in Chapter 4). Despite this, H_2 is characterised by peculiar chemical properties which affect its chemistry and the physics where it is involved. Being a homonuclear (symmetric) molecule, H_2 does not possess a permanent dipole moment and the rovibrational transitions in its ground state occur through quadrupole radiation. As a first consequence, the formation of H_2 by direct radiative association of ground state H atoms is negligible because there are no allowed stabilising transitions. In present-day molecular clouds, H_2 is formed by surface reactions on dust grains that can efficiently absorb the energy which must be released in the process. The only catalysts available to form H_2 in zero-metallicity cloud are electrons and protons, rather minor constituents of the post-recombination Universe.

Main formation routes. The two routes to H_2 formation via gas-phase reactions in a cloud of primordial composition are

$$\mathrm{H} + \mathrm{H}^+ \rightarrow \mathrm{H}_2^+ + h\nu, \tag{8}$$

$$\mathrm{H}_2^+ + \mathrm{H} \rightarrow \mathrm{H}_2 + \mathrm{H}^+ \tag{9}$$

(the H_2^+ channel) and

$$\mathrm{H} + e^- \rightarrow \mathrm{H}^- + h\nu, \tag{10}$$

$$\mathrm{H}^- + \mathrm{H} \rightarrow \mathrm{H}_2 + e^- \tag{11}$$

(the H^- channel). Of these two routes, the latter is generally the most efficient. However, since H^- has a smaller binding energy than H_2^+ (0.754 eV vs. 2.65 eV), it is more easily destroyed by any UV radiation released by Pop. III stars, minihalos and microquasars (Glover, 2007; Chuzhoy *et al.*, 2007; Coppola *et al.*, 2013). The H_2 formed by these routes is destroyed mainly by photodissociation

$$\mathrm{H}_2 + h\nu \rightarrow \mathrm{H} + \mathrm{H}, \tag{12}$$

and by collisional dissociation with H atoms

$$\mathrm{H}_2 + \mathrm{H} \rightarrow \mathrm{H} + \mathrm{H} + \mathrm{H}. \tag{13}$$

Photodissociation. In spite of its low binding energy (4.48 eV), H_2 cannot be easily photodissociated, since direct excitation to the vibrational continuum of the ground electronic state $X^1\Sigma_g^+$ is forbidden by the dipole selection rules (Field *et al.*, 1966). Direct excitation to the continua of higher electronic states is allowed, but requires absorption of a photon with energy larger than 14.7 eV, which is more likely to photoionise H (threshold 13.6 eV) or H_2 (threshold at 15.5 eV). Thus, direct photodissociation is very inefficient (Coppola *et al.*, 2011). Photodissociation of H_2 mainly occurs by a two-step process (also called *Solomon process*): absorption of the Lyman ($X^1\Sigma_g^+ \rightarrow B^1\Sigma_u^+$, threshold at 11.2 eV) or Werner ($X^1\Sigma_g^+ \rightarrow C^1\Pi_u$, threshold at 12.3 eV) band UV photons, followed by radiative decay into the vibrational continuum of the ground electronic state (Stecher and Williams, 1967; Glover, 2007). On average, $\sim$ 15% of all radiative decays following absorption in the Lyman–Werner bands results in dissociation. Since the Lyman–Werner bands arising from the lowest rotational levels in the ground state comprise several hundred absorption lines between 912 Å and 1100 Å, the photodissociation rate of H_2 depends mainly on the intensity of the UV radiation in this wavelength range, quantified in terms of the J_{21}.

Collisional dissociation. While photodissociation is the main destruction process of H_2 in the presence of a significant UV background around 1000 Å, providing an efficient feedback mechanism to prevent further cooling and collapse (Haiman *et al.*, 1997, 2000; Ciardi *et al.*, 2000; Machacek *et al.*, 2001; Mesinger *et al.*, 2006), collisional dissociation dominates at temperatures larger than 2000 K and in dense, shock-compressed regions (Lepp and Shull, 1983; Mac Low and Shull, 1986). Typically, H_2 is completely dissociated by (non-magnetic) shocks with velocities above $\sim$ 20 km s^{-1}. The main contribution to collisional dissociation at low-densities is the *dissociative tunneling.* This process involves quasi-bound states, i.e. excited states which are separated from the continuum by a rotational barrier, also called predissociative states. There is a high probability that molecules in these states can dissociate via tunneling instead to decay radiatively to a bound-state (Martin *et al.*, 1996).

Three-body reactions. At densities above $\sim 10^8$ cm^{-3}, efficient H_2 formation is provided by three-body reactions (Palla *et al.*, 1983),

$$\mathrm{H} + \mathrm{H} + \mathrm{H} \rightarrow \mathrm{H_2} + \mathrm{H}, \tag{14}$$

$$\mathrm{H} + \mathrm{H} + \mathrm{H_2} \rightarrow \mathrm{H_2} + \mathrm{H_2}, \tag{15}$$

the reverse processes of H_2 dissociation induced by collisions with H and H_2. For the impact of these reactions on the evolution of primordial clouds and the fragmentation process, see Turk *et al.* (2011a), Bovino *et al.* (2014b).

2.2. H_2^+ *chemistry*

In present-day molecular clouds, H_2^+ is formed by cosmic-ray ionisation of H_2 and rapidly transformed in H_3^+ by collision with other H_2 molecules. Conversely, in the primordial gas H_2^+ is formed by radiative association (reaction 8) in excited vibrational states, and is destroyed by the reverse reaction, photodissociation, at a rate strongly dependent on the distribution of excited levels (Coppola *et al.*, 2011). In its ground vibrational state, H_2^+ is photodissociated by photons with $h\nu > 2.65$ eV, or $\lambda < 4680$ Å, and therefore survives better than H^- at intense UV radiation fields. In those circumstances, H_2^+ represents the main source of H_2 via the proton exchange reaction (4), as, e.g., in the post-recombination Universe at redshifts $z \gtrsim 300$ (Saslaw and Zipoy, 1967; Galli and Palla, 1998). However, when the gas is sufficiently ionised, H_2^+ is promptly removed by dissociative recombination with electrons, a reaction enhanced by the presence of several low-energy resonances (Takagi, 2002; Schneider and Suzor–Weiner, 2002).

2.3. H^- *chemistry*

The hydrogen anion H^- (hydride ion) is the main catalyst for the formation of H_2 in the primordial gas. This species, a proton with two bound electrons, has a binding energy of only 0.754 eV and no bound excited states. In the present-day Universe, it provides the continuum opacity of the Sun and most other late-type stars via bound-free and free-free transitions; similarly, in the post-recombination Universe, it represents the main source of opacity for the cosmic microwave background radiation, although at levels presently undetectable (Black, 2006; Schleicher *et al.*, 2008). In the primordial gas, H^- is formed by radiative attachment (reaction 10) and removed by associative detachment (reaction 11). Recently, reaction (10) has been the object of intensive laboratory experiments and theoretical calculations that have largely reduced the uncertainties on the reaction rate of this crucial process (Bruhns *et al.*, 2010; Kreckel *et al.*, 2010; Miller *et al.*, 2011).

Photodetachment. In the presence of a significant flux of photons with $h\nu > 0.754$ eV ($\lambda < 16{,}500$ Å), H^- is efficiently destroyed by *photodetachment* (reverse of reaction 10). The cross-section of this process has been calculated by different groups with (Miyake *et al.*, 2010) and without (Wishart, 1979) resonant contributions (for photon energies greater than ~ 11 eV). The photodissociation rate is also strongly affected by the shape of the UV spectrum, as for instance a typical power-law radiation background produced by miniquasar can boost the rate by a factor of ~ 5 (Chuzhoy *et al.*, 2007). In general the net effect of the photodetachment process is to suppress the formation of H_2 reducing the efficiency of the H^- channel by a factor of approximately $1 + k_{ph}/(k_{ad}n_H)$, where k_{ph} is the photodetachment rate (reverse of action 10), k_{ad} the associative detachment rate (reaction 11), and n_H the number density of atomic hydrogen (see e.g. Chuzhoy *et al.* (2007); Miyake *et al.* (2010)). Recently it has also been suggested that trapped Ly-α cooling radiation emitted from collapsing gas (see Sec. 5.1.3) could contribute to the photodetachment of H^-. The rate for this process has been provided in Johnson and Dijkstra (2017) as a function of the cloud mass, the density, and the temperature of the gas.

3. High-energy chemistry

As mentioned in Sec. 2.1, the main cooling agent in the primordial gas, H_2, is efficiently photodissociated by photons with energies in the range of the Lyman–Werner bands (11.2–13.6 eV). Since these photons are not strongly absorbed by ambient H atoms, it is natural to expect the formation of the very first massive stars to be accompanied by the generation of a strong UV background radiation field (Sec. 1.2), which is able in principle to photodissociate any residual H_2 and suppress further star formation (see Sec. 2.1 and reference therein). Ionisation of H_2 by UV photons, on the other hand, is relatively inefficient because ionising photons with energy above the threshold energy of 15.4 eV only exist inside HII regions.

In contrast, X-rays ($h\nu > 1$ keV) and, to a larger extent, cosmic rays ($E > 1$ MeV) can penetrate protogalactic clouds and ionise both H and H_2, producing ions and excited molecules that can interact with the neutral atomic or molecular gas (see, e.g., Tielens (2005)). The ionisation of atomic hydrogen generates electrons that can catalyse further formation of H_2, compensating, or even overcoming the destruction of H_2 by the same process. These reactions carry away a significant amount of the available energy in the form of chemical heating, and contribute to the heating of the gas as much as the ionisation electrons themselves (Goldsmith and Langer, 1978; Glassgold *et al.*, 2012).

Possible sources of an X-ray background in the early Universe include high-redshift quasars, massive X-ray binaries, Bremsstrahlung and inverse Compton emission from Pop. III supernova remnants (Haiman *et al.*, 2000; Glover and Brand, 2003; Hummel *et al.*, 2015; Latif *et al.*, 2015). The latter are also the most likely sources of cosmic-ray particles, although more exotic processes like the decay of supermassive particles are also possible (Shchekinov and Vasiliev, 2004; Ripamonti *et al.*, 2007; Jasche *et al.*, 2007; Hummel *et al.*, 2016). In general, cosmic rays penetrate deeper into a cloud than X-rays, but their relative importance as ionising and heating agents depends on poorly constrained quantities like the intensity, distribution and spectral characteristics of the sources. Numerical models show that the effects of an X-ray background on the chemical and thermal evolution of protogalactic clouds could be significant at relatively low densities, while the high-density, star-forming gas remains largely unaffected (Hummel *et al.*, 2016).

4. Shielding effects

When considering a cloud irradiated by a radiation background it is very important to take possible *shielding* effects into account. The photodissociation rates and the cooling functions vary with the gas column density and only high energy photons and cosmic rays can penetrate deep into the cloud (where column densities are high). In present-day molecular clouds photons can be shielded both by the gas as well as dust, the latter being the main contribution for energies below the Lyman limit.

In a primordial environment the most important process is the H_2 *self-shielding*. At high enough densities H_2 molecules gather together creating a thick layer of material which can absorb energetic photons decreasing the photodissociation rate. This is usually taken into account by a suppression factor $f_{\rm sh}$, so that the rate is expressed as $k_{\rm ph}^{\rm thick} = f_{\rm sh} k_{\rm ph}^{\rm thin}$, where $f_{\rm sh} = 1$ is the "no-shielding" case (optically thin regime), while $f_{\rm sh} = 0$ is a complete shielding, i. e. radiation fully absorbed.

Evaluating the shielding is non-trivial as it usually requires the solution of the radiative transport equations and the computation of the gas column density. This depends on geometry, density, temperature, velocity gradients, and the presence of turbulence. Accurate calculations have been performed over the years and simple analytical formulae have been provided (Draine and Bertoldi, 1996; Wolcott–Green *et al.*, 2011; Safranek–Shrader *et al.*, 2012; Hartwig *et al.*, 2015c) based on approximate definitions of the column density. The latter is frequently expressed based on local properties as $N_i = n_i \ell$, with n_i being the number density of the considered species (in a primordial cloud usually H_2) and ℓ a characteristic length. Its choice however depends strongly on the physical problem, and this approximation can fail to reproduce accurate radiative transfer calculations under particular conditions. Detailed comparisons between the different characteristic lengths have been provided by several authors (Wolcott–Green *et al.*, 2011; Safranek–Shrader *et al.*, 2012; Hartwig *et al.*, 2015c).

When calculating the self-shielding one also has to take into account turbulence and thermal Doppler broadening (Richings *et al.*, 2014). The latter is important for instance in the study of collapsing gas where the self-shielding of H_2 molecules occurs only if relative velocities u_r are smaller than the thermal velocity, $u_r \ll v_{\rm th}$. An accurate method to account for this effect has been developed by Hartwig *et al.* (2015c).

To conclude, changes in the H_2 self-shielding treatment can introduce up to an order of magnitude differences on the final photodissociation rate, affecting in equal measure the outcome of numerical simulations of first cosmological objects.

5. Thermodynamics in primordial environments

A fundamental ingredient in any cosmological scenario for the formation of the first stars is the ability of primordial clouds to radiate energy efficiently during the phase of gravitational collapse. Clearly, the main cooling mechanisms at work in present-day clouds, namely CO and metal line emission or IR emission from dust grains are not available in a cloud of zero metallicity.

5.1. *Cooling processes*

In a H/He plasma in ionisation equilibrium, thermal kinetic energy can be removed from the gas by collisional excitation/ionisation, recombination, and radiation of free electrons scattering off free ions (free-free emission or Bremsstrahlung). The latter is in fact the primary source of cooling for $T_{\rm gas} \gtrsim 10^6$ K, when the gas is fully ionised, but its efficiency decreases with decreasing temperature. At $T \approx 10^5$ K, and $T \approx 5 \times 10^4$ K, the cooling is dominated by collisional excitation of He^+ and H, respectively, followed by radiative decay (mostly in the Lyman α line of neutral hydrogen). Below $T \approx 10^4$ K, the plasma recombines very rapidly. The very few residual electrons are unable to excite the first excited level of H, with an energy 10.2 eV above the ground state, and the cooling rate drops to zero. The cloud then relaxes to a stable thermal state with $T \approx 10^4$ K. While in a gas of nonzero metallicity, oxygen and nitrogen ions with the lowest energy levels just a few eV above the ground state can further cool the gas, the thermal evolution of a primordial cloud is controlled by the amount of H_2 and HD, and, to a lesser extent, H_3^+, formed by gas-phase reactions in a dust-free environment (Saslaw and Zipoy, 1967; Matsuda *et al.*, 1969).

5.1.1. H_2 *cooling*

The cooling of a metal-free cloud is dominated by the spontaneous emission of rovibrational levels of H_2 collisionally excited mainly by H and He atoms or other H_2 molecules. In the optically thin regime, the emitted photons can freely escape the cloud without being re-absorbed or scattered, transferring thermal energy outward at a rate proportional to the H_2 abundance. However, despite being the most abundant molecular species formed by gas-phase reactions in a zero-metal gas, H_2 is not an efficient coolant. Due to its lack of a permanent electric dipole moment, rovibrational levels of H_2 have small transition probabilities, and can be collisionally de-excited even at relatively low densities. In fact, above a *critical density* $n_{\rm crit} \approx 10^4$ cm^{-3}, the H_2 cooling rate reaches its local thermodynamic equilibrium (LTE) value, given by

$$\Lambda_{\rm LTE} = \sum_{u,l} n_u A_{ul} E_{ul}, \qquad (16)$$

where n_u is the population density of H_2 in upper energy level u, A_{ul} is the transition probability for spontaneous decay to energy level l, and $E_{ul} = h\nu_{ul}$. Note that typically, $n_{\rm crit} \approx 10^4$ cm^{-3} for the lower rotational states (Le Bourlot *et al.*, 1999) but the vibrational states can have a higher $n_{\rm crit}$ of about 10^8 cm^{-3}. However, high vibrational levels will have a negligible population already at 8000 K (only the first 2-3 levels will be populated but the ground will still dominate). Useful formulae for the H_2 cooling in the low-density and LTE conditions can be found in Glover and Abel (2008) and Coppola *et al.* (2012), respectively. The former reference also contains a thorough discussion of the H_2 ortho/para ratio in the typical conditions of primordial clouds, where it is generally different from the equilibrium value 3:1 expected at high densities and temperatures.

At densities above $n_{\rm thick} \approx 10^9$ cm^{-3}, H_2 becomes optically thick to its own radiation, and the emitted rovibrational photons can be efficiently scattered and re-absorbed, decreasing the overall cooling efficiency. However, velocity gradients can Doppler-shift photons outside the optically thick cores of the rovibrational lines, allowing them to leave the cloud (Yoshida *et al.*, 2006; Stacy and Bromm, 2013; Hartwig *et al.*, 2015b). In this case, one usually adopts the Sobolev approximation, assuming that each radiative transition probability A_{ul} is reduced by a factor β_{ul}, the so-called "escape probability", that depends on the velocity profile of the cloud (Yoshida *et al.*, 2008; Clark *et al.*, 2011a; Greif *et al.*, 2011, 2012).

5.1.2. HD *cooling*

In spite of a cosmological D/H ratio of $\sim 10^{-5}$, the role of HD as a cooling agent is not negligible. First, the HD molecule has a small permanent electric dipole moment and radiative transition probabilities ~ 100 times larger than those of H_2. Therefore, the critical density for the onset of LTE in the case of HD cooling is larger than that for H_2, of the order of $n_{\rm crit} \approx 10^6$ cm^{-3}. Second, because of its larger mass, the energy difference of the lowest rotational transition of HD corresponds to a temperature of 128 K, rather than 510 K as in the case of H_2, and therefore can efficiently cool a gas to lower temperatures than H_2. Third, since HD is mostly formed by the reaction

$$D^+ + H_2 \rightarrow HD + H^+, \tag{17}$$

which is exothermic by 462 K, its abundance relative to H_2 is boosted at low temperature as $[HD/H_2] \approx 2[D/H] \exp(462\ {\rm K}/T)$ (a process called "deuterium fractionation"). Thus, HD is able to cool the gas to temperatures below 200 K, possibly all the way to that of the cosmic background, which sets a lower limit to radiative cooling (Johnson and Bromm, 2006; Glover and Abel, 2008). Useful formulae for HD cooling for collisions with H are given by Lipovka *et al.* (2005) and Coppola *et al.* (2011) in the low-density and LTE limits, respectively.

In general, HD cooling plays a marginal role for the evolution of typical primordial halo masses ($M \approx 10^6\ M_\odot$), where the temperature never falls below ~ 200 K (Bromm *et al.*, 2002; Yoshida *et al.*, 2006). However, HD cooling can affect significantly the collapse of low-mass halos ($M \lesssim 3 \times 10^5\ M_\odot$), leading to a less massive stellar population compared to the case of halos dominated by H_2 cooling (Ripamonti *et al.*, 2007; McGreer and Bryan, 2008). In addition, HD cooling becomes important in all situations where the gas is initially significantly ionised, as in haloes formed behind strong shocks (Shapiro and Kang, 1987; Johnson and Bromm, 2006; Greif and Bromm, 2006; Nakauchi *et al.*, 2014) as in merger events (Prieto *et al.*, 2014; Bovino *et al.*, 2014b) or in relic HII regions (Mackey *et al.*, 2003; Johnson and Bromm, 2006).

A comparison between HD and H_2 cooling is reported in Fig. 3 together with H and He atomic cooling.

5.1.3. *Ly-α cooling*

The energy levels of hydrogen atoms mainly depend on two quantum numbers, the electronic quantum number n, and the orbital quantum number

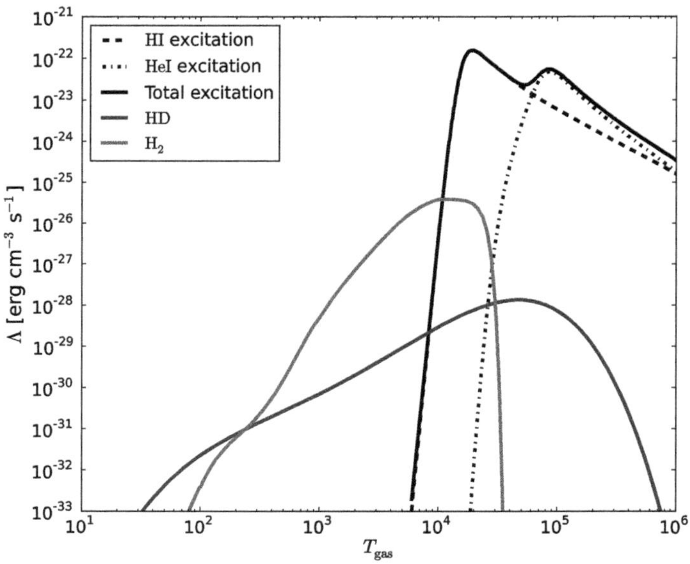

Fig. 3. Cooling functions due to hydrogen (HI), helium (HeI) excitations, molecular hydrogen H_2, and HD as a function of the gas temperature for a given density $n = 1\ \mathrm{cm}^{-3}$. The H_2, and HD cooling functions are computed at fixed species number fractions, i.e. $x_{H_2} = 10^{-4}$, and $x_{HD} = 10^{-7}$. The atomic cooling is instead calculated assuming collisional ionisation equilibrium (see Katz *et al.* (1996)).

l, with energy scaling as 13.6 eV/n^2. We restrict our discussion here to the first three electronic levels, termed $1s$ ($n = 0, l = 0$), $2s$ ($n = 1, l = 0$), and $2p$ ($n = 1, l = 1$). Spontaneous transitions between $2s$–$1s$ are forbidden by selection rules, while transitions involving the ground state $n = 1$ give rise to the Lyman series.[d]

The so called Ly-α cooling originates from the collisional excitation of hydrogen atoms from the ground electronic state to the $2p$ state, which in the subsequent emission process produces Ly-α photons at 10.2 eV (left sketch in Fig. 4). In metal-free cloud this contribution is crucial to bring the gas down to 8000 K. The cooling function is usually expressed as $\Lambda^{\mathrm{Ly}\alpha} = n_\mathrm{e} \sum_i \sum_u n_i E^i_{ul} \langle \sigma^i_{ul} u_\mathrm{e} \rangle$, with n_e the electron density, n_u and n_l the density of the upper and lower levels, E_{ul} the energy difference between the two levels, σ_{ul} the excitation cross-section, and u_e the electron velocity. Note

[d]When the final state is $n = 3$, or $n = 2$ we refer to the Paschen and Balmer series, not of interest in this chapter.

that the cross-section is averaged over a Maxwellian distribution (Shapiro and Kang, 1987).

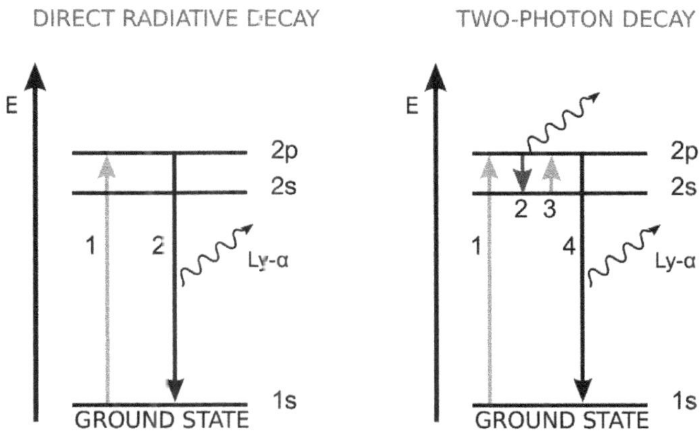

Fig. 4. (Color online) Sketch of the hydrogen Ly-α cooling and the two-photon process. We only show the first three hydrogen levels. Numbers represent the sequence of events. The grey arrows indicate the excitation by collisions with electrons; the black arrow the photon emission that produces Ly-α photons; the blue arrow the decay from $2p$ to $2s$.

5.1.4. *Two-photon decay*

Another process which can contribute in part to the cooling of the gas producing Ly-α photons is the so called two-photon decay (Spitzer and Greenstein, 1951; Omukai, 2001; Glover, 2007). Even though the $2s$–$1s$ transition is forbidden there is a chance that hydrogen atoms (after being excited through collisions with electrons into the $2p$ state) decay to the $2s$ state and are again excited to the $2p$ followed by the decay to the ground state (see sketch in Fig. 4, right). The sum of the two emitted photons is a net Ly-α transition at 10.2 eV. This process is in general $\sim 10^8$ orders of magnitude slower than the Ly-α direct radiative decay and depends on different parameters as for instance the Ly-α optical depth of the gas. An analytical formula of this cooling is given for example in Omukai (2001).

5.1.5. *High-density cooling*

When densities increase, H_2 reaches LTE and its capability to cool the gas is highly reduced. Three-body reactions then trigger the cooling at density between 10^8–10^{10} cm^{-3}, efficiently contributing to the conversion of hydrogen atoms to H_2 molecules. When densities reach $\sim 10^{14}$ cm^{-3} the gas becomes optically thick to H_2 and the H_2 cooling is completely inefficient. In the high density regime between 10^{14}–10^{17} cm^{-3} other processes contribute to the thermodynamics of the gas, among these the collisional induced emission. In the following we discuss the main contributions in the density regime where H_2 cooling becomes inefficient, focusing in particular on the cooling produced by chemical reactions and by continuum processes.

Chemical cooling. A chemical reaction proceeds in time along its reaction path going from reactants to products through different mechanisms. We can distinguish between two main classes of reactions: (i) *exothermic*, and (ii) *endothermic*. The former ones proceed fast and release into the medium a given amount of energy equal to the enthalpy of the reaction ΔH. On the contrary, endothermic reactions need to absorb energy from the medium to take place. The outcome of endothermic reactions is a net cooling effect, with cooling function proportional to the energy needed for the reaction to occur (ΔH), the abundances of the reactants (i, j), and the reaction rate $k(T)$, which regulates the speed of the entire process.[e] The chemical heating $\Lambda_{\rm chem} \propto n_i n_j k(T) \Delta H$ is expressed in erg cm^{-3} s^{-1}.

In a primordial environment the main source of chemical cooling comes from the destruction of H_2 via three-body reactions (reaction 15) and collisional dissociation (reaction 13), and results in 4.48 eV carried away from the medium. These processes have an impact on the thermal evolution of a collapsing gas at density larger than 10^8 cm^{-3}.

Continuum processes. The continuum cooling is usually expressed as $\Lambda_{\rm cont} = 4\sigma T^4 \kappa_{\rm gas} \beta_{\rm esc}$, i.e. photon emission based on the Stefan–Boltzmann law (with σ the Stefan–Boltzmann constant) but attenuated by the opacity of the gas $\kappa_{\rm gas}$ and the escape probability $\beta_{\rm esc}$ introduced in Sec. 5.1.1. The main processes which contribute to the opacity of the gas at high-densities are bound-free absorption of H, He, H^-, and H_2^+ ; free-free absorption of H^- and H; collision-induced absorption of H_2–H_2 and H_2–He; Rayleigh scattering of H; and Thomson scattering of electrons (Lenzuni *et al.*, 1991).

The main contribution is provided by collisionally-induced absorption (emission) of H_2 (Lenzuni *et al.*, 1991; Ripamonti and Abel, 2004; Yoshida *et al.*, 2008; Hirano and Yoshida, 2013). At high densities the number of collisions is highly enhanced; in particular H_2 can inelastically collide with other H_2 molecules or helium atoms inducing the formation of a transient

[e]For reactions involving a third body the expression should read as $n_i n_j n_k k(T) \Delta H$.

supramolecule with a nonzero dipole moment, with a high probability of emitting a photon. This is the last radiative cooling channel available during the collapse of a primordial cloud. Above 10^{16} cm^{-3} the gas becomes completely opaque to the continuum radiation and also collisionally-induced emission becomes inefficient (Yoshida *et al.*, 2008).

5.2. *Heating processes*

There are different ways to heat the gas in the ISM, i.e. to transfer energy into the gaseous medium: (*i*) via photons/electrons, (*ii*) via chemical reactions, (*iii*) via dust grains, and (*iv*) via dynamical processes. In this section we will focus on the microscopic processes relevant for a primordial environment, mainly the chemical heating and the photoheating. However, other processes at macroscales can strongly contribute to the heating, e.g. dynamical compression (first term in Eq. 1) during gravitational collapse, shocks and turbulence, mechanical heating produced by stellar winds and supernova explosion, and viscous heating (e.g. ambipolar diffusion). For an overview on these dynamical processes we refer to Tielens (2005).

5.2.1. *Chemical Heating*

Exothermic reactions, i.e. processes which proceed very fast, often without a barrier, and with an excess of energy, can contribute to the heating of the gas. In this class of reactions, collisions between atoms and molecules are usually converted into (*i*) translation energy of the newly formed molecule, and (*ii*) rotational/vibrational excitations. In the former case the heating is directly released into the gas, in the latter the gas is heated up via subsequent de-excitations.

In a primordial gas chemical heating is mainly due to processes which form molecular hydrogen, in particular the H^- channel, the H_2^+ channel, and three-body reactions. The heat deposited per formed molecular hydrogen is defined similarly to the chemical cooling as $\Gamma_{\mathrm{chem}} = \epsilon n_i n_j k(T)\Delta H$, where $\Delta H = 4.48$ eV, 3.53 eV, and 1.83 eV respectively for the reactions cited above. At variance with chemical cooling, chemical heating is weighted by a critical density factor $\epsilon = (1 + n_{\mathrm{crit}}/n)^{-1}$ depending on the critical density n_{crit}, i.e. the ratio between the Einstein coefficient and the collisional de-excitation rate (Hollenbach and McKee, 1979). This critical density factor keeps into account the amount of energy which is radiated away and does not contribute to the heating. We can define two extreme cases: if (*i*) $n \gg n_{\mathrm{crit}}$, then $\epsilon \to 1$, the gas is heated up by ΔH; (*ii*) if

$n \ll n_{\rm crit}$, then $\epsilon \to 0$, and the energy ΔH is fully lost by radiative emission. The cases in between contribute to the heating of the gas to the extent of an amount of energy smaller than ΔH.

5.2.2. *Photoheating*

The photodissociation and photoionisation induced by UV radiation generate an excess of energy equal to $h\nu - E_0$ which can go into heating, where $h\nu$ is the energy of the incident photon and E_0 the dissociation/ionisation energy threshold for the given species. Photoheating is mainly caused by (*i*) photoionisation of atoms in HII regions ($h\nu > 13.6$ eV), (*ii*) photoionisation of large molecules and small dust grains in HI regions ($h\nu < 13.6$ eV), and (*iii*) photoionisation of molecules in molecular regions. In a primordial environment the main heating source comes from photodissociation of H_2 and H_2^+, and under the presence of ionising radiation, from the photoionisation of H, and He. The heating is usually expressed as $\Gamma_{\rm ph} = H_{\rm ph} n_i$, where

$$H_{\rm ph} = \frac{4\pi}{h} \int_{E_0}^{\infty} \frac{J(E)\sigma(E)}{E} (E - E_0)\eta(E)\, dE \qquad (18)$$

is the photoheating rate in erg s^{-1} for the given species. Its value depends on the radiation flux per energy $J(E)/E$, the cross-section $\sigma(E)$, the energy excess $E - E_0$, and the efficiency factor $\eta(E)$ that determines the amount of energy released into the gas.

The heating produced by the photodissociation of H_2 discussed in Sec 2.1 is called *UV pumping heating* and it is the sum of two different contributions (see sketch in Fig. 5). The first (releasing 0.4 eV) is the heating produced by the dissociation of H_2 which directly goes into kinetic energy and has an efficiency $\sim$ 10%–15%, as only this fraction of molecules dissociates following this path. The second contribution (which releases 2.2 eV) is due to the de-excitation processes which involve vibrational excited states of the H_2 electronic ground state, and depend on the critical density already introduced in Sec. 5.2.1. Photoheating can be the dominant contribution to the thermodynamics at low-densities where shielding effects are negligible and radiation can penetrate efficiently.

6. Chemical uncertainties and dynamical implications

Although the chemistry of the early Universe is bound to a few elements (H, He, and D), and simple quantum molecular systems (H_2, HD), some of

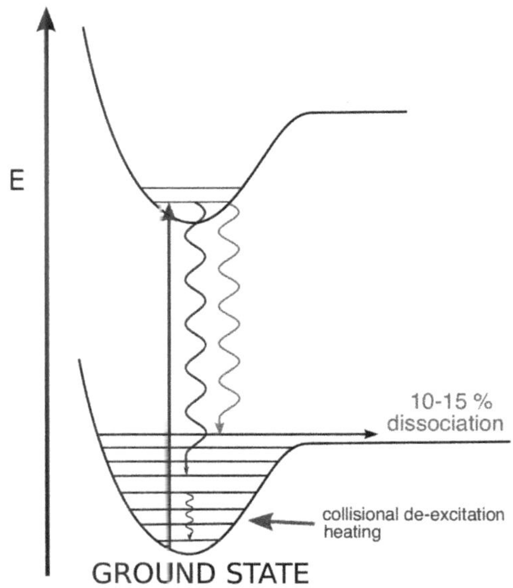

Fig. 5. H_2 UV pumping heating sketch. Dissociation produces direct heating corresponding to 0.4 eV, while collisional de-excitations release up to 2.2 eV with an efficiency proportional to the critical density factor ϵ introduced in Sec. 5.2.1.

the ingredients leading the gas kinetics are affected by uncertainties. This in turn can alter the outcome of computational simulations of the formation and evolution of first stars and primordial black holes.

Particularly relevant are the uncertainties in studying the dynamical evolution of primordial halos irradiated by UV photons. The fate of the gravitational collapse is tied to the formation/destruction of H_2 and its capability to cool the gas down to $\sim$ 200 K. The devoid of H_2 can strongly modify the outcome of the gravitational process and lead for example to the formation of super-massive black holes (Omukai, 2001; Shang *et al.*, 2010; Latif *et al.*, 2014a). As discussed in this chapter, the presence of a strong extragalactic UV background suppresses H_2 directly via photodissociation or indirectly through the photodetachment of H^-. The minimum LW radiation flux required to efficiently destroy H_2 and suppress its cooling is denoted $J_{\rm crit}$. Besides the chemical uncertainties on some of the rate coefficients, it has been shown (Wolcott–Green *et al.*, 2011; Sugimura *et al.*, 2014; Visbal *et al.*, 2014; Agarwal and Khochfar, 2015) that the spectrum shape, and the treatment of the self-shielding are very important for determining the critical flux. Several works reported values of $J_{\rm crit}$ in the range 20–10^5

(see Omukai (2001); Shang *et al.* (2010); Inayoshi and Omukai (2011); Latif *et al.* (2014a, 2015); Sugimura *et al.* (2014)) and large discrepancies have been found between simplified zero-dimensional model and more realistic three-dimensional hydrodynamical simulations (Latif *et al.* (2014a)). These differences span orders of magnitudes which are much more relevant than the factor of 2 uncertainties within the chemical model itself, as outlined e.g. in Glover (2015a,b).

The subsequent chapters will explore the role of chemistry in different environments. Particularly, Chapter 4 discusses the role of primordial chemistry and H_2 cooling during primordial star formation. In Chapter 5, the main emphasis turns to the formation of direct collapse black holes, forming in a primordial gas in the presence of atomic line cooling due to strong radiation backgrounds. This chapter also includes a detailed analysis on the determination of $J_{\rm crit}$ both due to chemistry and the dynamics. How the latter depends on the radiation spectrum is further outlined in Chapter 6. The local chemical conditions, depending on the cosmological context, are further quantified within Chapter 9.

Chapter 4

Formation of the first stars

Ralf Klessen

Institut für Theoretische Astrophysik, Albert-Ueberle-Str. 2, 69120, Heidelberg, Germany

klessen@uni-heidelberg.de

From studying the cosmic microwave background, we know our Universe started out very simple. It was by and large homogeneous and isotropic, with small fluctuations that can be described by linear perturbation theory. In stark contrast, the Universe today is highly structured on a vast range of length and mass scales. In the evolution towards increasing complexity, the formation of the first stars marks a primary transition event. The first generation of stars, the so-called Population III (or Pop. III) build up from truly metal-free primordial gas. They have long been thought to live short, solitary lives, with only one massive star forming per halo. However, in recent years this simple picture has undergone substantial revision, and we now understand that stellar birth in the early Universe is subject to the same complexity as star formation at present day. In this chapter, I review the current state of the field. I begin by introducing the basic concepts of star-formation theory and by discussing the typical environment in which Pop. III stars are thought to form. Then I argue that the accretion disk that builds up in the center of a halo is likely to fragment, resulting in the formation of a cluster of stars with a wide range of masses, and I speculate about how this process may be influenced by stellar feedback, the presence of magnetic fields, the energy input from dark matter annihilation, and the occurrence of large-scale streaming velocities between baryons and dark matter. Finally, I discuss direct and indirect constraints on Pop. III star formation from high-redshift observations and from the search for extremely metal-poor stars in the Milky Way and its satellites.

1. Introduction

The appearance of the first stars marked a primary transition in cosmic history. Their light ended the so-called 'dark ages', and they played a key role in cosmic metal enrichment and reionization, thereby shaping the galaxies and their internal properties as we see them today. Understanding high-redshift star formation is thus important for many areas of modern astrophysics. However, studying stellar birth in the primordial Universe is a relatively young area of astrophysical sciences. Only with the advent of new numerical methods and powerful supercomputers did the numerical modeling of early star formation become feasible. As a consequence, there is still little consensus on the physical processes that govern stellar birth at high redshifts. This chapter aims at providing a concise overview of the current state of the field.

The first generation of stars, the so-called Population III (or Pop. III) built up from truly metal-free primordial gas. They have long been thought to live short, solitary lives, with only one extremely massive star forming in each dark matter halo with about 100 solar masses or more (Omukai, 2001; Abel *et al.*, 2002; Bromm *et al.*, 2002; O'Shea and Norman, 2007). The idea was that first star formation is simple and one only needs to know the initial Gaussian density perturbations of material at very high-redshift which are very well understood, e.g. from measuring the cosmic microwave background, see the Planck Collaboration *et al.* (2016) and the growth of cosmological structures as described in Chapter 2 and the heating and cooling processes in the primordial gas described in Chapter 3 (again, which are thought to be very simple because of the lack of elements heavier than hydrogen and helium). In addition, the early numerical models had considerable limitations. First, they only followed the collapse of the very first object in the first halo to form, since the hydrodynamic timestep decreases as the gas collapses, and so the simulations effectively grind to a halt once the protostar has formed. Second, most calculations did not include the effects of protostellar feedback, which may substantially reduce the accretion rate onto the protostar by dumping energy and/or momentum into the infaling gas.

The ever increasing capabilities of modern supercomputers allow us now to perform significantly improved numerical simulations of early star formation that reveal a much more complex picture. The introduction of numerical techniques to first star formation studies that have been standard repertoire in present-day star formation lead to the conclusion that fragmentation is a wide-spread phenomenon in first star formation (Clark *et al.*, 2011a; Greif *et al.*, 2012), and that Pop. III stars form as members of multiple stellar systems with separations as small as the distance between the Earth and the Sun, see Turk *et al.* (2009); Clark *et al.* (2011b); Greif *et al.* (2011); Smith *et al.* (2011); Stacy and Bromm (2013). Studies that do include radiative feedback (Hirano *et al.*, 2014, 2015; Hosokawa *et al.*, 2016), magnetic fields (Machida *et al.*, 2006, 2008; Schleicher *et al.*, 2009, 2010a; Sur *et al.*, 2010, 2012; Turk *et al.*, 2012; Schober *et al.*, 2012a,b; Bovino *et al.*, 2013b), dark matter annihilation (Smith *et al.*, 2012; Stacy *et al.*, 2014), as well as the primordial streaming velocities (Tseliakhovich and Hirata, 2010; Greif *et al.*, 2011; Maio *et al.*, 2011b; Stacy *et al.*, 2011) add to this complexity. All of these processes are relevant and need to be included in any realistic model. There is agreement now that primordial star formation is equally dynamic and difficult to understand as stellar birth at present days.

For further reading on primordial star formation, we refer to the reviews of Bromm and Larson (2004); Yoshida *et al.* (2012); Bromm (2013), or to the book by Loeb (2010). A good overview with specific focus on the astrochemistry of first star formation is provided by Glover (2005, 2013). The current numerical frontier of high-redshift star formation is summarized by Greif (2015). For further reading on the transition to the second generations of stars and build up of the first galaxies we recommend Bromm and Yoshida (2011) or the textbook by Loeb and Furlanetto (2013). As we calibrate our understanding of primordial star formation with what we know about stellar birth at present days, we also recommend the reviews by Mac Low and Klessen (2004); McKee and Ostriker (2007); Zinnecker and Yorke (2007); Padoan *et al.* (2014); Krumholz (2015); Klessen and Glover (2016) on different aspects of this subject.

In this chapter, we try to provide a brief summary of primordial star formation theory, we discuss some of the difficulties we face in this context and speculate about possible observational constraints from the high-redshift Universe as well as from our local neighborhood in the Galaxy. Next, in Section 2 we introduce some of the basic concepts that are behind any theory of star formation. The environment of Pop. III star formation and the thermodynamic behavior of primordial gas are discussed in Section 3. Section 4 illustrates the complexity of stellar birth in the early Universe. We start with simple one-dimensional collapse models, but then argue that the accretion disks around the first stars are highly susceptible to fragmentation. We also speculate about how stellar feedback, magnetic fields, possible dark matter annihilation, and large-scale streaming velocities between baryons and dark matter may influence the stellar mass spectrum. In Section 5 we discuss indirect constraints on the properties of the first stars from high-redshift observations and from stellar archeology in the solar neighborhood, and we speculate about the possible detection of genuine low-mass Pop. III stars in the Universe today. We summarise and conclude in Section 6.

2. Basic concepts

The mean density of the Sun is $1.4\,\mathrm{g\,cm^{-3}}$, and the numbers for other stars are very similar (Kippenhahn *et al.*, 2012). In contrast to that, the mean density of the gas in the Milky Way at the solar radius is $\sim 10^{-23}\,\mathrm{g\,cm^{-3}}$ (Ferrière *et al.*, 2007), and the mean baryon density of the Universe at present day is with $5 \times 10^{-31}\,\mathrm{g\,cm^{-3}}$ even less (Planck Collaboration *et al.*, 2016). The formation of stars therefore requires a density increase by many orders of magnitude. The only way nature can achieve such enormous density contrasts is by gravitational collapse. All other fundamental forces are either short range, as the strong and weak nuclear forces, or cancel out on average. The electromagnetic force plays no role on global scales because the Universe as a whole is charge neutral and so there is no net force between positive and negative charges. And so it is by far the weakest of the four fundamental forces, gravity, that dominates the large-scale dynamics, simply because it is the only one that is purely attractive and has infinite reach.

2.1. *Stability of spherical gas clouds — Jeans criterion*

As argued above, any theory of star formation needs to be based on the competition between gravitational attraction and a large number of dispersing processes ranging from gas pressure, to the coupling between matter and radiation, magnetic fields, turbulence, cosmic rays to name some of the most important ones (Mac Low and Klessen, 2004; McKee and Ostriker, 2007). In its most basic form, a criterion for gravitational collapse goes back to Jeans (1902), who studied the stability of self-gravitating isothermal gas spheres. He found that these systems have a critical mass,

$$
\begin{aligned}
M_{\mathrm{J}} &= \frac{\pi^{5/2}}{6}\left(\frac{1}{G}\right)^{3/2} \rho^{-1/2} c_{\mathrm{s}}^{3} \qquad (1)\\
&= \frac{\pi^{5/2}}{6}\left(\frac{k}{G}\right)^{3/2}\left(\frac{1}{\mu m_{\mathrm{H}}}\right)^{2} n^{-1/2} T^{3/2}\\
&\approx 50\,\mathrm{M}_{\odot}\,\mu^{-2}\left(\frac{n}{1\mathrm{cm}^{-3}}\right)^{-1/2}\left(\frac{T}{1\mathrm{K}}\right)^{3/2},
\end{aligned}
$$

where the proportionality factor depends on Boltzmann's constant k, on the gravitational constant G, on the proton mass $m_{\rm H}$, and on the mean molecular weight of the gas particles μ. Recall that $\mu \approx 1.22$ for primordial atomic gas, and that $\mu \approx 2.33$ for molecular gas. If this critical mass is exceeded, gravity overwhelms pressure gradients and the system collapses. If the mass is smaller, then gas pressure wins and the sphere expands. The Jeans mass $M_{\rm J}$ depends inversely on the square root of the density ρ and on the third power of the sound speed $c_{\rm s}$. Because of $\rho = \mu m_{\rm H} n$ and $c_{\rm s} = (kT/\mu m_{\rm H})^{1/2}$, we can also write this expression in terms of the number density n and the temperature T as $M_{\rm J} \propto n^{-1/2}T^{3/2}$. In summary, the larger the density the smaller the critical mass for collapse, and the higher the temperature the more stable the system.

The concept of the Jeans mass (1) can be extended to include additional physics by introducing an 'effective' density (say when dealing with multi-component fluids) and an 'effective' temperature or sound speed. For example when the gas is turbulent on scales much smaller than the dynamical scales of interest (Chandrasekhar, 1951a,b; von Weizsäcker, 1951) then the velocity dispersion σ can be simply added to the sound speed,

$$c_{\rm s,eff} = (c_{\rm s}^2 + \sigma^2)^{1/2} \,. \tag{2}$$

Similar has been proposed for the presence of magnetic fields, then we add half of the Alfvén velocity squared, $v_{\rm A}^2 = B^2/4\pi\rho$, in Equation (2), see e.g. Federrath and Klessen (2012). In order to trigger star formation in an otherwise stable medium, either the density needs to increase (say, due to an external compression) or the temperature needs to decrease (by some cooling process). Consequently, a significant fraction of this chapter focuses on a discussion of the various astrophysical processes that can lead to this behavior.

We note that the Jeans criterion can also be formulated in terms of energies. If the absolute value of the potential energy W is more than twice the integral over the pressure U, which is closely related to the internal energy of the system, then collapse occurs. We get

$$\eta_{\rm vir} = \frac{|W|}{U} = \frac{\frac{1}{2}\int d^3x \rho(x)\phi(x)}{\frac{3}{2}\int d^3x P(x)} > 1/2 \,, \tag{3}$$

where ϕ is the potential and P is the pressure. This is simply the scalar virial theorem. Another alternative is to look at typical timescales. If the free-fall time, $\tau_{\rm ff}$, is smaller than the sound crossing time, $\tau_{\rm s}$, then the system collapses. Otherwise sound waves travel fast enough to wipe out inhomogeneities and the system is stable. Both times are defined as

$$\tau_{\rm ff} = \left(\frac{3\pi}{32G\rho}\right)^{1/2} \quad \text{and} \quad \tau_{\rm sound} = \frac{R}{c_{\rm s}}\,, \tag{4}$$

where R is the radius of the sphere.

We can obtain a crude estimate of the accretion rate onto the center of the halo from combining Equations (1) and (4) as

$$\dot{M} = \zeta \frac{M_{\rm J}}{\tau_{\rm ff}} \propto \frac{c_{\rm s}^3}{G}\,, \tag{5}$$

where the factor ζ can take values of up to several tens depending on the actual density profile and on how much the gas mass M exceeds the Jeans mass $M_{\rm J}$, for the classical studies, see Shu (1977); Larson (1969); Penston (1969); Whitworth and Summers (1985). Note that the accretion rate for isothermal collapse only depends on the gas temperature.

2.2. *Stability of rotating disks — Toomre criterion*

In systems that are supported by rotational motions, like protostellar accretion disks or spiral galaxies, the criterion for gravitational instability takes a slightly different form. Besides gas pressure the stabilizing effect of shear in the disk also needs to be taken into account now (Toomre, 1964). The criterion for instability reads

$$Q = \frac{\kappa c_{\rm s}}{\pi G \Sigma} \lesssim 1\,, \tag{6}$$

with the surface density Σ and the epicyclic frequency κ, which for Keplerian disks takes the simple form $\kappa = \Omega$, where Ω is the rotational frequency, see the review by Kratter and Lodato (2016). The criterion was originally derived for infinitesimally thin disks, but it can be extended to thick disks with multiple components (Rafikov, 2001; Elmegreen, 2002) which introduces correction factors of order unity to Equation (6). Again, there are two main pathways towards disk fragmentation. If the mass load onto the disk by accretion from the enclosing gas cloud exceeds its capability to transport material inwards by viscous stresses, Σ increases beyond the critical value and the disk becomes unstable. First, spiral arms form and speed up the inward transport. If this is not enough, these arms become non-linear and interact with each other with run-away collapse occuring in the interception regions. By a similar token, also in the absence of accretion an initially stable disk will become unstable if it is able to cool rapidly enough compared to the viscous time scale, see Gammie (2001).

For accretion disks that are continuously fed by infalling gas from an extended envelope, in our case from the gas that is further out in the halo, the Toomre criterion can be extended. Kratter *et al.* (2010), for example argue that two dimensionless parameters can be used to address the stability of mass-loaded disks. The primary parameter,

$$\xi = \frac{\dot{M}_{\rm in}}{\dot{M}_{\rm disk}} = \frac{c^3_{\rm s,halo}/G}{c^3_{\rm s,disk}/G} = \frac{T^{3/2}_{\rm halo}}{T^{3/2}_{\rm disk}}\,, \tag{7}$$

compares the effective sound speed, or equivalently the effective temperature, in the halo with the corresponding value in the disk. Effective means here that the normal thermal sound speed could be increased by the presence of microturbulence or by the presence of magnetic fields, see Equation (2). The effective temperature is then simply $T_{\rm eff} = \rho c^2_{\rm s,eff}/k$. Equation (7) makes use of the fact that in isothermal collapse models the accretion rate scales as $c^3_{\rm s}/G$, see Equation (5). This relation can also be used to characterize the mass flow through disks in steady state (Kratter and Lodato, 2016). The interpretation of (7) is simple, if the mass load onto the disk from the halo (numerator) exceeds the accretion from the disk onto the central protostar (denominator), then the disk becomes Toomre unstable and will fragment. With other words, instability is likely to occur if the halo gas is effectively hotter than the disk material, say because of the presence of strong microturbulence in the halo or because the disk is so dense that additional cooling processes can work there, see the reviews Glover (2005, 2013). Numerical simulations indicate critical values of $\xi \gtrsim 3$ for fragmentation to set in (Offner *et al.*, 2010).

A secondary parameter,

$$\Gamma = \frac{\dot{M}_{\rm in}}{M_{\rm disk}\Omega}\,, \tag{8}$$

compares the infall timescale $M_{\rm disk}/\dot{M}_{\rm in}$ to the orbital timescale Ω in the disk. If $\Gamma \ll 1$ then the disk radius is governed by the angular momentum of the infalling material and not by viscous spreading. Roughly speaking the parameter Γ compares the relative strength of gravity versus rotation in the halo.

If these concepts are applied to the accretion disks around the first stars, it is found that fragmentation is a widespread phenomenon and that first stars typically form as members of multiple systems with a wide range of masses (see Section 4.2).

3. Environment of first star formation

Before we discuss further details of protostellar collapse and the evolution of the accretion disks around the first stars let us briefly review the large-scale environment and the thermodynamic properties of the primordial gas.

3.1. *Cosmological context*

The formation of the first stars in the Universe occurs in systems which have reached sufficiently large masses so that gas cooling becomes important and baryons can go into run-away collapse within a dark matter halo. In the current ΛCDM paradigm (Planck Collaboration *et al.*, 2016), gravitationally bound objects form in a hierarchical fashion with smaller objects decoupling earlier from the Hubble flow of cosmic expansion (see also Chapter 2). Calculations of the growth of density perturbations in an expanding Universe (Barkana and Loeb, 2001) show that the corresponding Jeans mass scales with redshift z as well as cosmological matter and baryon density parameters, $\Omega_{\rm m}$ and $\Omega_{\rm b}$, as

$$M_{\rm J} \approx 5 \times 10^3 {\rm M}_\odot \left(\frac{\Omega_{\rm m} h^2}{0.14}\right)^{-1/2} \left(\frac{\Omega_{\rm b} h^2}{0.022}\right)^{-3/5} \left(\frac{z+1}{10}\right)^{3/2} . \tag{9}$$

Here h is the value of the Hubble parameter in units of $100\,{\rm km\,s^{-1}\,Mpc^{-1}}$, and the parameters are normalized to the 2015 Planck data. However, the criterion for gas to be bound is not sufficient. It also needs to be able to cool for gravitational collapse to set in and lead to first star formation. Considering the various cooling channels of primordial gas described in Chapter 3, see also Glover (2005, 2013), this leads to another critical mass scale of

$$M_{\rm cool} \approx 6 \times 10^5 {\rm M}_\odot\, h^{-1} \Omega_{\rm m}^{-1/2} \left(\frac{\mu}{1.22}\right)^{-3/2} \left(\frac{z+1}{10}\right)^{3/2} , \tag{10}$$

where again the mean molecular weight μ of primordial gas enters the equation.

We note that below a redshift of $z \approx 40$ the critical mass from cooling $M_{\rm cool}$ becomes larger than $M_{\rm J}$ and so we expect a significant number of halos that are not able to form stars, because their gas is not able to cool sufficiently fast (that is within a Hubble time). We also note that there are several processes that can increase the critical mass for collapse even further. For example the relative streaming velocity between baryons and dark matter (Tseliakhovich and Hirata, 2010; Fialkov *et al.*, 2014) is able to add a turbulent contribution to the effective sound speed in Equation (1) and so can delay collapse (Section 4.5). Similar holds for the slow build up of background radiation in the Lyman–Werner band with photon energies from 11.2 eV to 13.6 eV or cosmic reionization at later times, where the additional heat input or change of chemical composition may again delay the gravitational collapse of gas in halos with masses that exceed $M_{\rm cool}$. Clearly, this is most extreme in the case of halos that lie in close vicinity to sites of first star formation, where this external feedback may lead to the formation of supermassive stars (Agarwal *et al.*, 2012, 2016b), which in turn could be the progenitors of the supermassive black holes that we observe in the centers of some galaxies at high redshift (Mortlock *et al.*, 2011; Wu *et al.*, 2015). The latter will be discussed in more detail in Chapter 5.

Altogether, first star formation is likely to start at a redshift $z \gtrsim 30$ in rare high-sigma fluctuations, then rapidly becomes possible in more and more halos, and reaches a peak rate at redshifts $z \sim 20 - 15$. Although the overall cosmic star formation rate continues to increase (Madau *et al.*, 2014), the rate at which metal-free stars form declines again. Regions in the Universe that are not enriched by supernova ejecta from massive stars become increasingly rare. This is the transition to so-called Population II star formation (see Section 4.7). When exactly the formation of genuine Population III stars ends is difficult to assess, and different models and numerical simulations give vastly different answers. It is conceivable that some very low-density regions in voids in the cosmic web have not yet been polluted by metals and so the formation of genuine Population III stars is still possible.

A typical example of the formation rate of metal-free stars predicted for regions similar to the Local Group is shown in Figure 1. It shows comoving Pop. III star formation rate densities from Magg *et al.* (2018). The semi-analytic calculations are based on a suite of 30 Local Group simulations (Griffen *et al.*, 2016) using four different assumptions about stellar feedback and metal enrichment. Panel (a) gives the fiducial model with radiative feedback and metal enrichment from Pop. III supernovae. The star formation rate has two peaks, one at $z \approx 17$ and one at $z \approx 7$. Panel (b) demonstrates that the metal enrichment from Pop. III supernovae in neighboring halos delays the transition to Pop. II star formation. This model considers only yields from stars within the same halo. Altogether more Pop. III stars are forming. The largest impact comes from switching off ionizing feedback, as illustrated in panel (c). This means that individual halos are able to form Pop. III stars for a longer time and this model leads

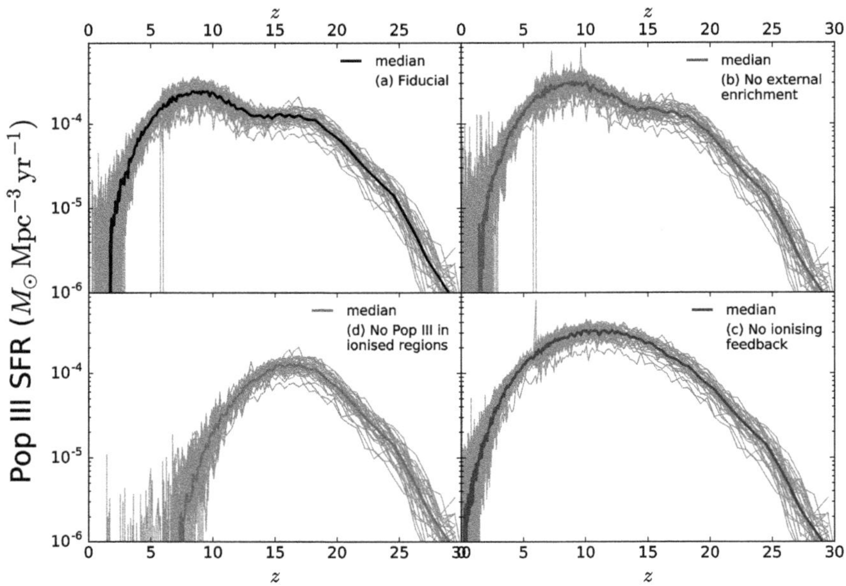

Fig. 1. Comoving Pop. III star formation rate densities in units of $M_\odot\,\mathrm{Mpc}^{-3}\,\mathrm{yr}^{-1}$ and their medians. The values are calculated from a suite of 30 Local Group simulations named Caterpillar, see Griffen *et al.* 2016 using four different assumptions about stellar feedback and metal enrichment. Full details are provided in the main text. Adopted from Magg *et al.* (2018), reproduced by permission of Oxford University Press/on behalf of the RAS.

to the highest Pop. III star formation rates at low redshifts. Panel (d) depicts the model with the most extreme feedback, where star formation is completely shut off inside an ionized region, even in the most massive halos. This model does not allow for Pop. III star formation below a redshift of $z \approx 7$. Overall, Figure 1 demonstrates the impact of environmental conditions on the overall rate of Pop. III star formation and provides some estimate for the uncertainties in the current modeling efforts.

3.2. *Thermodynamic behavior of primordial gas*

As discussed above, the ability of gas to collapse and form stars depends on its thermodynamic behavior as determined by the competition between the available heating and cooling functions. Despite its chemical simplicity several cooling and heating channels exist for zero-metallicity gas, and become important in different density and temperature regimes. All cooling

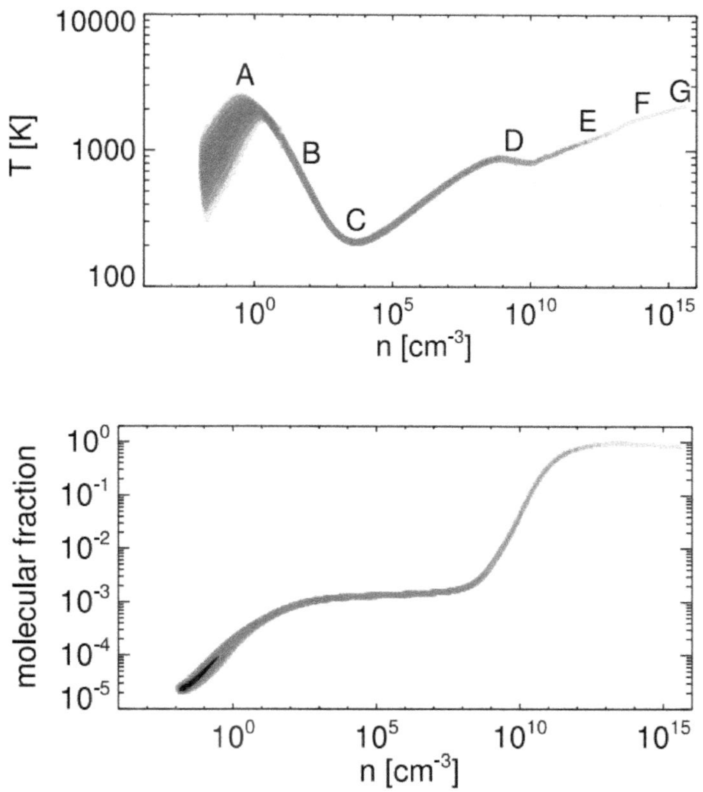

Fig. 2. *Top panel:* Temperature T of metal-free gas as a function of number density n. The labels are discussed in the main text. *Bottom panel:* H_2 fraction as a function of n. Adopted from Yoshida *et al.* (2006), ©AAS. Reproduced with permission.

processes are related to hydrogen, either in atomic or molecular form. At high temperatures, collisions can populate the first excited electronic states of H which then de-excite by emitting Ly-α photons. This process is most efficient around temperatures of $\sim 10^4\,$K. To reach lower temperatures, molecular hydrogen is needed. Since H_2 is a symmetric molecule, it has no permanent dipole moment, and so only quadrupole or higher order transitions are permitted. The corresponding rates are relatively small. In addition, H_2 exists in two states with either parallel nuclear spins (para-hydrogen) or anti-parallel spins (ortho-hydrogen). Transitions between para- and ortho-states are forbidden, so that the lowest allowed transition occurs between the $J = 2$ and $J = 0$ rotational levels in the vibrational ground state of para-hydrogen. The transition energy corresponds to a temperature of $\sim 512\,\mathrm{K}$. The high-velocity tail in the thermal Maxwell–Boltzmann velocity distribution allows the gas to cool down to about 200 K, see Greif (2014). The temperature can drop even further, if cooling by deuterated hydrogen takes over. The HD molecule has a nonzero dipole moment and it is not separated into para- and ortho-states and so its lowest energy transition from the ground state is between the $J = 1$ to $J = 0$ rotational levels, corresponding to a temperature of $\sim 128\,\mathrm{K}$. In practice, HD cooling becomes significant only in regions with enhanced fractional ionization, for example in very massive or in externally irradiated halos. In most sites of Pop. III star formation this process is not important. For a more detailed account of the cooling and heating processes in primordial gas, see Chapter 3 or the reviews by Glover (2005, 2013).

At low densities, $\mathrm{H_2}$ forms mostly by reacting with $\mathrm{H^-}$ (McDowell, 1961; Peebles and Dicke, 1968), which itself requires free electrons to form. The reactions are:

$$\mathrm{H} + \mathrm{e}^- \rightarrow \mathrm{H}^- + \gamma\,, \tag{11}$$

$$\mathrm{H}^- + \mathrm{H} \rightarrow \mathrm{H}_2 + \mathrm{e}^-\,. \tag{12}$$

There is also a contribution from the interaction with $\mathrm{H^+}$ as intermediary molecule (Saslaw and Zipoy, 1967):

$$\mathrm{H} + \mathrm{H}^+ \rightarrow \mathrm{H}_2^+ + \gamma\,, \tag{13}$$

$$\mathrm{H}_2^+ + \mathrm{H} \rightarrow \mathrm{H}_2 + \mathrm{H}^+\,. \tag{14}$$

These chanels lead to typical molecular fractions of about 10^{-3}. At high particle densities above $\sim 10^9\,\mathrm{cm}^{-3}$, the three-body reaction becomes important (Palla *et al.*, 1983):

$$\mathrm{H} + \mathrm{H} + \mathrm{H} \rightarrow \mathrm{H}_2 + \mathrm{H}\,. \tag{15}$$

As a result all atomic hydrogen is converted into $\mathrm{H_2}$ once particle densities of $n \approx 10^{11}\,\mathrm{cm}^{-3}$ are reached. This is illustrated in the bottom panel of Figure 2, which is adopted from Yoshida *et al.* (2006). The top panel of this figure shows the corresponding equilibrium temperature T as function of n. The labels indicate important phases of the collapse. (A) As the gas begins to flow into the potential well of the dark matter halo, it is compressionally heated to the virial temperature of the system. Once sufficient molecular hydrogen is formed the gas goes into a run-away cooling phase (B), which brings it down to the minimum temperature of $\sim 200\,\mathrm{K}$ (C). At this stage of the collapse cold gas can be accumulated in the center of the halo. Bromm *et al.* (2002) call this a 'loitering' phase. When enough gas is accumulated the collapse proceeds and gas slowly heats up again. Eventually three-body H_2 formation becomes important (D) and the gas turns fully molecular. The gas remains roughly isothermal over several decades in density. As n increases further, the cloud slowly becomes optically thick (E) and the temperature rises again. At densities $n \approx 10^{14}\,\mathrm{cm}^{-3}$ (F) collisional induced emission is an important coolant. When two molecules come close to each other, van der Waals forces can induce a temporary dipole which allows for efficient dipole emission during the interaction time interval, for details see Omukai and Nishi (1998); Ripamonti and Abel (2004). Finally, at temperatures around 2000 K collisional dissociation of H_2 sets in (G), and the gas becomes molecular again. Altogether, we note that over more than ten orders of magnitude in density, from $n \approx 10^4\,\mathrm{cm}^{-3}$ to $n \approx 10^{16}\,\mathrm{cm}^{-3}$, the temperature only rises by a factor of 10 at most. The gas roughly follows a polytropic equation of state, $P \propto n^\gamma$ with effective index $\gamma \approx 1.08$ (Omukai *et al.*, 2005). This close to isothermal behavior is essential for allowing the accretion disk in the center of the halo to fragment efficiently (Section 4.2). We note that Figure 2 was calculated in the absence of additional heat sources, the situation may change if radiative feedback from newly formed stars (Section 4.3) or the possible energy input from dark matter annihilation is considered (Section 4.6).

4. Stellar birth in the halo center

In this section we investigate how star formation progresses once the combination of gravitational collapse and cooling leads to a strong accretion flow into the center of a high-redshift halo, and we discuss the most important physical processes that govern stellar birth on small scales. We argue that the accretion disk that forms around the first object is likely to fragment, which typically results in the formation of a cluster of stars with a wide range of masses rather than the build-up of a single high-mass object. We then speculate about how stellar feedback, the presence of magnetic fields, the potential energy input from dark matter annihilation and the possible existence of large-scale streaming velocities between baryons and dark matter may influence this picture.

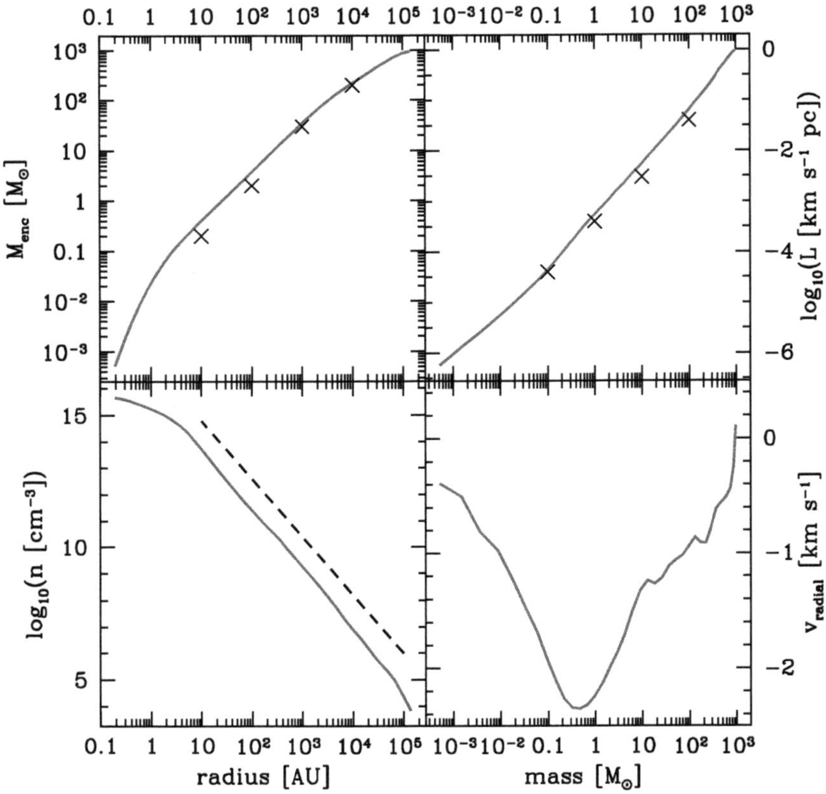

Fig. 3. Physical properties of the inner $\sim 0.1\,\text{pc}$ of a star-forming halo, plotted just before the formation of the first star in the center. *Upper left:* Enclosed mass as function of distance from the center. *Lower left:* Radial number density profile. *Upper right:* Specific angular momentum plotted as a function of the enclosed mass. *Lower right:* Radial inflow velocity as function of enclosed mass. The black crosses represent the corresponding values in the simulation presented by Abel *et al.* (2002). The dashed line represents the power-law behavior of the density, $n \propto r^{-2.2}$, typically found in simulations of Pop. III star formation. Figure from Clark *et al.* (2011b).

4.1. *Simple spherical collapse*

The most simple model we can construct for describing primordial star formation is the spherically symmetric continuation of the run-away collapse that sets in when a halo exceeds the critical mass for cooling $M_{\rm cool}$ as defined in Equation (10) above. This was modeled extensively in the early 2000's, as discussed by Abel *et al.* (2002), Bromm *et al.* (2002), and Yoshida *et al.* (2003). Although the numerical simulations were fully three-dimensional, the halos considered were relatively round and so assuming spherical symmetry and only considering radial profiles was a very good approximation during the early stages of collapse, see also Yoshida *et al.* (2006, 2008) or Bromm and Larson (2004). Figure 3 adopted from Clark *et al.* (2011b) shows enclosed mass and number density as function of radius *(left)*, as well as the specific angular momentum and the radial inflow velocity as function of enclosed mass *(right)*. The gas temperature is $\sim 1500\,\mathrm{K}$ and we can use Equation (5) to obtain a good estimate of the measured accretion rate of $\dot{M} \approx 2 \times 10^{-3}\,\mathrm{M}_\odot\,\mathrm{yr}^{-1}$.

These early calculations typically stopped when the object in the center reached number densities of $n \approx 10^{16}\,\mathrm{cm}^{-3}$ or more, because then the computational timestep became prohibitively small. At this time the protostar has a mass of only $\sim 10^{-3}\,\mathrm{M}_\odot$. At such early stage of evolution the accretion disk has only a small mass and it is strongly sub-Keplerian, meaning that the rotational velocity is smaller than the circular velocity corresponding to centrifugal support, due to the stabilizing influence of large pressure gradients and, if magnetic fields are taken into account, due to the additional support from magnetic pressure and tension. Consequently, this early disk shows no sign of fragmentation at the time these simulations have been stopped. Similar holds for the gas further out in the infalling envelope. The authors of these studies concluded that the same should be true for the entire protostellar accretion history, and so they argued that all the inflowing mass would end up in one single high-mass star, see also Tan and McKee (2004). Clearly this supposition needed to be tested, in particular, because the idea that primordial stars only form in isolation is in tension with present-day star formation where fragmentation is ubiquitous and young stars are typically found in clusters and aggregates Lada and Lada (2003).

4.2. *Disk fragmentation*

In the Universe today, the process of stellar birth is controlled by the intricate interplay between the self-gravity of the star-forming gas and various opposing agents, such as supersonic turbulence, magnetic fields, radiative and mechanical feedback, gas pressure, and cosmic rays. In particular turbulence has been identified to play a key role, see the reviews by Mac Low and Klessen (2004); McKee and Ostriker (2007); Klessen and Glover (2016). On global scales it provides support, while at the same time it can promote local collapse. This process is modified by the thermodynamic response of

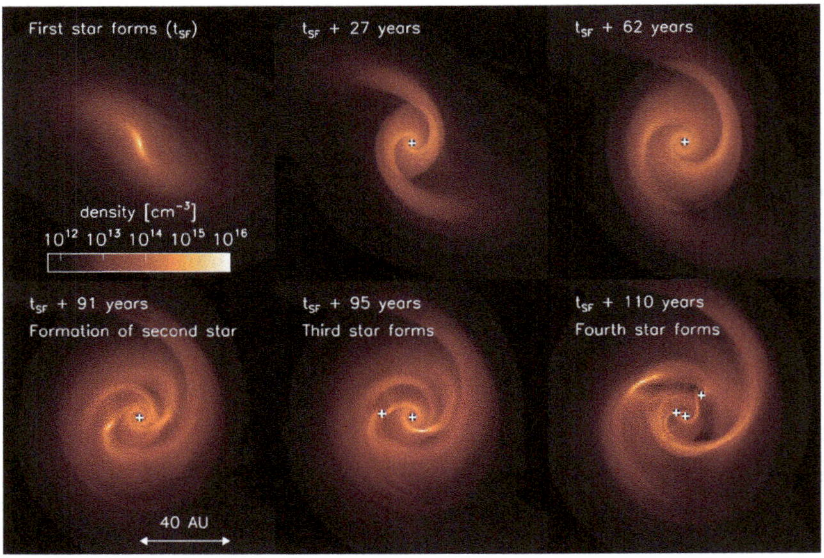

Fig. 4. Density evolution in a 120 AU region around the first protostar, showing the build-up of the protostellar disk and its eventual fragmentation. The prominent two-arm spiral structure is caused by the gravitational instability in the disk, and the resulting gravitational torques provide the main source of angular momentum transport that allows disk material to accrete onto the protostar. Eventually, as mass continues to pour onto the disk from the infalling envelope, the disk becomes so unstable that regions in the spiral arms become self-gravitating in their own right. The disk fragments and a multiple system is formed. Figure adopted from Clark *et al.* (2011b).

the gas, which is determined by the balance between various heating and cooling processes. These in turn depend on the chemical composition of the material. Here, clearly, differences between present-day and primordial star formation are expected (see Section 3.2).

The introduction of techniques that had been standard repertoire in present-day star formation studies, in particular the use of sink particles, allowed continuation of the numerical simulations beyond the first occurrence of a protostar. In this approach, a contracting high-density region is replaced by a single Lagrangian particle when the resolution limit is reached. The particle inherits the mass as well as the linear and angular momentum of the original region, and it continues to accrete mass that falls in at later times, see Bate *et al.* (1995); Krumholz *et al.* (2004); Jappsen *et al.* (2005); Federrath *et al.* (2010); Bleuler and Teyssier (2014); Howard *et al.* (2014); Sormani *et al.* (2017). This technique enables calculations that model the build-up and long-term evolution of the accretion disk, and that follow the entire protostar mass growth history. These studies unambiguously show that primordial accretion disks are highly prone to fragmentation. They indicate that the standard pathway of Pop. III star formation leads to a stellar cluster with a wide distribution of masses rather than the build-up of one single high-mass object. An example is illustrated in Figure 4, adopted again from Clark *et al.* (2011b), which shows the evolution of the accretion disks and the build-up of altogether four protostars within only about hundred years after the formation of the first object.

Disk fragmentation on various spatial and temporal scales is also reported by Machida *et al.* (2008); Greif *et al.* (2011, 2012); Dopcke *et al.* (2013); Susa (2013); Susa *et al.* (2014); Hosokawa *et al.* (2016); Stacy and Bromm (2013); Stacy *et al.* (2016); Turk *et al.* (2009). The reason is always the same. Under typical conditions of Pop. III star formation the mass load onto the disk by accretion from the infalling envelope exceeds its capability to transport this material inwards by gravitational or magnetoviscous torques, that is by spiral arms (Binney and Tremaine, 1987) or by the magnetorotational instability (Balbus and Hawley, 1998). The mass of the disk grows, and it quickly becomes Toomre unstable (Section 2.2). This preferentially occurs at the outer edge of the disk. The instability region moves outwards as the disk grows larger by accretion of higher angular momentum material. And so, fragmentation and the formation of new protostars occur at larger and larger radii as the evolution progresses, as visible in Figure 4.

This has important consequences for the overall accretion history and the resulting mass spectrum. As matter flows through the disk towards the center, it first encounters the Hill volume of secondary protostars further out and preferentially gets swallowed by these objects. Otherwise, this gas would have continued to move inwards and would eventually be accreted by the central object. Clearly, disk fragmentation limits the mass growth of the primary star in the center, and so this process has been termed 'fragmentation-induced starvation' in the context of present-day star formation (Kratter and Matzner, 2006; Peters *et al.*, 2010; Girichidis *et al.*, 2011, 2012b,a). Some of these protostars may get ejected by dynamical encounters with other protostars or fragments while some may move inwards to get accreted by the central object (Clark *et al.*, 2011a; Greif *et al.*, 2012; Smith *et al.*, 2012; Stacy *et al.*, 2016). As a result of these highly unpredictable and stochastic events, the mass spectrum of Pop. III stars is expected to cover a wide range of masses, possibly reaching down into the substellar regime. Possible implications of this are discussed in Section 5.

We note that some numerical studies indicate that fragmentation may also occur on larger scales in the halo, on scales of the star-forming cloud as a whole (Turk *et al.*, 2009; Stacy *et al.*, 2010; Clark *et al.*, 2011b). It may happen when the initial turbulence in the halo gas, that is always present at some low subsonic or transsonic level, gets amplified during gravitational collapse and induces density fluctuations that can go into run-away growth in their own right. This is then very similar to the turbulence-driven mode of star formation that is dominant at present days, for reviews see Mac Low and Klessen (2004); McKee and Ostriker (2007); Krumholz (2015); Klessen and Glover (2016). The importance of this process is enhanced in atomic cooling halos, where cold streams of gas bring dense material towards the center with supersonic velocities (with respect to the cold gas), thereby strongly raising the level of turbulence in the halo (Greif *et al.*, 2008; Wise and Abel, 2007a; Wise *et al.*, 2008). Similar is also expected in halos that are strongly affected by streaming velocities between baryons and dark matter (Section 4.5). It has been suggested that the primordial IMF may be different under more turbulent conditions (McKee and Tan, 2008; Clark *et al.*, 2011b; Maio *et al.*, 2011a; Stacy *et al.*, 2011), however, overall the results are not fully conclusive.

4.3. *Radiative feedback*

Since the protostellar Kelvin–Helmholtz contraction time decreases rapidly with increasing stellar mass, massive stars enter the hydrogen burning main sequence while still accreting (Zinnecker and Yorke, 2007; Maeder and Meynet, 2012). The properties of the resulting star hereby depend very strongly on the details of the mass growth history. First stars with accretion rates below values of $\dot{M} \approx 10^{-3}\,\mathrm{M}_{\odot}$, as we typically expect in low-mass minihalos at high redshift (Section 4.1), are compact and very hot at their surface (Hosokawa and Omukai, 2009; Hosokawa *et al.*, 2010, 2012b). These stars emit copious amounts of ionizing photons (Schaerer, 2002) that can significantly influence their birth environment, as we discuss below. On the other hand, for rates of $\dot{M} \gtrsim 10^{-2}\,\mathrm{M}_{\odot}$ stellar evolution calculations suggest that stars remain bloated and fluffy, and because of the large radius the surface temperature is relatively low. Although very luminous, these stars do not emit much ionizing radiation and could be able to maintain this high accretion flow for a very long time. In the right environment this could possibly lead to the formation of supermassive stars (Hosokawa *et al.*, 2013; Haemmerlé *et al.*, 2016; Umeda *et al.*, 2016; Woods *et al.*, 2017).

Compact and hot Pop. III stars create HII regions which are likely to break out of the parental halo and affect stellar birth in neighboring halos. Many aspects of this problem have been addressed, for example by Kitayama *et al.* (2004), Whalen *et al.* (2004); Alvarez *et al.* (2006); Abel *et al.* (2007); Yoshida *et al.* (2007); Greif *et al.* (2008); Wise *et al.* (2012b,a); Jeon *et al.* (2014a). Here we focus on the impact of radiative feedback on the immediate birth environment of the star and on the question of how this influences the fragmentation behavior of the disk and the resulting stellar mass spectrum.

As before, we can seek guidance from models of present-day star formation. Radiation hydrodynamic simulations in two and three dimensions that take both non-ionizing and ionizing radiation into account (Yorke and Sonnhalter, 2002; Krumholz *et al.*, 2009; Kuiper *et al.*, 2010, 2011; Peters *et al.*, 2010, 2011; Commerçon *et al.*, 2011; Rosen *et al.*, 2016) demonstrate that once a protostellar accretion disk has formed, it quickly becomes gravitationally unstable and so material in the disk midplane flows inwards along dense filaments, whereas radiation escapes through optically thin channels above the disk. Even ionized material can be accreted, if the accretion flow is strong enough (Keto, 2007; Peters *et al.*, 2010). Radiative feedback is thought not to be able to shut off the accretion flow onto massive stars. Instead it is the dynamical evolution of the disk material that controls the mass growth of individual protostars. Accretion onto the central object is shut off by the fragmentation of the disk and the formation of lower-mass companions which intercept inward-moving material as argued in Section 4.2. This requires three-dimensional simulations, because this fragmentation process is not properly captured in two dimensions.

Due to the lack of metals and dust, protostellar accretion disks around Pop. III stars can cool less efficiently and are much hotter than disks at present days. Similarly, the stellar radiation field couples less efficiently to the surrounding because the opacities are smaller. It is thus not clear how well the above results can be transferred to the primordial case. Insight can be gained from the two-dimensional radiation hydrodynamic simulations of Pop. III star formation by Hosokawa *et al.* (2011); Hirano *et al.* (2014, 2015). They follow disk formation and the long-term accretion history of the central object and find that radiative feedback can indeed stop stellar mass growth and blow away the remaining accretion disk, leading to final stellar masses from a few $10\,M_\odot$ up to about $1000\,M_\odot$. However, these calculations cannot capture disk fragmentation and the formation of multiple stellar systems. Three-dimensional calculations have been reported by Stacy *et al.* (2012); Susa (2013); Susa *et al.* (2014); Hosokawa *et al.* (2016). These studies find widespread fragmentation, again with a wide range of stellar masses down to $\sim 1\,M_\odot$. These simulations have their own limitations and shortcomings, either in terms of resolution or in the number of physical processes included. And so, any firm conclusions about the resulting mass spectrum of Pop. III stars in the presence of radiative feedback is premature at this stage. This is clearly one of the frontiers of current research in primordial star formation.

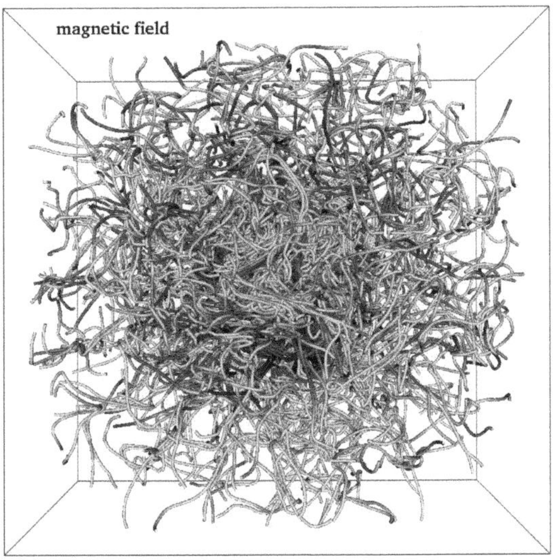

Fig. 5. Magnetid field configuration in the center of a collapsing halo. The highly tangled field lines illustrate nicely the complexity of the field structure in the presence of a turbulent dynamo. The figure is adopted from Federrath *et al.* (2011b), ©AAS. Reproduced with permission.

4.4. *Magnetic fields*

The presence of dynamically important magnetic fields could significantly alter the picture presented so far. We know that the current Universe is highly magnetized on all scales (Beck *et al.*, 1996) and that this influences the birth of stars and the evolution of the interstellar medium, for the extreme viewpoint of magnetically mediated star formation, see Shu *et al.*

(1987). The properties of the magnetic fields observed today are well explained by a combination of small-scale and large-scale dynamo processes (Brandenburg and Subramanian, 2005). In contrast, our knowledge of magnetic fields at high redshifts is very sparse. Theoretical models predict that magnetic fields could be produced in various ways, for example via the Biermann battery (Biermann, 1950), the Weibel instability (Lazar *et al.*, 2009; Medvedev *et al.*, 2004), or thermal plasma fluctuations (Schlickeiser and Shukla, 2003). Other theories place their origin in phase transitions that occur during cosmic inflation (Sigl *et al.*, 1997; Grasso and Rubinstein, 2001; Banerjee and Jedamzik, 2003; Widrow *et al.*, 2012). The resulting fields are thought to be orders of magnitudes too weak to have any dynamical impact, and so, magnetohydrodynamic effects have typically been neglected in numerical simulations of primordial star formation however, see the analytic models by Pudritz and Silk (1989); Tan and McKee (2004); Silk and Langer (2006).

This situation has changed, when people started to realize that the small-scale turbulent dynamo is efficiently amplifying even extremely small primordial seed fields to the saturation level, and that this process is very fast, acting on timescales much shorter than the dynamical free-fall time. An analytic treatment is possible in terms of the Kazantsev model (Kazantsev, 1968; Subramanian, 1998; Schober *et al.*, 2012a,b) for low as well as large Prandtl numbers. This theory describes how the twisting, stretching, and folding of field lines in turbulent magnetized flows (see Figure 4.3) lead to an exponential growth. The amplification timescale is comparable to the eddy-turnover time on viscous or resistive length scales. This is the kinematic regime. Once backreactions become important, the growth rate slows down, and saturation is reached within a few large-scale eddy-turnover times (Schekochihin *et al.*, 2002, 2004; Schober *et al.*, 2015). Depending on the properties of the turbulent flow the magnetic energy density at saturation is thought to lie between 0.1% and a few 10% of the kinetic energy density (Federrath *et al.*, 2011b, 2014).

Magnetic fields with that strength can strongly affect the evolution of protostellar accretion disks. They remove angular momentum from the star-forming gas (Machida *et al.*, 2008; Machida and Doi, 2013; Bovino *et al.*, 2013b; Latif *et al.*, 2013b, 2014d), drive protostellar jets and outflows (Machida *et al.*, 2006), and reduce the level of fragmentation in the disk (Turk *et al.*, 2011b; Peters *et al.*, 2014). We expect Pop. III clusters to have fewer members with somewhat higher masses than predicted by purely hydrodynamic simulations (as discussed in Section 4.2). However, the details of how magnetic fields influence the overall star-formation process in primordial gas, and how they affect the resulting IMF are not well understood. This remains to be an area of very active research.

4.5. *Relative streaming velocities between baryons and dark matter*

Prior to recombination, baryons are tightly coupled to photons resulting in a standing acoustic wave pattern (Sunyaev and Zel'dovich, 1970). In turn, this leads to oscillations between baryons and dark matter with relative velocities of about $30\,\mathrm{km\,s^{-1}}$ and coherence lengths of several 10 Mpc to 100 Mpc (Silk, 1986) at $z \approx 1000$. After recombination, baryons are no longer tied to photons, their sound speed drops to $\sim 6\,\mathrm{km\,s^{-1}}$, and the velocity with respect to the dark matter component becomes supersonic with Mach numbers of $\mathcal{M} \approx 5$. This has been first described by Tseliakhovich and Hirata (2010). In the subsequent cosmic evolution, the relative streaming velocity decays linearly with z, and reaches $\sim 1\,\mathrm{km\,s^{-1}}$ at the onset of first star formation at $z \approx 30$. It is comparable to the virial velocity of the first halos to cool and collapse, and so it has been suggested that this velocity offset strongly influences the star formation process (Tseliakhovich *et al.*, 2011; Fialkov *et al.*, 2012).

Simulations that include streaming velocities indeed suggest that their presence reduces the baryon overdensity in low-mass halos, delays the onset of cooling, and leads to a larger critical mass for collapse to set in (Greif *et al.*, 2011; Stacy *et al.*, 2011; Maio *et al.*, 2011a; Naoz *et al.*, 2012, 2013; O'Leary and McQuinn, 2012; Latif *et al.*, 2014b; Schauer *et al.*, 2017b). They may also have substantial impact on the resulting 21 cm emission (Fialkov *et al.*, 2012; McQuinn and O'Leary, 2012; Visbal *et al.*, 2012). Gas in halos which are located in regions of large streaming velocities will be more turbulent than in more quiescent systems, and so we expect more fragmentation and a bias towards smaller stellar masses (Clark *et al.*, 2008). However, this process has not yet been modeled with sufficient resolution, and so a reliable prediction of the IMF in regions with large streaming velocities is still outstanding.

4.6. *Potential impact of dark matter annihilation*

Despite its importance for cosmic evolution and structure formation, the true physical nature of dark matter is still unknown. Many models introduce a new class of weakly interacting massive particles, so called WIMPs, as they naturally occur in supersymmetry theories. The lightest supersymmetric particle is expected to be stable and to have properties that agree with the phenomenological requirements on dark matter (Bertone *et al.*, 2005). Specifically, they are self-annihilating Majorana particles, that is they are their own antiparticles, and they interact only weakly with baryons.

In most environments the dark matter density is much too low for this to be significant. However, it has been suggested to be different in star-forming halos in the early Universe. Here the collapse of baryons may drag along dark matter particles. This process is called 'adiabatic contraction' (Blumenthal *et al.*, 1986), and it can lead to a density increase of several orders of magnitude in the very center of the halo. As the annihilation rate scales quadratically with the density the corresponding energy input and ionization rate may become large enough to influence the dynamics of the gas. Spolyar *et al.* (2008); Freese *et al.* (2009) realize that this process could potentially overcome the cooling provided by H_2. They speculate that this could halt further collapse and lead to the formation of a dark star, an object of a few AU in size that is powered by dark matter annihilation rather than by nuclear fusion. If dark matter particles also scatter weakly on baryons this could lead to a structure that is stable for a long time. These dark stars are thought to be much larger than normal Pop. III stars, have lower surface temperatures, and are more luminous (Freese *et al.*, 2008; Iocco, 2008; Iocco *et al.*, 2008; Yoon *et al.*, 2008; Hirano *et al.*, 2011).

There are several problems with this scenario. First, it is not clear whether collapse stalls once the energy input from dark matter becomes comparable to the cooling rate. Ripamonti *et al.* (2010), for example argue that this may not be the case because the larger heating rate catalyzes further formation of H_2 and is compensated by the corresponding larger cooling rate. Second, the inherent assumption of perfect alignment between dark matter cusp and gas collapse is most likely violated in realistic star formation conditions. There is always some degree of anisotropy in the halo and there is always some level of turbulence in the infalling gas that make it highly unlikely for the dark matter cusp and the collapsing gas to overlap perfectly. This problem was studied by Stacy *et al.* (2012, 2014), who indeed found that non-axisymmetric perturbations lead to a separation between dark matter cusp and collapsing gas, rendering the annihilation energy input insignificant for dark star formation. A similar report was made by Smith *et al.* (2012), who also found no evidence for a dark stellar phase in their calculations, but instead formed more normal Pop. III stars. These authors, however, also reported that dark matter annihilation was able to influence the dynamics of the accretion disk and that the energy input associated with this process led to a suppression of disk fragmentation. They concluded that dark matter annihilation may reduce the multiplicity of metal-free stars and increase their overall mass. But, as before, the existing simulations are still too premature to draw firm conclusions about the Pop. III IMF.

4.7. *Second generation star formation*

Second generation stars, sometimes termed Population II.5 stars, have formed from material that has been enriched from the debris of the first stars. Unlike the very first stars, for which we have no direct detections yet (Section 5.1), members of the second generation have already been found in surveys looking for extremely metal-poor stars in our Milky Way and neighboring satellite galaxies (Section 5.3). There are two competing models for their origin based on different low-metallicity cooling channels that determine the ability of the gas to collapse and fragment, and so set the stellar mass spectrum.

Unlike purely primordial gas, where cooling is solely provided by hydrogen in various forms (Section 3.2), metal-enriched gas has access to a wide range of different cooling processes. Atomic fine-structure lines from alpha elements such as carbon or oxygen can provide sufficient cooling at number densities around $10^4\,\mathrm{cm}^{-3}$ above a critical metallicity of about 10^{-3} of the solar value (Smith *et al.*, 2009). Details of the element abundance in the Sun are discussed by Asplund *et al.* (2009); Caffau *et al.* (2011b). Up to $n \approx 10^2\,\mathrm{cm}^{-3}$ and metallicities of $\sim 10^{-2}$ solar the cooling rate is dominated by molecular hydrogen (Jappsen *et al.*, 2007, 2009; Glover and Clark, 2014). This can be estimated by calculating the amount of heavy elements required to produce a cooling rate equal to the rate of adiabatic compression heating for given halo properties Bromm and Loeb (2003); Santoro and Shull (2006). Frebel *et al.* (2009) combined the available abundance measures and introduced a transition criterion, with low-mass stars forming only above a certain threshold. This proposition is challenged by the discovery of SDSS J1029151+172927, a truly primitive star in the constellation of Leo, that falls below this value (Caffau *et al.*, 2011a). The star has elemental abundances in the range 10^{-5} to 10^{-4} of the solar value for all of the elements measured in its spectrum, setting it apart from other extremely metal-poor stars which typically have enhanced CNO abundances despite being very iron poor.

Another line of reasoning considers dust cooling as the primary agent for fragmentation, which leads to a lower critical metallicity in the range 10^{-5} to 10^{-6} (Omukai *et al*., 2005; Schneider *et al.*, 2006, 2012; Chiaki *et al.*, 2013b,a). The existence of SDSSJ1029151+172927 provides evidence for the validity of the dust-induced fragmentation model (Schneider *et al.*, 2012; Klessen *et al.*, 2012; Bovino *et al.*, 2016). The analytical models, based on comparing the strength of various cooling and heating processes in a metal-poor environment, are supported by numerical simulations (Tsuribe and Omukai, 2006, 2008; Clark *et al.*, 2008; Dopcke *et al.*, 2013; Chiaki *et al.*, 2014, 2016). These suggest that indeed dust is responsible for a transition to a star formation mode similar to the one observed at present days. This transition occurs at a metallicity of about 10^{-5} solar and leads to an IMF peaking below $1\,M_\odot$ with a functional form similar to the one inferred for the solar neighborhood (Kroupa, 2002; Chabrier, 2003).

5. Observational constraints

Determining the properties of the first population of stars is a difficult task. Most of our current knowledge of Pop. III stars comes from theoretical model building and numerical simulations. To a large degree, this is due to the fact that, unlike in present-day star formation, stringent observational constraints are rare and extremely difficult to obtain. Here we provide a brief overview of possible indirect constraints from the primordial as well as the present-day Universe, and we speculate about possible detections of genuine low-mass Pop. III stars in Galactic archeological surveys.

5.1. *Indirect constraints from high redshift observations*

Unfortunately, Pop. III stars are very hard to see directly in the high-redshift Universe. They are too faint (Schaerer, 2002) to be within reach of the next generation of telescopes. This holds in space for the James Webb Space Telescope (Gardner *et al.*, 2006), Euclid (Laureijs *et al.*, 2011), or the Wide-Field Infrared Survey Telescope (Spergel *et al.*, 2015), and it is true for ground-based observatories such as the European Extremely Large Telescope (Tamai and Spyromilio, 2014) or the Thirty Meter Telescope (Skidmore *et al.*, 2015). Constraints on the IMF from high-z observations therefore are indirect at best. For example we can search for the supernova explosions that accompany the end of massive Pop. III stars. Numerical simulations indicate that the light-curves of Pop. III pair-instability supernovae and even less energetic core collapse supernovae will be within easy reach of JWST or E-ELT out to very high redshifts (Hummel *et al.*, 2012; Kasen *et al.*, 2011; Pan *et al.*, 2012; Whalen *et al.*, 2013b,a). However, because of their narrow field of view, and for the space missions also because of the long slewing time, these telescopes are not suited for large area surveys and will have difficulties to find these supernovae. Lists of suitable target candidates are needed. Here, gravitational lensing might help and bring some of the supernova events above the sensitivity limit of large-area surveys such as planned with LSST, WFIRST or Euclid, see also (Oguri and Marshall, 2010, Rydberg *et al.*, in prep.). Once a candidate is found, JWST or E-ELT can be used for detailed follow-up observations. The detection rates in large surveys furthermore can contribute to distinguish between different cosmological models (Magg *et al.*, 2016).

Studying the high-redshift Universe will only allow us to address the high mass end of the IMF. Information about the low-mass Pop. III stars is completely inaccessible via this route. Similarly this is true for potential detections of gravitational wave emission from merging Pop. III black hole binaries. The recent discoveries of gravitational waves from three systems with masses of $62\,\mathrm{M}_\odot$, $21\,\mathrm{M}_\odot$, and $49\,\mathrm{M}_\odot$ (Abbott *et al.*, 2016a,b, 2017b) have demonstrated that this might indeed be possible. Mergers of very massive Pop. III remnants could contribute to the statistical background (Inayoshi *et al.*, 2016; Sasaki *et al.*, 2016) or may even be detected directly (Belczynski *et al.*, 2017; Hartwig *et al.*, 2016b; Nakama *et al.*, 2017). More details on detecting mergers of black hole binaries from Pop. III stars via gravitational wave emission will be given in Chapter 13. We should also mention studies by Kulkarni *et al.* (2013, 2014) who investigate the impact of the chemical enrichment from Pop. III stars on damped Lyman-α systems and on cosmic reionization, and find that this could be a sensitive probe of the primordial IMF at high redshift.

At the extreme end of the mass spectrum, we note that the discovery of very massive quasars with several $10^9\,\mathrm{M}_\odot$ at redshifts of $z \approx 6$ or above (Mortlock *et al.*, 2011; Wu *et al.*, 2015) provides support for the existence of supermassive stars in the early Universe. We note that stellar evolution calculations with very high accretion rates suggest that stars could be stable up to several $10^5\,\mathrm{M}_\odot$ before the general relativistic instability (Iben, 1963; Chandrasekhar, 1964) leads to the collapse into a black hole of the same mass (Hosokawa *et al.*, 2013; Umeda *et al.*, 2016; Woods *et al.*, 2017). These could be the seeds for the observed very massive quasars when able to maintain high accretion rates close to the Eddington limit down to $z \approx 6$. The formation of such objects via gravitational collapse will be described in Chapter 5, and their subsequent evolution in Chapter 8.

5.2. *Indirect constraints from observations in the Local Group*

More stringent limits on the IMF are likely to come from the study of nearby stars in the Local Group. Current Galactic archeological surveys (Beers and Christlieb, 2005; Caffau *et al.*, 2013; Frebel, 2010) in the halo and bulge of our Milky Way or the analysis of stars in nearby satellite dwarf galaxies (Koch *et al.*, 2013; Salvadori *et al.*, 2015; Kirby *et al.*, 2015; Roederer *et al.*, 2016; Ji *et al.*, 2016a,b) can contribute to our knowledge of primordial stars in several ways. There is a large body of work that uses the abundance pattern determined in extremely metal-poor stars to infer the properties of the progenitor stars which provided the enrichment (Heger and Woosley, 2002, 2010). Assuming that the oldest and most metal-poor stars in the Galaxy have been supplied with heavy elements by only one or at most two supernova explosions (Chan and Heger, 2017), it turns out that the measured relative abundances of heavy elements are most consistent with core collapse supernovae from Pop. III stars in the mass range $20 - 40\,\mathrm{M}_{\odot}$ (Aoki *et al.*, 2014; Bonifacio *et al.*, 2015; Caffau *et al.*, 2012; Cooke and Madau, 2014; Frebel *et al.*, 2005; Iwamoto *et al.*, 2005; Joggerst *et al.*, 2009, 2010; Keller *et al.*, 2014; Lai *et al.*, 2008; Norris *et al.*, 2013). Together with the fact that no genuine signatures of pair-instability supernovae from massive stars in the range of ~ 130 to $250\,\mathrm{M}_{\odot}$ have been found, this places strong constraints on the high-mass end of the primordial IMF.

5.3. *Possible direct detections of low-mass Pop. III stars*

The theory of stellar evolution (Kippenhahn *et al.*, 2012) tells us that any low-mass Pop. III star with $0.8\,\mathrm{M}_{\odot}$ or smaller must have survived until the present day. If these stars ever existed, then there is a chance to directly detect some of them in current or future stellar archeological surveys. Even non-detections allow us to put stringent limits on the low-mass end of the Pop. III IMF. For example Hartwig *et al.* (2015a) estimate the expected numbers of low-mass Pop. III stars in the Galactic halo based on semi-analytic models of the early star formation history in Milky Way-like halos. They conclude that if no such object is found in surveys with sample sizes of 4 million stars then we can exclude the existence of low-mass Pop. III stars with masses below $0.8\,M_\odot$ with a confidence level of 68%. This limit may soon be reached.

On similar grounds, Salvadori *et al.* (2007, 2010) develop a detailed model of the metallicity distribution function of metal-poor stars in the Galactic halo and suggest that Pop. III stars should be more massive than $0.9\,M_\odot$ when matching their predictions with the observational data. On the other hand, Tumlinson (2006, 2010) argues that the current abundance measurements are not really sufficient to distinguish between different models of the primordial IMF based on studying the chemical evolution during the early build-up of the Galaxy. However, he also suggests that characteristic masses of a few $10\,M_\odot$ provide a better fit to the available data than masses of $100\,M_\odot$ or above, consistent with the supernova yields mentioned above.

Mapelli *et al.* (2006) focus on the density of Galactic intermediate-mass black holes. They take them as the relics of higher-mass Pop. III stars and derive an upper limit by comparing their model to the non-detection of ultra-luminous x-ray sources in the Galaxy. This constraint, however, does not affect the low-mass end of the primordial IMF. The approach has been extended by de Bennassuti *et al.* (2014) who also conclude that the Galactic metallicity function and the abundances measured in extremely metal-poor stars in the halo are inconsistent with pair-instability supernovae. Their study indicates that the early enrichment of the Galaxy by Pop. III stars was dominated by core collapse supernovae, and they suggest an IMF that goes all the way down to $0.1\,M_\odot$. Altogether, the prospects of finding surviving low-mass first stars in our immediate neighborhood are highly exciting, but at present no clear conclusions are possible.

6. Summary

In this chapter of the book we aimed at providing a concise overview of our current understanding of the physical processes that govern stellar birth in the primordial Universe. As the first stars are too faint to be directly detectable at high redshift, the available observational data give little input and provide very indirect constraints at best. And so, most of the current progress in this rapidly evolving field of astrophysical research has been driven by numerical simulations and theoretical model building.

While early models predicted that the first stars formed in isolation with only one single high-mass star in the center of a halo, the community now acknowledges that the birth of stars in truly metal-free gas is subject to the same complexity and stochasticity that is well known in present-day star formation. Probably the most important development in the past decade is the realization that the accretion disks that build up around the first stars are highly susceptible to fragmentation. The natural formation pathway therefore leads to Population III stars that are members of multiple stellar systems and clusters. This has important consequences for the stellar mass spectrum. Numerical simulations of first star formation that are able to resolve the evolution of the accretion disk and the mass growth of individual protostars with high resolution predict a wide range of masses, with a relatively flat distribution that spans the substellar regime up to several hundred solar masses, and with the most likely values being around a few tens of solar masses. This is different from the IMF observed at present day, which shows a clear peak at $\sim 0.3\,M_\odot$ and a power-law fall-off towards higher masses. So, one of the key questions is when and where did the transition between the primordial and the current mode of star formation occur, and what is its primary physical cause.

A convincing answer to this question is still outstanding. Besides the lack of a good observational base, the primary reason for this is that all existing calculations have severe shortcomings. And so still very little is actually known about the mass spectrum of Pop. III stars. The current numerical simulations either lack sufficient resolution, or they cover only a short fraction of the overall star formation timescale. Most of them ignore important physical processes, such as protostellar feedback or magnetic fields. It may also be that more speculative phenomena such as heating from dark matter annihilation or the relative streaming velocity between baryons and dark matter are important for the star formation process in the center of a halo. The theoretical and numerical treatment of these processes is still in its infancy. Furthermore, some simulations are two-dimensional, whereas a full three-dimensional approach is required to properly capture gas fragmentation or to fully assess the dynamical impact of stellar radiation on the infalling gas. So, correctly modeling stellar birth in high-redshift halos with high physical fidelity with strong predictive power is still an unsolved problem. It is a true frontier of computational astrophysics.

It is probably fair to say that most researchers in the field agree that the genuine Pop. III stars have formed in binary or higher-order multiple stellar systems and that they most likely had a wide range of masses. Whether the mass scale reached down to low-mass stars or even the sub-stellar regime depends on physical processes that we only now start to consider properly, such as radiative feedback, the level of turbulence in halo gas and accretion disk, properties of the dark matter, or the larger scale environment of the star forming halo. Since low-mass Pop. III stars must have survived until the present days, they should be detectable in stellar archeological surveys in the Milky Way and neighboring satellite galaxies, and so there are good chances that we shall soon have stringent constraints on the low-mass end of the Pop. III IMF. The abundance patterns we observe in extremely metal-poor stars nearby are all consistent with enrichment by core collapse supernovae with progenitor masses of a few tens of solar masses. This is in agreement with the most likely stellar masses from fragmentation calculations of primordial accretion disks. The fact that no signatures of enrichment by pair-instability supernovae are seen, suggests that Pop. III stars with masses above $100\,M_\odot$ must have been extremely rare. How far up the mass range of primordial stars extends is not clear yet. Theoretical limits from including general relativistic instabilities in the stellar structure equations predict an upper mass limit of several $10^5\,M_\odot$. Indeed, such stars could be the seeds for the extremely massive quasars with billions of solar masses we observe at redshifts of $z \approx 6$ and higher. However, the density of these objects is only a few per Gpc^3, and so our statistical base is scarce.

In this chapter, we have focused on what could be called the standard pathway to Pop. III star formation. We have considered cooling and collapse of truly pristine gas that has not been affected by any form of feedback from earlier episodes of star formation either within the same halo or in neighboring halos. The temperature and composition of the material has not been altered by strong Lyman–Werner or ionizing fluxes, and there was no enrichment from the supernovae of previous stars. These effects are discussed in detail in other chapters of this book. Particularly, Chapter 5 will discuss the formation of very massive black holes via direct collapse, in situations where the cooling is suppressed by radiation backgrounds. An alternative formation channel via collisions in dense stellar clusters will be presented in Chapter 7, and the evolution of supermassive stars is discussed in Chapter 8. Predictions for gravitational wave emission from black hole binary mergers, including remnants from Pop. III stars, are summarised in Chapter 13.

7. Acknowledgements

Writing this text would not have been possible without the input and the support of many friends and colleagues in Heidelberg and around the world.

I would also like to thank Christoph Federrath, Simon Glover, and Naoki Yoshida for carefully reading this manuscript and for providing very helpful feedback. I also acknowledge financial support by the European Research Council via the ERC Advanced Grant *STARLIGHT: Formation of the First Stars* (project number 339177), and by the Deutsche Forschungsgemeinschaft (DFG) in the Collaborative Research Centre (SFB 881) *The Milky Way System* (subprojects B1, B2, and B8) and in the Priority Program SPP 1573 *Physics of the Interstellar Medium* (grant numbers KL 1358/18.1, KL 1358/19.2).

Chapter 5

Black hole formation via gas-dynamical processes

Muhammad A. Latif
Physics Department, College of Science,
United Arab Emirates University, UAEU
malatif@uaeu.ac.ae

Understanding the formation of earliest supermassive black holes is a question of prime astrophysical interest. In this chapter, we focus on the formation of massive black holes via gas dynamical processes. The necessary requirement for this mechanism are large inflow rates of about 0.1 solar mass per year. We discuss how to obtain such inflow rates via an isothermal collapse in the presence of atomic hydrogen cooling, and the outcome of such a collapse from three-dimensional cosmological simulations in Subsection 2.2. Alternatives to an isothermal direct collapse are discussed in Subsection 3 which include trace amounts of metals and/or molecular hydrogen. In the end, we briefly discuss future perspectives and potential detection of massive black hole seeds via upcoming missions.

1. Introduction

Understanding the formation of supermassive black holes remains an open and fascinating issue, as outlined in detailed reviews on this topic by Volonteri (2010); Volonteri and Bellovary (2012); Haiman (2013); Latif and Ferrara (2016). The three main pathways for the formation of supermassive black holes are: (1) stellar remnants, (2) seed black holes forming in dense stellar clusters via dynamical processes, (3) monolithic collapse of protogalactic gas cloud into a massive black hole so-called direct collapse scenario. In case of stellar remnants, the most promising pathway is likely in the context of massive Population III stars, as introduced previously in Chapter 4. Such a mechanism, if it was to produce the observed supermassive black holes, would certainly require a few episodes of super-Eddington accretion, as described in Chapter 11, which may however also be present in other scenarios. Black hole formation via collisions in stellar clusters will be discussed in Chapter 7, and the observed masses and constraints on the supermassive black hole population at high redshift are reported in Chapter 12. In this chapter, the main focus will be on black hole formation via a monolithic collapse.

The idea for the formation of a massive black hole directly via the gas dynamical processes was conceived in the pioneering work of Martin Rees (Rees, 1984). The expectation is that gas in the low spin halos collapses on the viscous time scale, forms a rotationally supported compact disk which later may lead to the formation of a massive black hole (BH) (Loeb and Rasio, 1994; Eisenstein and Loeb, 1995). Similarly, models proposed in the cosmological context suggest that conditions for the formation of a massive black hole are ideal in high redshift protogalaxies with the lowest angular momentum gas (Koushiappas *et al.*, 2004; Volonteri and Rees, 2005; Lodato and Natarajan, 2006). It is proposed that gas can rapidly lose angular momentum via 'bars within bars' instabilities and may lead to rapid formation of a self-gravitating core supported by gas pressure. Such a core catastrophically cools by thermal neutrino emission and contracts to potentially form a massive BH (Begelman *et al.*, 2006). The first smoothed particle hydrodynamical simulations starting from idealised initial conditions showed that a low spin metal free halo collapses into a single clump while higher spin halo formed a binary (Bromm and Loeb, 2003).

The key requirement for this scenario is that gas should rapidly collapse by efficiently transporting angular momentum and avoiding fragmentation. The goal is to bring large inflows ($\geq 0.1\ \mathrm{M_{\odot}/yr}$, see discussion below) of gas into the halo centre within a short time scale of about ~1 Myr and rapidly build up a massive object of $10^4 - 10^6\ \mathrm{M_{\odot}}$. Such large inflow rates can be obtained thermodynamically by keeping the gas warm as the inflow rate is $\propto T^{3/2}$ and also via dynamical processes such as 'bars within bars' instabilities (Shlosman *et al.*, 1989; Begelman *et al.*, 2006) or in the aftermath of galaxy mergers (Mayer *et al.*, 2010). Depending on the time evolution of mass inflow rates, the central object may form a supermassive star/quasi-star (details are mentioned below) or directly collapse into a massive black hole.

2. Black hole formation in primordial atomic gas

The mass inflow rate ($\dot{M}$) of collapsing gas is related to its thermodynamical properties, as $\dot{\mathrm{M}} \sim \mathrm{c_s^3/G} \sim 0.1\ \mathrm{M_{\odot}/yr} \left(\frac{\mathrm{T}}{8000\ \mathrm{K}}\right)^{3/2}$ where c_s is the sound speed and T is the gas temperature. The higher the sound speed the larger the mass inflow rate. Therefore, the thermodynamical requirement for getting large inflows is that gas should not cool down to lower temperatures, otherwise it will fragment and form ordinary stars. The cooling ability of the gas strongly depends on its chemical composition. In the presence of a trace amount of dust/metals, the gas cools down to a few tens of Kelvin by radiating away its thermal energy and forming stars. Even in primordial gas, molecular hydrogen cooling can bring the gas temperature down to $\sim$ 200 K and induces star formation. In the absence of molecular hydrogen, primordial gas remains in the atomic phase, cools mainly via atomic line radiation and the gas temperature remains around 8000 K. In atomic primordial gas, collapse is expected to proceed isothermally with T $\sim$ 8000 K and large mass inflow rates of the order of 0.1 $M_{\odot}$/yr can be achieved easily. Therefore, the conditions for forming a massive object are ideal in massive primordial halos with $T_{vir} \geq 10^4$ K cooled only via atomic lines. Such halos have masses of a few times $\geq 10^7$ $M_{\odot}$, formed at $z > 10$ and their gravitational potentials are sufficiently deep to allow the rapid collapse. Therefore, massive primordial halos deprived of H_2 cooling are the potential cradles for the formation of massive black holes.

The prerequisites for the formation of massive black holes in atomic cooling halos are that they should be metal-free and the formation of molecular hydrogen remains suppressed. In the next sub-section, we discuss in detail how to quench the molecular hydrogen formation which could be detrimental for forming massive black holes via isothermal direct collapse (DC).

2.1. *Conditions to keep the gas primordial and atomic*

Our understanding of structures in the cosmos is based on the hierarchical paradigm of structure formation according to which minihalos of $10^5 - 10^6$ $M_{\odot}$ were formed first at $z \sim 30 - 40$ which later merged to form larger halos. According to the Big Bang theory, the gas in the Universe initially has a primordial composition of predominantly hydrogen and helium. The first stars, so-called Population III stars, will then form out of a primordial gas in these minihalos. As structure formation proceeds, depending on their mass, these stars end their lives as supernovae and pollute the ambient medium with metals. At earlier cosmic times, a few hundred Myrs after the Big Bang, the metal enrichment is expected to be patchy and also found from the large scale numerical simulations (Trenti *et al.*, 2009; Maio *et al.*,

2011a; Ritter *et al.*, 2015; Pallottini *et al.*, 2014; Habouzit *et al.*, 2016b). Therefore, some of the halos may remain pristine until their masses reach the atomic cooling regime (above 10^7 $M_\odot$). Indeed, there is observational evidence that pockets of metal free gas can exist down to z=7 (Simcoe *et al.*, 2012) and even at z=3 (Fumagalli *et al.*, 2011). The estimates about the fraction of metal free halos computed from cosmological simulations including self-consistently star formation and supernova feedback suggest that about 40 % of halos remain metal free down to z=10 (Latif *et al.*, 2016a; Habouzit *et al.*, 2016a), see Figure 1.

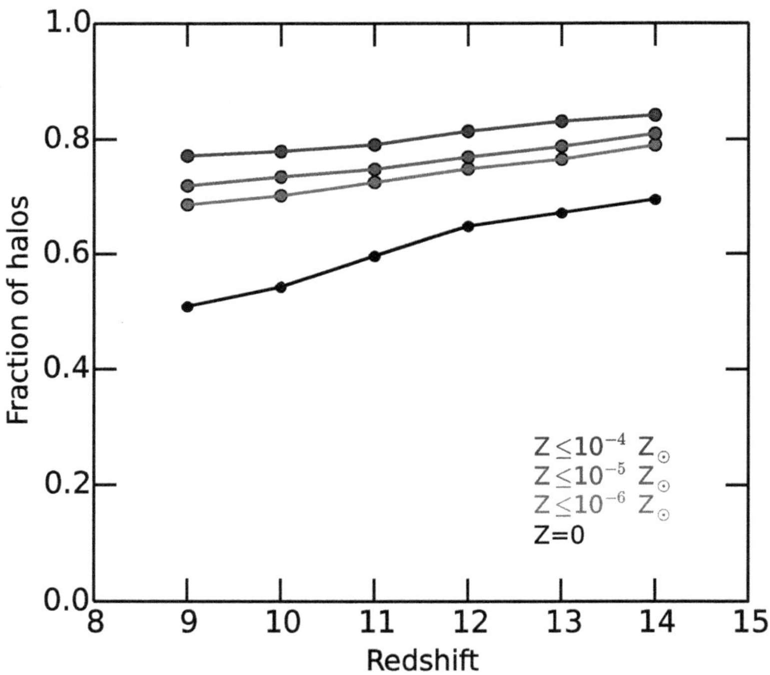

Fig. 1. Fraction of halos with metallicity below the given value in the figure legend and masses between $2 \times 10^7 - 10^8$ $M_\odot$. Adopted from Latif *et al.* (2016a), ©AAS. Reproduced with permission.

As we discuss below, due to the requirement of a strong UV flux, a DC halo may form in the surroundings of an intensely star forming galaxy. This may affect the abundance of DC halos by metal pollution from supernova winds from a nearby star forming halo (Dijkstra *et al.*, 2014; Habouzit *et al.*, 2016b). Even such pollution can be avoided in a synchronised pair of halos where the halo which is the radiation source forms first while a rapid collapse in the DC halo helps in avoiding the metal pollution (Visbal *et al.*, 2014). Moreover, cosmological hydrodynamical simulations starting from first principles show that metal ejection preferentially occurs in the low density regime (Ritter *et al.*, 2015; Pallottini *et al.*, 2014) and neighboring halos may remain metal free. In a nutshell, metal pollution is not a bottleneck in the formation of DCBHs.

The second main constraint for DCBHs is that the formation of H_2 should remain suppressed in DC halos. Trace amount of H_2 can be formed via gas phase reactions where a residual fraction of electrons from the recombination epoch acts as a catalyst. As discussed in the previous chapter on the chemistry of the early universe, the main pathway for the formation of H_2 is the following:

$$\mathrm{H} + \mathrm{e}^- \rightarrow \mathrm{H}^- + \gamma \tag{1}$$

$$\mathrm{H} + \mathrm{H}^- \rightarrow \mathrm{H}_2 + \mathrm{e}^-. \tag{2}$$

The H_2 can be dissociated either directly or indirectly by UV radiation depending on the stellar spectra. The low energy photons with 0.75 eV can photo-detach H^- which is the main channel for the formation of H_2. The photons with energy between 11.2–13.6 eV, so-called Lyman–Werner (LW) photons directly photo-dissociate molecular hydrogen via the Solomon process. The dissociation processes are described by the following reactions:

$$\mathrm{H}_2 + h\nu \rightarrow \mathrm{H} + \mathrm{H} \tag{3}$$

$$\mathrm{H}^- + h\nu \rightarrow \mathrm{H} + \mathrm{e}^-. \tag{4}$$

The competition between the formation and dissociation timescales defines the critical value of UV flux (hereafter J_{21}^{crit}) above which the formation of H_2 remains quenched. Pop. III stars with $T_{rad} = 10^5$ K produce more high energy photons and are very effective in directly dissociating H_2 while normal stars with $\mathrm{T_{rad}} = 10^4$ K photo-detach H^-. The previous studies employed idealised spectra to compute the J_{21}^{crit} and found that it varies from 30–300 for $\mathrm{T_{rad}} = 10^4$ K and 1000 for $\mathrm{T_{rad}} = 10^5$ K (Omukai, 2001; Shang *et al.*, 2010; Latif *et al.*, 2014a; Johnson *et al.*, 2014). However, recent estimates of J_{21}^{crit} for realistic spectra of the first galaxies suggest that it has no single value but it depends on the stellar age, metallicity and mode of star formation (Sugimura *et al.*, 2014; Agarwal and Khochfar, 2015; Agarwal *et al.*, 2016b). Moreover, such spectra can be mimicked with $T_{rad} = 2 \times 10^4 - 10^5$ K. The values of J_{21}^{crit} from cosmological simulations for realistic spectra of first galaxies found that it further depends on the properties of the host halos and varies between 20,000–50,000 considering a uniform background UV flux (Latif *et al.*, 2015) or anisotropic flux (Regan *et al.*, 2014, 2016a). X-rays catalyse H_2 formation by boosting electron fraction and consequently the value of J_{21}^{crit} may further get enhanced (Latif *et al.*, 2015; Inayoshi and Tanaka, 2015; Regan *et al.*, 2016b), see Figure 2. The accurate treatment of H_2 self-shielding is also necessary to calculate J_{21}^{crit} (Wolcott–Green *et al.*, 2011; Hartwig *et al.*, 2015c).

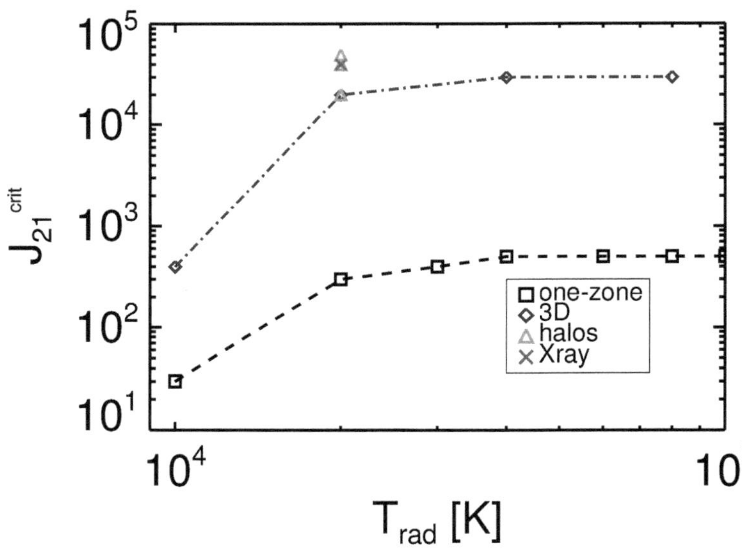

Fig. 2. Estimates of the critical value of the UV flux ($J_{21}^{\rm crit}$) both from one zone models and 3D simulations including variations from halo to halo, dependence on the radiation spectra and the impact of X-ray ionization. Adopted from Latif *et al.* (2015), reproduced by permission of Oxford University Press / on behalf of the RAS.

Estimates of J_{21}^{crit} and the fraction of metal free halos are crucial to estimate the number density of direct collapse black holes and to assess the feasibility of this scenario. A detailed assessment of its dependence on the stellar spectra reflecting our present understanding is therefore provided in Chapter 6.

2.2. *Outcome of an isothermal collapse*

In this section, we discuss the outcome of an isothermal collapse in massive primordial halos with $T_{vir} \geq 10^4$ K illuminated by a strong LW flux above the critical strength. Numerical cosmological simulations performed under these conditions confirm that the gas collapses isothermally in dark matter potentials with T $\sim$ 8000 K in a self-similar way and the density profile follows an R^{-2} behaviour (Bromm and Loeb, 2003; Wise *et al.*, 2008; Regan and Haehnelt, 2009; Latif *et al.*, 2011b, 2013a; Inayoshi *et al.*, 2014; Becerra *et al.*, 2015; Latif *et al.*, 2016b). The gas continues to cool by Lyman alpha radiation and keeps collapsing. At densities of 10^8 cm^{-3}, cooling due to H^- comes into play and brings the gas temperature down to 5000 K (Van Borm *et al.*, 2014; Latif *et al.*, 2016b). Above 10^{16} cm^{-3}, the gas cloud becomes optically thick to both Lyman alpha and H^- cooling, and the temperature starts to rise as shown in Figure 3. The collisional ionisation cooling becomes important and maintains the gas temperature close to 10^4 K. Eventually, at densities higher than 10^{20} cm^{-3}, the gas cloud becomes completely opaque and collapses adiabatically.

Under isothermal conditions, the gas is expected to collapse monolithically without fragmentation. Previous low resolution studies employing a fixed Jeans resolution of four cells confirmed this hypothesis (Regan and Haehnelt, 2009; Latif *et al.*, 2011b). However, recent work employing a detailed chemical model, resolving the collapse to unprecedentedly high densities of 10^{21} cm^{-3} with a Jeans resolution of 32 cells, shows that fragmentation occasionally occurs depending on the properties of host halos, but does not prevent the formation of a massive central object (Becerra *et al.*, 2015; Latif *et al.*, 2016b). The clumps forming due to the fragmentation on small scales quickly migrate inwards and merge with the central object (Inayoshi and Haiman, 2014; Latif and Schleicher, 2015a). Analytical models of primordial disks around the central object suggest that in the presence of large inflows and rapid rotation, viscous heating stabilises the disk and helps in the formation of a massive central object (Latif and Schleicher, 2015b; Schleicher *et al.*, 2016). Similarly, magnetic fields amplified via the so-called small scale dynamo reach the equipartition field strength within a dynamical timescale and provide a support against gravity. Such strong fields further help in suppressing the fragmentation in atomically cooled halos (Schober *et al.*, 2012a; Schleicher *et al.*, 2010a; Latif *et al.*, 2013b, 2014d). On larger scales, the angular momentum in these halos gets transferred due to the gravitational torques exerted by the

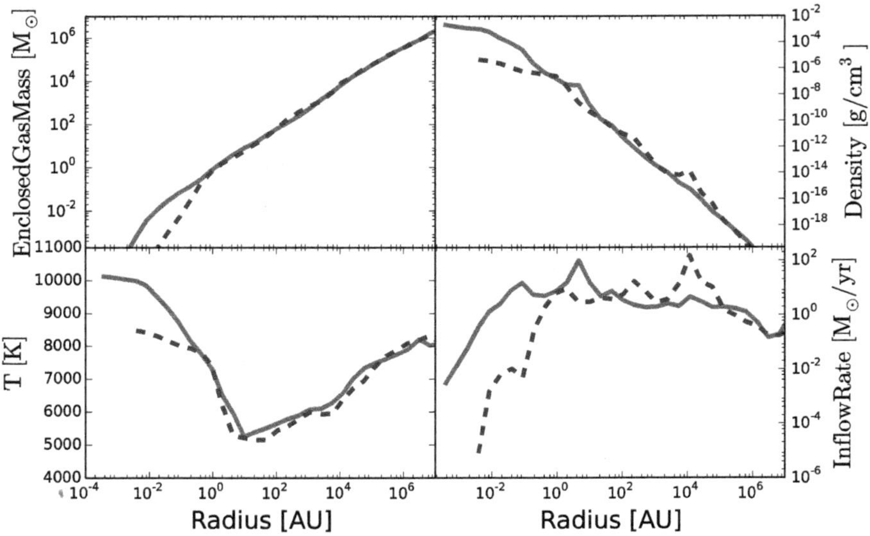

Fig. 3. Spherically averaged and radially binned profiles of density, temperature, mass and mass inflow rates are depicted here. The simulated halos are metal free and illuminated by a strong LW flux above the J_{21}^{crit}. Consequently, H_2 formation remains suppressed and cooling proceeds via atomic lines. Adopted from Latif *et al.* (2016b), reproduced by permission of Oxford University Press / on behalf of the RAS.

triaxility of DM haloes via 'bars within bars' instabilities (Choi *et al.*, 2013, 2015).

One of the salient features of forming DCBHs through isothermal collapse is that large inflow rates of $0.1 - 1$ M$_\odot$/yr are easily available, see bottom right panel of figure 3. If these inflow rates can be sustained for about one Myr then the formation of a massive object of $\sim 10^4 - 10^5$ M$_\odot$ is feasible. In fact, it has been confirmed from cosmological numerical simulations that a massive central object of 10^5 M$_\odot$ can be formed within a Myr after the initial collapse (Latif *et al.*, 2013c; Shlosman *et al.*, 2016), though these simulations did not employ the feedback from a supermassive proto-star. In fact, stellar evolution calculations show that for large mass accretion rates $\dot{m}$ of $\geq 0.1\ M_\odot/yr$, the radius of the star monotonically increases with mass because of the short accretion time compared to the Kelvin–Helmholtz contraction timescale (Hosokawa *et al.*, 2012a; Schleicher *et al.*, 2013). These stars have low surface temperatures of about 5000 K like supergiants and do not produce strong UV flux. Due to the short accretion time in comparison with the nuclear burning time, the core of

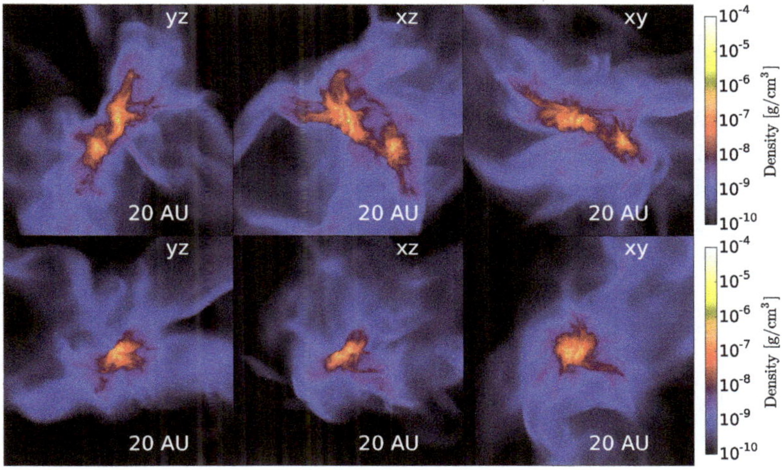

Fig. 4. Averaged density along the line of sight in the central 20 AU for two halos (top and bottom). The simulated halos are metal free and illuminated by a strong LW flux above the J_{21}^{crit}. Therefore, cooling proceeds via atomic lines and an isothermal collapse occurs. Adopted from Latif *et al.* (2016b), reproduced by permission of Oxford University Press/on behalf of the RAS.

such a supermassive star may collapse into a BH provided that sufficiently large accretion rates can be retained (Begelman *et al.*, 2008; Begelman, 2010). Such stars with a BH at their centre are called quasi-stars and it has been found that $\dot{m} > 0.14\ M_{\odot}/yr$ is required for their formation (Schleicher *et al.*, 2013). These supermassive/quasi-stars may collapse into a BH via general relativistic instabilities (Volonteri and Begelman, 2010; Ball *et al.*, 2011; Johnson *et al.*, 2014; Ferrara *et al.*, 2014) and are the potential embryos of massive black holes forming via direct collapse. A more detailed picture on the evolution of such supermassive stars is provided in Chapter 8.

Most of the previous works studying the evolution of a supermassive star used constant mass accretion rates. However, recent studies employing time dependent accretion rates show that if the time interval between two consecutive episodes of accretion (Δt_{acc}) is longer than 1000 years then the star can sufficiently contract and produces strong UV flux which may halt further accretion (Sakurai *et al.*, 2015). It is expected that accretion onto the supermassive proto-star will be episodic due to the possible fragmentation of proto-stellar disk and clumps inward migration. Under isothermal conditions, Δt_{acc} is expected to be much shorter in 1000 years

(Sakurai *et al.*, 2016; Latif and Schleicher, 2015b; Latif *et al.*, 2016b). Therefore, intermittent accretion does not halt the formation of a potential supermassive star. As mentioned earlier, due to the requirement of a strong LW flux, the host halos of DCBHs are expected to form in the close vicinity of actively star forming galaxy and some of these DC halos get tidally disrupted by the nearby massive halos and cannot host a DCBH (Chon *et al.*, 2016). Recently, radiation hydrodynamical simulations investigating the impact of ionising radiation on the formation of DCBHs found that rich structures such as filaments and clumps between DC halo and star forming galaxy shield it from ionising photons (Chon and Latif, 2017). So, the formation of DCBHs is expected to continue under these conditions.

2.3. *Outcome of collapse under less idealised conditions*

Various alternatives to isothermal direct collapse mediated by a strong LW flux have been proposed. For large columns ($\geq 10^{20}$ cm^{-2}) of neutral hydrogen, the escape time for Lyman alpha photons becomes longer than the cloud collapse time scale. Consequently, Lyman alpha photons get trapped inside the cloud, the equation of state gets stiffened, the temperature of the gas cloud starts to increase and eventually collapse proceeds adiabatically (Spaans and Silk, 2006). However, later studies found that cooling still proceeds via 2s-1s transition and collapse becomes isothermal, identical to the Lyman alpha cooling case (Schleicher *et al.*, 2010b; Latif *et al.*, 2011a). Similarly, large baryonic streaming motions (3 σ fluctuations) produced prior to the epoch of recombination naturally increase the critical mass for H_2 cooling until it reaches the atomic cooling limit. Such motions may also collisionally dissociate H_2 molecules and avoid metal enrichment by suppressing in-situ star formation (Tanaka *et al.*, 2013; Tanaka and Li, 2014). However, it was found that streaming motions are not effective in quenching H_2 formation and may require the ubiquity of a strong LW flux (Latif *et al.*, 2014b).

The large inflow rates of the gas required to assemble massive back holes can be obtained dynamically via 'bars within bars' instabilities, and fragmentation even in metal rich gas may be suppressed via supersonic turbulence (Begelman and Shlosman, 2009). However, studies of contemporary star formation show that supersonic turbulence locally compresses the gas and induces star formation in molecular clouds (Federrath *et al.*, 2010, 2011a). During galaxy mergers, gravitational torques drive large inflows of about $10^3 - 10^4\ M_\odot/yr$ to the halo centre and form a compact circum-nuclear disk which later may coalesce into a massive BH (Mayer *et al.*, 2010). The cooling timescale of such a disk is very short, of the order of 100 years, and hence the resulting mass scale is still controversial (Ferrara *et al.*, 2013). In a recent study, improving on their previous work of galaxy mergers, Mayer *et al.* (2015) found that the gas becomes optically thick to cooling radiation and the central stable core may directly collapse into a massive BH via general relativistic instabilities. In this scenario, the central core collapse must avoid the stellar phase otherwise it will blow away most of the mass via stellar winds as the mass loss from star is directly proportional to the metallicity of the gas.

So far, we have mainly focused on the isothermal direct collapse with emphasis on keeping the halo metal free and devoid of H_2. The studies of primordial star formation suggest that although the proto-stellar disk forming in natal primordial clouds fragments, most of the clumps migrate inward, leading to intermittent accretion and merging with the central proto-star. The typical accretion rates in primordial minihalos are $10^{-2} - 10^{-4}$ $M_\odot$/yr which is about 3–4 orders of magnitudes larger than the present day star formation in molecular clouds. Moreover, the accretion rate in the bursty mode exceeds 10^{-2} $M_\odot$/yr, which keeps the stellar envelope bloated up and consequently the supermassive star produces weak UV feedback. Thus, even in the presence of H_2, massive stars up to 1000 $M_\odot$ or even higher may form (Hirano *et al.*, 2014; Latif and Schleicher, 2015a; Hosokawa *et al.*, 2016).

The numerical experiments exploring collapse in massive primordial halos with $T_{vir} \geq 10^4$ K irradiated by moderate strength of LW flux found that a trace amount of molecular hydrogen forms in the halo centre surrounded by warm gas with $T \sim 8000$ K. It was found that fragmentation occasionally occurs but clumps migrate inward and merge with the central star. Large inflow rates of ~ 0.1 $M_\odot$/yr are generated in halos which have recently gone through a major merger (Latif *et al.*, 2014c; Latif and Volonteri, 2015). Under these conditions, the central star may reach 10^4 $M_\odot$ within 1 Myr even if in-situ star formation occurs in these halos. Estimates about the mass of the central object for various strengths of LW flux below J_{21}^{crit} are shown in figure 5. These findings suggest that a massive central object/star of $10^3 - 10^4$ $M_\odot$ can be formed within 1 Myr after the initial collapse for moderate LW flux. Even if in-situ star formation occurs, the gas flow to the central star may still continue until stars reach the main-sequence and blow away the gas. This study was performed assuming that

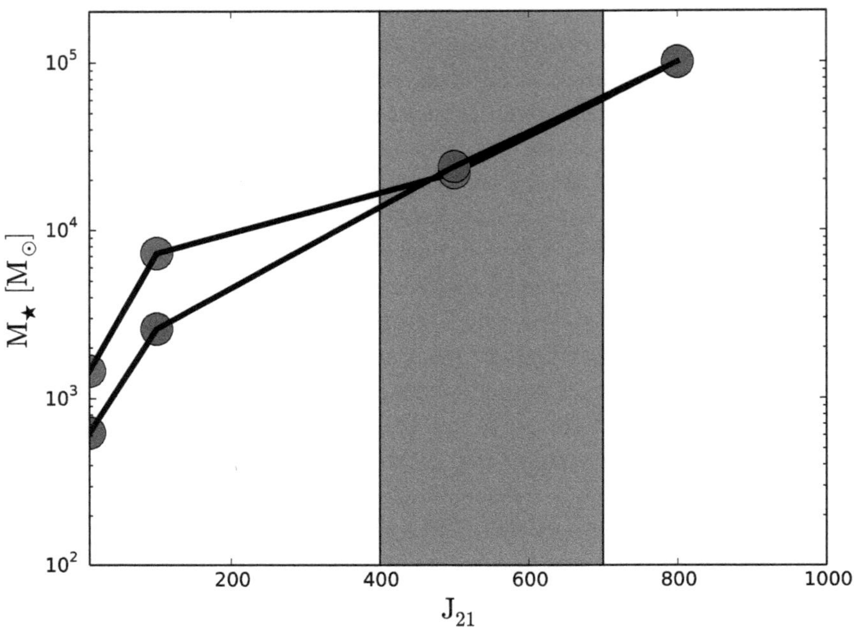

Fig. 5. Expected stellar masses for different strengths of background LW flux. The dashed vertical shaded region represents the range of the critical values. The blue and red spheres are two representative halos. Adopted from Latif *et al.* (2014c), ©AAS. Reproduced with permission.

the spectra of Pop. II stars are soft with $T_{\rm rad} = 10^4$ K. However, similar results are expected for realistic spectra of Pop. II stars.

In the presence of a strong LW flux, star formation within the halo remains suppressed but the halo may be polluted by supernova winds from a nearby star forming galaxy. It has been found that for a trace amount of metals/dust as low as $Z = 10^{-6} Z_\odot$, where Z is the metallicity of the halo and $Z_\odot$ is the solar metallicity, dust cooling becomes important at densities above $10^8 - 10^{12}$ cm^{-3} and lowers the gas temperature to a few hundred K (Omukai *et al.*, 2008; Latif *et al.*, 2016a). For very low metallicities $\leq 10^{-5}$ and a strong LW flux, cooling remains confined to the central 10 AU and large inflows of gas are available. The density structures for two halos illuminated by a strong LW flux and polluted by a trace amount of metals are shown in figure 6. It has been found that for $Z/Z_\odot = 10^{-4}$, due to the efficient dust cooling, a filamentary structure emerges and gravitationally bound clumps form. Instead, for $Z/Z_\odot = 10^{-5}$, the density structure remains spherical and gravitationally unbound sub-solar clumps are observed. The clumps for the cases with $Z/Z_\odot \leq 10^{-5}$ may migrate inwards and merge with the central object. In the case of efficient fragmentation, a dense stellar cluster is expected to form as fragmentation occurs in the very inner part of the halo. It is expected that the run-away collisions in such a dense cluster may lead to the formation of a very massive central star which may later collapse into a massive BH of about a thousand solar masses. So, the conditions for growing a massive object in very metal poor halos with $Z \leq 10^{-5}\ Z_\odot$ and illuminated by a strong UV flux are still favourable (Latif *et al.*, 2016a).

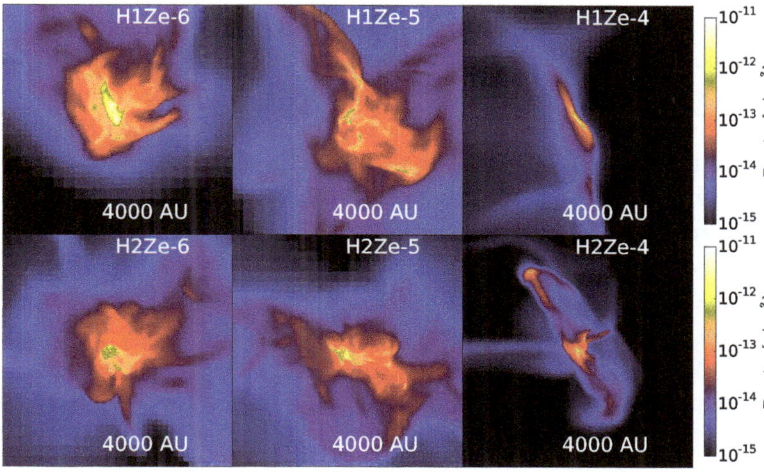

Fig. 6. Average gas density along x-axis for metallicities of $Z/Z_\odot = 10^{-6}, 10^{-5}, 10^{-4}$ and is shown for the central 4000 AU of a halo. Each row represents a halo (halo 1 on top and halo 2 on bottom) and each column represents metallicity (increasing from left to right). Adopted from Latif *et al.* (2016a), ©AAS. Reproduced with permission.

3. Future outlook

Despite the tremendous progress made regarding our understanding of massive black hole formation during the past decade, there are still many open questions. It is yet not clear what is the final outcome of an isothermal collapse, if it is always a supermassive star or a quasi-star or a direct massive black hole. The final result depends on the long term sustainability of the mass accretion rates and the amount of angular momentum retained during the end stages of gravitational collapse. Both of these quantities remain uncertain and are not fully comprehended due to the numerical constraints. In the future, numerical simulations starting from ab initio initial conditions should be performed to assess the feasibility of both parameters and to determine the ultimate fate of direct collapse. Similarly, stellar evolution calculations of supermassive proto-star formation ignored the role of rotation which may impede the collapse and may shut further mass accretion.

Some studies suggest that rapid accretion may launch strong winds leading to large mass loss (Dotan *et al.*, 2011; Fiacconi and Rossi, 2016). Also, preliminary work exploring the role of rotation during the quasi-stellar phase indicates that the quasi-stellar phase may be skipped for objects massive than 10^5 $M_\odot$ (Fiacconi and Rossi, 2017). However, detailed three-dimensional radiation hydrodynamical computations are necessary to support/reject this finding. The dynamical ways to obtain large inflow rates look promising, but it is not completely clear whether such inflows can be maintained down to AU scales. Also, whether metal-enriched gas remains optically thick or radiates away thermal energy to form stars remains to be determined. Detailed investigations exploring this channel will thus be necessary.

One of the biggest uncertainty in assessing the feasibility of direct collapse black holes is their number density which sensitively depends on the value of $J_{21}^{\rm crit}$. The variation in $J_{21}^{\rm crit}$ by a factor of a few changes the abundance of direct collapse black holes by an order of magnitude or even larger. Our current understanding suggests that values of $J_{21}^{\rm crit}$ vary by orders of magnitude and may also change with the age of stars and metallicity. In the future, more realistic estimates of $J_{21}^{\rm crit}$ are required from 3D radiation hydrodynamical simulations by properly modelling the SED of the source galaxy for different modes of star formation, as also discussed in Chapter 6.

Observational evidence is necessary to constrain these models of black hole formation. The first observations of CR7, the brightest Lyman alpha emitter at z=6.6, revealed that it shows strong Lyman alpha and He-1640 Å emission and no signatures of metal lines were observed (Sobral *et al.*, 2015). The drivers of strong Lyman alpha and He-1640Å can be either a cluster of primordial stars with 10^7 $M_\odot$ or a direct collapse black hole of $10^6 - 10^7$ $M_\odot$. However, due to the metal enrichment, forming such a young massive cluster of primordial stars at redshifts as low as z= 6 seems infeasible and a black hole of 10^6 $M_\odot$ possibly forming via DC is a more likely source of CR7 (Pallottini *et al.*, 2015; Hartwig *et al.*, 2016a;

Agarwal *et al.*, 2016a; Dijkstra *et al.*, 2016a; Smidt *et al.*, 2016; Smith *et al.*, 2016). The recent deep observations of CR7 show doubly ionised oxygen (OIII) emission (Bowler *et al.*, 2017) but the current photometry is still consistent with a mild amount of pollution DCBH site with metals (Hartwig *et al.*, 2016a; Pacucci *et al.*, 2017b; Agarwal *et al.*, 2017b). It is expected that JWST[a] will be able to directly probe the observational signatures of DCBHs in high-redshift galaxies.

The future X-ray space observatory ATHENA[b] is expected to detect a few hundred low luminosity active galactic nuclei (AGN) in the early universe with X-ray luminosities of $L_X \geq 10^{43}$erg/s. These observations will provide direct constraints on the masses and luminosity functions of the first AGN. Also future 21 cm experiments such as SKA[c] and LOFAR[d] will probe the proximity zones of high redshift quasars and help in better understanding their formation and growth mechanisms (Whalen *et al.*, 2017). In the local universe the deep observation of dwarf galaxies may help in tracing the seed BHs forming at high redshift (Reines and Comastri, 2016). The recent detections of BH-BH mergers with LIGO[e] have opened a new window of gravitational waves astronomy and the upcoming European space mission eLISA[f] will be able to detect the merging of massive black holes and help in understanding their formation mechanisms.

In the next chapters, we will clarify uncertainties in the value of $J_{21}^{\rm crit}$ (Chapter 6), which is crucial to quantify the expected number of black holes. Black hole formation through mergers in stellar clusters will then be explored in Chapter 7, and the evolution of supermassive stars is described in Chapter 8. The growth of seed black holes, potentially forming from the first stars, will be discussed in Chapters 11 and 10. Statistical predictions will be provided in Chapter 9. A comparison with current observations of high-redshift quasars is given in Chapter 12, while predictions on gravitational wave emission are provided in Chapter 13, and expectations for future observations are outlined in Chapter 14.

[a]https://jwst.nasa.gov
[b]http://www.the-athena-x-ray-observatory.eu
[c]http://skatelescope.org
[d]http://www.lofar.org/
[e]https://www.ligo.caltech.edu
[f]https://www.elisascience.org

Chapter 6

Primordial gas collapse in the presence of radiation: direct collapse black hole or Population III star?

Bhaskar Agarwal

Universität Heidelberg, Zentrum für Astronomie, Institut für Theoretische Astrophysik, Albert-Ueberle-Str. 2, D-69120 Heidelberg, Germany, bhaskar.agarwal@uni-heidelberg.de

The first billion years in the evolution of the Universe mark the formation of the first stars, black holes and galaxies. The radiation from the first galaxies plays an important role in determining the final state of primordial gas collapsing in a neighboring halo. This is due to the fact that the primary coolant for primordial gas is molecular hydrogen, which can be dissociated into atomic hydrogen by Lyman-Werner photons in the energy range $11.2 - 13.6$ eV. While cooling by molecular hydrogen leads to Pop. III star formation, cooling by atomic hydrogen can lead to the formation of a supermassive star (or quasi-star) which results in the formation of a massive $10^{4-5}\ M_{\odot}$ black hole, or a direct collapse black hole. The spectrum of this radiation field is critical in order to determine whether a primordial gas cloud forms a Pop. III star or a very massive black hole. We will in the following explore this scenario and discuss how the radiation spectrum influences the outcome of the collapse.

1. Introduction

1.1. *First stars or direct collapse black holes*

As discussed in Chapter 3, the primary coolant in the early Universe is H_2, which can cool the gas down to ~ 200 K. Cooling of primordial gas by molecular hydrogen leads to Pop. III star formation (Chapter 4), thus we must understand the formation and destruction of H_2 in the context of primordial star formation. Partial dissociation of molecular hydrogen can delay Pop. III star formation in a minihalo, as the halo needs to grow and regain its H_2 fraction to efficiently cool and collapse into stars. As discussed in Chapter 5, in the complete absence of H_2, atomic hydrogen assumes the role of the primary coolant which can cool the gas to ~ 8000 K. Comparing the Jeans mass, M_J, at a number density of $n = 10^3$ cm^{-3}, H_2 cooling at 200 K leads to a $M_J \sim 283\,M_\odot$, while atomic hydrogen (H) cooling at 8000 K leads to a $M_J = 10^5\,M_\odot$. The Jeans mass resulting from H cooling is readily available in atomic cooling halos, i.e. halos with $T_{vir} \sim 10^4$ K, and thus the idea of supermassive stars that undergo runaway collapse into black holes (BHs) was born (Rees, 1978b; Omukai, 2001). The black holes formed through this scenario are commonly referred to as direct collapse black holes (DCBH). To outline the basic thermodynamical framework, the density-temperature digram for gas undergoing collapse and cooling via H_2 (leading to Pop. III stars, dashed line) vs. cooling via H (leading to DCBH, solid line) is shown in Fig. 1. Thus, any channel that leads to the formation or destruction of H_2 becomes critically important in the gas collapse process.

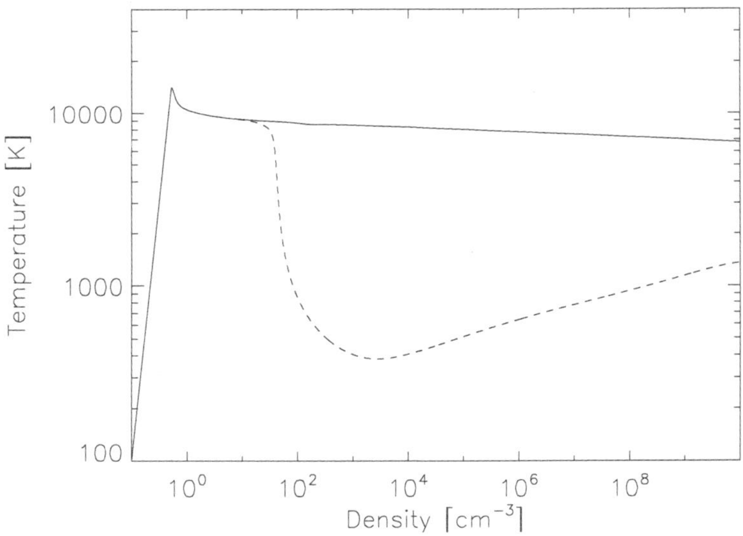

Fig. 1. Density-temperature diagram for primordial gas collapse. Cooling via atomic H leads to a DCBH (solid), while cooling via molecular hydrogen leads to Pop. III stars (dashed). Figure taken from Agarwal *et al.* (2016b), reproduced by permission of Oxford University Press/on behalf of the RAS.

1.2. *Soft-ultraviolet radiation feedback*

As soon as the first population of stars appears in the Universe, subsequently leading to the formation of the second generation of stars, the effects

Abbreviation	Full name
BH	black hole
DCBH	direct collapse black hole
LW	Lyman-Werner
SED	spectral energy distribution
SFR	star formation rate
SF	star formation

of stellar radiation become relevant for primordial gas collapse. Lyman-Werner (LW) photons in the energy range 11.2–13.6 eV can photo-dissociate H_2 into H via the reaction

$$H_2 + h\nu \rightarrow H + H. \tag{1}$$

Thus LW radiation from the first stars (and galaxies) can dissociate H_2 and delay Pop. III star formation in primordial minihalos (Haiman *et al.*, 1996; Ciardi *et al.*, 2000). This LW radiation is often parametrised in the literature in the form of a specific intensity, J_{LW}, in units of 10^{-21} erg/s/cm^2/Hz/sr. In order to form Pop. III stars a primordial minihalo in the presence of LW radiation, the mass of the halo needs to be above a critical threshold (Machacek *et al.*, 2001; O'Shea and Norman, 2008), M_{th}, or in other words

$$M_{th} \sim 4\left(1.25 \times 10^5 + 8.7 \times 10^5 (4\pi J_{LW})^{0.47}\right) M_\odot. \tag{2}$$

Setting $J_{LW} = 1$ in the above equation sets the threshold mass to $10^7\, M_\odot$ or the atomic cooling limit. Thus, one might argue from that DCBH formation should ensue in a minihalo exposed to $J_{LW} > 1$. However, detailed studies (Shang *et al.*, 2010; Latif *et al.*, 2014a) show that in the presence of a LW background, only the molecular hydrogen in the outer regions of minihalos is dissociated and a considerable fraction of molecular hydrogen (0.001) still exists in the central region of the halos. In order to completely suppress Pop. III star formation, it is essential to lower the H_2 fraction to 10^{-8} which can be achieved by extremely high values of LW specific intensity as discussed in Sec. 2.1.

1.3. *Importance of H^- photo-detachment*

While H_2 is an obvious determinant of the fate of primordial gas collapse, another important molecular species is that of H^-. At low densities an important pathway for the formation of H_2 is via H^- where

$$\mathrm{H} + \mathrm{e} \rightarrow \mathrm{H}^- + \mathrm{e} \tag{3}$$

$$\mathrm{H}^- + \mathrm{H} \rightarrow \mathrm{H}_2 + \mathrm{e}. \tag{4}$$

Thus it is important to understand the role of H^- in pristine gas collapse as well. A stellar radiation field can help in destroying H^- via photodetachment by photons in the energy range $h\nu \geq 0.754$ eV,

$$\mathrm{H}^- + \mathrm{h}\nu \rightarrow \mathrm{H} + \mathrm{e}. \tag{5}$$

Due to the importance of this species, it has only recently been argued (Wolcott-Green and Haiman, 2012) that besides UV, the infrared (IR) component of radiation is also critical to primordial gas collapse. This highlights the role of low mass stars in the early Universe which are the primary producers of IR photons. We refer here to Chapter 3 for a detailed discussion on the chemical processes regarding H^-.

2. Modeling gas collapse under the influence of a stellar radiation field

The photodestruction of the two most relevant species, i.e. photodetachment of H^-, and the photodestruction of H_2 is parametrised as

$$\mathrm{k_{di}} = \mathrm{C_{di}}\alpha \mathrm{J_{LW}}\ \mathrm{s}^{-1}, \tag{6}$$

$$\mathrm{k_{de}} = \mathrm{C_{de}}\beta \mathrm{J_{LW}}\ \mathrm{s}^{-1}, \tag{7}$$

where $C_{di} = 1.38 \times 10^{-12}$ and $C_{de} = 1.1 \times 10^{-10}$ are rate constants, α and β are rate parameters that depend on the spectral shape of the irradiating source, and J_{LW} is the specific intensity in the LW as defined earlier. In order to better understand the rate parameters, we first define the quantity L_n with the spectral energy distribution (SED) normalised to be 1 at

$h\nu = 13.6$ eV, in units of K_{21}. Then, α is defined as

$$\alpha = \frac{\int_{\nu_{0.76}}^{\nu_{13.6}} \frac{4\pi L_n}{h\nu}\sigma_\nu d\nu}{1.1 \times 10^{-10}}, \tag{8}$$

where $\nu_{0.76}$ & $\nu_{13.6}$ correspond to the frequency limits of 0.76 and 13.6 eV, respectively. The cross section denoted by σ_ν is written as

$$\sigma_\nu = 10^{-18}\lambda^3(\frac{1}{\lambda} - \frac{1}{\lambda_o})^{1.5}\sum_{n=1}^{6} C_n[\frac{1}{\lambda} - \frac{1}{\lambda_o}]^{\frac{n-1}{2}} \text{ cm}^2, \tag{9}$$

where $\lambda_o \approx 1.6\mu$m (0.76eV) and the wavelength is in μm. Lastly, the parameter C_n is of a tabulated form (John, 1988; de Jong, 1972; Wishart, 1979; Tegmark *et al.*, 1997; Abel *et al.*, 1997).

The parameter β on the other hand is defined with a much simpler form

$$\beta = \frac{1}{\Delta\nu_{LW}} \int_{\nu_{11.2}}^{\nu_{13.6}} L_n d\nu, \tag{10}$$

where $\nu_{11.2}$ & $\nu_{13.6}$ correspond to the frequency limits 11.2 and 13.6 eV respectively (i.e. the LW band), and $\Delta\nu_{LW} = \nu_{13.6} - \nu_{11.2}$.

It is now clear that for a given level of flux (or specific intensity) the assumed spectral shape of the irradiating source, that determines the spectral parameters, can affect the rates of destruction of H_2 and H^-.

2.1. *First galaxies as single temperature black bodies*

The first attempts (Omukai, 2001) to model stellar radiation attributed a single temperature black body (BB), or a power-law spectrum to the first galactic populations. A Pop. III stellar cluster was modeled using a T= 10^5 K (T5) black body, and Pop. II stellar clusters with a T= 10^4 K black body (T4). The presence of such an external radiation field led to the conclusion that primordial gas can withstand collapse into Pop. III type protostellar cores for a high level of incident flux. In the absence of H_2, where molecular hydrogen has been dissociated into H due to the incident radiation, the gas can only cool down to $\sim$ 8000 K, thereby yielding a Jeans mass of $\sim 10^6\,M_\odot$ at $n \sim 10^3$ cm^{-3}. Such high gas masses are readily available in atomic cooling haloes with a virial temperature $T_{vir} \sim 10^4$ K, thus giving rise to the idea of the formation of a single supermassive star (or quasi-star) that can *directly* collapse into a massive $10^{4-5}\,M_{\odot}$ BH.

The first study (Omukai, 2001) that explored the minimum value of the LW flux at which the gas undergoes isothermal collapse at 8000 K, or J_{crit}, was done using a 1D model, that set the flux for a power-law type spectrum to $J_{crit} = 10^5$ and a T4 spectrum to $J_{crit} = 10^3$. Exposing the gas to fluxes higher than J_{crit} leads to a similar chemo-thermodynamical state as seen at J_{crit}. A decade later, another study (Shang *et al.*, 2010) revisited the idea and using their 3D gas collapse simulations found a significantly lower $J_{crit} \sim 30 - 100$ for a T4 type spectrum, while for a T5 spectrum they found $J_{crit} = 10^{4-5}$. Besides an updated methodology, they attributed this difference to the softer shape of the T4 spectrum, which leads to a much more effective destruction of H^- at lower densities.

In other words, one can understand this difference in terms of the reaction rates, k_{de} and k_{di} defined earlier. Using Eqs. (8) and (10), for a T4 type spectrum we obtain $\alpha \sim 2000$, $\beta \sim 0.9$. Thus for a $J_{crit} = 30$ the rates are evaluated as

$$k_{de} = 1.1 \times 10^{-10} \times 2000 \times 30 = 6 \times 10^{-6}\ s^{-1}, \tag{11}$$

$$k_{di} = 1.38 \times 10^{-12} \times 0.9 \times 30 = 3.3 \times 10^{-11}\ s^{-1}. \tag{12}$$

Similarly for a T5 type spectrum, we obtain $\alpha \sim 0.1$, $\beta \sim 3$, and for a $J_{crit} = 10^4$ the rates are

$$k_{de} = 1.1 \times 10^{-10} \times 0.1 \times 10^4 = 1 \times 10^{-7}\ s^{-1}, \tag{13}$$

$$k_{di} = 1.38 \times 10^{-12} \times 3 \times 10^4 = 4.14 \times 10^{-8}\ s^{-1}. \tag{14}$$

When compared with the typical formation rate of H_2 at a density of $n = 10^3 cm^{-3}$, both the above pairs of values for k_{de} and k_{di} are able to dissociate enough molecular hydrogen into atomic hydrogen leading to isothermal collapse of the pristine gas (Shang *et al.*, 2010). Note that the values of k_{de} for both the T4 and T5 spectra are similar $\sim 10^{-7}\ s^{-1}$, which can be understood in terms of the interplay between the rate parameter α and J_{crit} for the respective spectra. The H_2 photodissociation rate however varies by 3 orders of magnitude between the two spectra. This requires investigation of the relative importance of k_{de} and k_{di} for a given value of J_{LW}.

2.2. *First galaxies from realistic spectra*

Attributing a single blackbody to either a Pop. III or Pop. II type stellar population is a strong approximation. One can better understand the behaviour of J_{crit} for *realistic* galaxy samples by first studying the values of the rate parameters. Attributing single stellar populations, extracted from STARBURST99 (Leitherer *et al.*, 1999), to the first galaxies (Pop. II type), one can obtain the values of α and β for a range of stellar parameters such as the stellar mass ($M_\star$), star formation rate (SFR), metallicity (Z), and age ($t_\star$). For SEDs computed from STARBURST99 (or from Agarwal *et al.* (2017a) binary stellar synthesis models) using a Salpeter IMF, the rate parameters are plotted in Fig. 2, where star formation (SF) is modelled either as a single burst of 10^6 $M_\odot$ that is then allowed to age, or in the form of a continuous SFR of 1 $M_\odot$/yr. The rate parameters vary over several orders of magnitude for the burst mode, and less than an order of magnitude for a continuous mode of SF (Agarwal and Khochfar, 2015). This suggests that approximating stellar populations as single temperature black bodies where the rate parameters are constant for any given star formation history is an inaccurate way of capturing the effects of their radiation on pristine gas collapse.

It is now understood that the radiation constraint for isothermal collapse of pristine gas into a BH cannot be understood as a single value of J_{crit}, but rather in terms of the interplay between the relevant photo-destruction rates.

A more robust formulation of the critical LW radiation field required for DCBH formation comes in the form of a critical curve in the k_{de} and k_{di} phase space (Agarwal *et al.*, 2016b; Wolcott–Green *et al.*, 2017). One can draw a curve in the k_{de}–k_{di} parameter space to differentiate the region of DCBH formation from that of fragmentation into Pop. III stars. The curve represents the minimum value of k_{di} needed for a given value of k_{de} at which the gas collapses isothermally. The curve itself does not change for a given irradiating source-DCBH halo arrangement and can be viewed as a fundamental relation. However, an irradiating galaxy can occupy any part of this phase space depending on its spectral shape and separation from the target halo (or the site of DCBH), thereby determining the chemo-thermodynamical evolution of the pristine gas in the target halo.

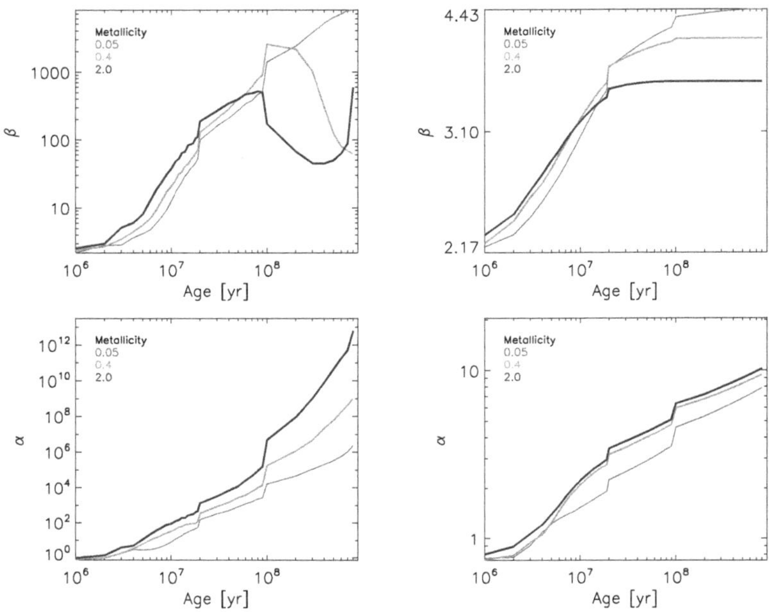

Fig. 2. The rate parameters for stellar populations modelled as STARBURST99 SEDs. The metallicity is quoted in solar units with $Z_\odot = 0.02$. In the left panels a single burst of $10^6\,M_\odot$ is used to model the SFR, whereas in right panels, a continuous star formation rate of $1\,M_\odot/\text{yr}$ is employed (Figure reproduced Agarwal and Khochfar (2015) by permission of Oxford University Press/on behalf of the RAS).

3. Other radiation effects

3.1. *The Escape Fraction of LW photons (Kitayama* et al., *2004; Schauer* et al., *2015; Schauer* et al., *2017a)*

Although the effects of radiation on gas collapse are broadly understood, the escape of radiation from the host haloes of stellar components is still under investigation. The escape fraction of LW photons from the first generation of stars forming in massive haloes is close to zero, and varies from $0-85\%$ in smaller haloes, depending on the stellar mass. Higher mass Pop. III stars ($> 10\,M_\odot$) in low mass haloes ($\sim 10^{5-6}\,M_\odot$) generally lead to larger escape fractions of up to $\sim 80\%$. On the other hand, escape fractions of LW photons from the first galaxies in the Universe, primarily composed of the second generation or Population II stars, are generally higher than 60%. Thus, the stellar populations of the first galaxies can assume their role as the main contributors to the LW radiation field in the early Universe.

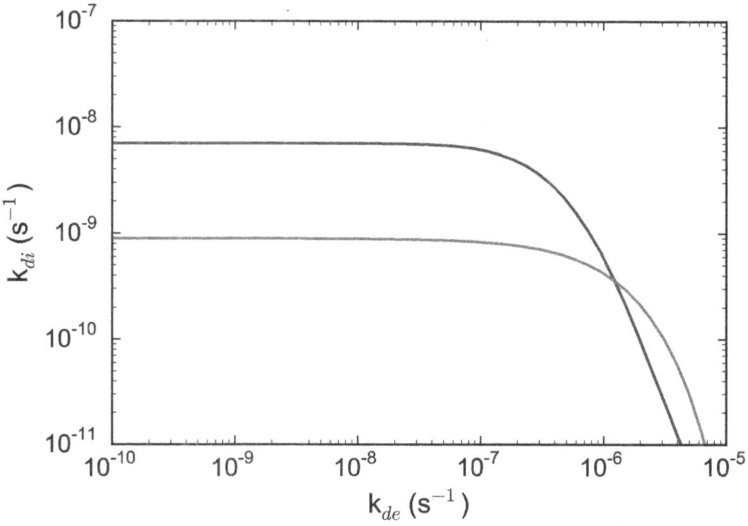

Fig. 3. The critical curve in the k_{de}–k_{di} phase space above which the gas undergoes isothermal collapse into a DCBH, and below which the gas cools by H_2 leading to the formation of Pop. III stars. The difference between the curves is mostly due to the difference in the Jeans length and self-shielding of H_2, where the upper curve (Agarwal *et al.*, 2016b) is obtained with a Jeans length that is twice that of the one used in the lower curve. Figure adopted from Wolcott–Green *et al.* (2017), reproduced by permission of Oxford University Press/on behalf of the RAS.

3.2. *Effects of H ionising radiation (Johnson* et al.*, 2014; Chon and Latif, 2017)*

Photons in the energy range $h\nu > 13.6$ eV are capable of ionising atomic H and can thus heat up the gas in the first haloes, preventing subsequent collapse. If the gas in the haloes is self-shielded to ionising photons, the effects of ionising radiation are minimal and DCBH formation in atomic cooling haloes can proceed as usual. However if the gas densities are low (e.g. in the early stages of a halo's evolution), such that self-shielding is not efficient, ionising photons can delay collapse of gas in atomic cooling haloes by $\sim 20 - 30$ Myr. Furthermore, if the collapsing halo is close enough to the radiation source the gas in it can even get photo-evaporated, thus halting the collapse entirely. However, these calculations are heavily dependent on the modelling of the propagation of the ionising photons, or in other words the inter-galactic-medium (i.e. gas between the ionising source and the collapsing halo) and the inter-stellar-medium (i.e. gas within the ionising source).

3.3. *Effects of X-rays (Inayoshi and Omukai, 2011; Latif* et al.*, 2015; Inayoshi and Tanaka, 2015; Glover, 2016; Regan* et al.*, 2016b)*

X-rays can prove to be critical to the scenario of DCBH formation as they can penetrate dense gas and ionise H, thereby increasing the fraction of free electrons which can lead to the formation of H_2 via the reactions discussed in Sec. 1.3. Thus, presence of X-ray photons can thus lead to the J_{crit} that is higher than the original one, i.e. in the absence of X-ray photons. The X-ray specific intensity is defined as J_x in the same units as J_{LW} which facilitates an easy comparison and investigation of the relative importance of the two quantities. The current consensus on the topic is that for values of $J_x < 0.01$, the revised J_{crit} needed for DCBH formation is at most a factor of two higher than the original one. The J_{crit} needed for larger values of J_x is even higher and can be up to 10 times higher than the original one if $J_x \sim 1$. The field still lacks a full investigation of this topic in the form of detailed 3D hydrodynamical simulations, and the results discussed here still employ BB type spectra to represent stellar populations.

With the presented chapter, we conclude the discussion regarding the uncertainties due to microphysical processes in black hole formation scenarios. In Chapter 7 which follows, we will investigate the role of collisional processes in the formation of massive black holes, while the evolution of supermassive stars is described in Chapter 8. Growth and feedback, statistical predictions and a comparison with current and future observations are given in subsequent chapters.

Chapter 7

Black hole formation in the first stellar clusters

Harley Katz
University of Oxford,
Department of Physics, Clarendon Laboratory,
Parks Road, Oxford OX1 3PU, Great Britain,
hk380@ast.cam.ac.uk

The early Universe was composed almost entirely of hydrogen and helium, with only trace amounts of heavy elements. It was only after the first generation of star formation that the Universe became sufficiently polluted to produce a second generation (Population II) of stars which are similar to those in our local Universe. Evidence of massive star cluster formation is nearly ubiquitous among the observed galaxy population and if this mode of star formation occurred at early enough epochs, the higher densities in the early Universe may have caused many of the stars in the cluster to strongly interact. In this scenario, it may be possible to form a very massive star by repeated stellar collisions that may directly collapse into a black hole and form a supermassive black hole seed. In this chapter, we will explore this scenario in detail to understand the dynamics which allow for this process to ensue and measure the probability for this type of seed to represent the supermassive black hole population observed at $z > 6$.

1. Introduction

In previous chapters, we have explored the conditions for black holes to form via a direct, monolithic collapse (Chapters 5 & 6). The latter is however not the only possible pathway for black hole formation. In fact, as we saw in Chapter 4, the first stars in the Universe are expected to form as clusters, and even denser and more compact clusters may form after trace amounts of metal enrichment. The first Population II star clusters to form in the Universe may be ideal laboratories for dynamically building supermassive black hole seeds.

In these star clusters, a very massive star (VMS) may form through repeated stellar collisions and mergers which may collapse directly to an intermediate mass black hole (IMBH) if the star becomes massive enough (Heger *et al.*, 2003a). Because the galaxies which formed in the early Universe were significantly more dense than those we observe in the local Universe, it is very likely that the first Population II star clusters which formed were also significantly more dense compared to those observed nearby. Stellar densities in some of the most extreme systems in our local neighborhood can reach values of $\sim 10^5 M_\odot$/yr (Espinoza *et al.*, 2009). At these densities, stellar interactions are expected to be frequent (Portegies Zwart *et al.*, 1999). The core collapse and relaxation time scales of the systems are very short compared to the age of the Universe (at the time these systems were likely to first form) as well as the main sequence lifetimes of the most massive stars[a] in the system. Therefore, if these stellar densities are commonplace in the first star clusters, it is expected that the mass of the most massive star to form in the cluster is not set by the fragmentation properties of the gas, but rather by the number of mergers or collisions between high mass stars which build a VMS dynamically.

The general dynamical process for forming an IMBH seed from collisional runaway proceeds as follows:

(1) A dense Population II star cluster forms in a metal enriched gas cloud in the early Universe with total mass M_c, radius R_c, total number of stars N_*, and stellar density n_*. This star cluster will have a given stellar IMF, ξ, virial ratio, Q, which determines whether the initial velocities of the star cluster are dynamically cold, hot, or virialized, a binary fraction, b, a fractal degree, D, as well as a degree of mass segregation S. There is no guarantee that the initial conditions of the star cluster are spherical, but in the absence of gas, direct N-body experiments show that the star cluster will become spherical rather quickly (Katz *et al.*, 2015) and the shape of the cluster in the initial conditions only weakly affects the final outcome.

(2) Once the star cluster has formed, if the IMF is not a delta function, the massive stars sink to the center of the cluster (if not already segregated at the center) due to dynamical friction. This increases the collision/merger probability of stars in the cluster as the cross

[a]The main sequence lifetime of very massive stars is expected to plateau at ~ 3 Myr for high mass ($\geq 50\ M_\odot$).

sections of these stars are much larger than their lower mass counterparts. Once a collision or merger has occurred, the cross section of the newly formed star increases, thus enhancing the probability for a future collision or merger. This process is clearly unstable as the star with the largest cross section is most likely to dominate all subsequent collisions or mergers in the cluster.

(3) After the collision/merger time scale in the cluster becomes larger than the main sequence lifetime of the VMS, the star cluster will continue to evolve normally without many more stellar collisions or mergers. The VMS then evolves off the main sequence and depending on its mass and metallicity, it may directly collapse to an IMBH with minimal mass loss (Heger *et al.*, 2003a).

Conceptually, this mechanism for forming a SMBH seed at high redshift is rather simple. Furthermore, unlike two of the other scenarios which remain popular in the literature (Population III star formation and direct collapse), the conditions needed to initiate the process of collisional runaway have been observed in the local Universe — stellar densities high enough for multiple stellar interactions have been observed in the Milky Way. Similarly, the remnants of stellar collisions may have already been observed — it is believed that stellar collisions may be responsible for forming blue stragglers (Sanders, 1970; Hills and Day, 1976; McNamara and Sanders, 1976; Lombardi *et al.*, 1996), X-ray binaries (Fabian *et al.*, 1975), and millisecond pulsars (Lyne *et al.*, 1987, 1988). For this reason, collisional runaway remains worthwhile to explore the probability of this scenario being the dominant seeding mechanism for high redshift SMBHs.

Unfortunately, the evolution of these types of star clusters is highly nonlinear and thus, properly modeling this scenario requires extremely accurate direct N-body integrators (as these systems are highly chaotic). Furthermore, the processes of stellar evolution (single and binary, including that of a merger remnant), stellar mergers/collisions, the gas dynamics and the gravitational effects due to the larger galaxy within which the star cluster is embedded must be modeled correctly. Finally, the initial conditions of this type of star cluster at $z > 6$ are unknown. Our current best estimates for their properties rely on detailed cosmological hydrodynamic simulations which can only resolve the scales needed to address this issue in single objects, limiting our ability to make general statements. In what follows below and throughout this chapter, we discuss the current, state-of-the-art understanding of the collisional runaway scenario, beginning with the probability of forming a dense star cluster in the early Universe, continuing to the star cluster dynamics and stellar evolution which govern the formation of a VMS, and finishing with evaluating the probability that this scenario is the dominant mechanism for SMBH seed formation.

2. Forming a high redshift birth cloud

Unfortunately, there is no expectation that the birth clouds in which these star clusters might form at high redshift will be observed any time in the near future. Therefore, our entire understanding of the initial conditions of the star clusters relies on cosmological simulations and analytic modeling. The two conditions required for the formation of a dense, bound star cluster are a metal enriched birth cloud and a high star formation efficiency. The first criterion ensures that the birth cloud will fragment efficiently into a Population II star cluster with many stars, while Population III stars tend to form in smaller groups (Bromm *et al.*, 2002; Turk *et al.*, 2009; Stacy *et al.*, 2010) which may prohibit collisional runaway. The estimated metallicity for efficient fragmentation is $Z \gtrsim 10^{-4} Z_{\odot}$ or $Z \gtrsim 10^{-6} Z_{\odot}$ in the presence of dust (Omukai *et al.*, 2005; Schneider *et al.*, 2012). The latter criterion ensures that a bound star cluster actually forms which is a necessary requirement for collisional runaway to ensue. The properties of a star cluster which forms in a birth cloud with an arbitrary set of properties is an unsolved problem, even at low redshift. However, the fact that old, dense star clusters such as globular clusters exist means that at high redshift, dense birth clouds can form with a high efficiency of gas to star conversion.

3. Metal enriched birth clouds in a cosmological context

Devecchi and Volonteri (2009) used an analytic model to consider the case where metal enriched birth clouds form at the centers of atomic cooling haloes at high redshift. In this model, they assumed that Population III star formation occurs in the low mass mini-halos which merge to form the more massive object, therefore setting the global metallicity of the halo above the critical value needed for efficient fragmentation. They assume that a gaseous disk forms as the baryons condense from the halo and that the Toomre parameter is sufficiently high so that the disk does not fragment efficiently. This allows a massive star cluster to form in the center of the disk which may undergo collisional runaway. This is put into a cosmological context by using theoretical halo mass functions as a function of redshift (Press and Schechter, 1974), and assuming a metallicity dependent on redshift. Finally, using a model to populate the halos with star clusters, they check whether the system has a core-collapse time scale less than 3 Myr (i.e. the time at which the first stars are likely to explode as SN) which would be sufficient to develop a runaway instability. The metal enrichment histories are highly uncertain so they provide multiple different models to understand the effect of this parameter (Devecchi and Volonteri, 2009).

Katz *et al.* (2015) attempted a different approach where they aimed to form the birth clouds of high redshift Population II star clusters from first principles in high resolution cosmological hydrodynamical simulations. The advantage of this approach compared to the analytic models is that one can directly measure the properties of the gaseous birth clouds from the simulation with limited additional assumptions. The disadvantage of this method is that the physical scales required to model the dynamics of the birth cloud are small enough that this method is only practical for few numbers of objects which prohibits making general statements. Nevertheless, the two approaches are complementary as the former provides global statistics while the later can focus on individual properties of the cloud.

Rather than focusing on nuclear star clusters formed at the center of disk galaxies in atomic cooling haloes, Katz *et al.* (2015) considered the scenario of the pairwise collapse of two high redshift mini-halos. If the collapse of the two mini-halos is slightly offset in time by a few Myrs, the first collapsing object will undergo Population III star formation. The main sequence lifetimes of the average mass Population III star (Hosokawa *et al.*, 2011) is expected to be short ($\sim$ 3.9 Myrs (Schaerer, 2002)) and metals ejected in the supernova of these stars will enrich the gas which will fall in the second object. If the second object is enriched to a level above the critical metallicity for efficient fragmentation, it will form Population II stars where collisional runaway may occur.

In particular, Katz *et al.* (2015) ran a zoom-in simulation on a pair of mini-haloes with halo masses at collapse of $M_h \sim 3 \times 10^5$ $M_\odot$. The second halo collapsed 12.9 Myr after the first (at $z = 28.9$) at a distance of 117pc which is more than enough time for the first object to form Population III stars and for the metals from the supernova ejecta to reach the second halo (Ritter *et al.*, 2012). The evolution of the gas in the second halo was followed in great detail (at a resolution of 0.9 comoving pc h^{-1} which corresponds to a physical resolution of 0.046 pc at $z = 30$). The simulations demonstrated that the mass in the center of the second mini-halo can grow to $M > 10^4$ $M_\odot$ in less than 10 Myr within a radius of 1 pc. The exact structure of the central cloud is subject to the resolution in the simulation but is likely to be fragmented, flattened, and ellipsoidal (Katz *et al.*, 2015). While the structure differs greatly depending on the simulation resolution, the mass contained within a fixed radius is very well conserved indicating that birth clouds with $M > 10^4$ $M_\odot$ and a radius of 1 pc are robustly predicted to form in high-redshift mini-halos. This type of birth cloud may provide the necessary conditions needed to form a dense star cluster which may undergo collisional runaway.

4. From birth cloud to star cluster

As described earlier, how a birth cloud with an arbitrary set of initial parameters forms a star cluster remains an open problem in modern astrophysics. This issue is even harder to constrain at high redshift as there is very limited observational data at the metallicities and densities which are common in the early Universe. Of particular interest are properties such as the star formation efficiency (SFE, ϵ) and the stellar initial mass function (IMF). In the local Universe, the most massive and dense cores within a molecular cloud often have embedded star clusters which form with an SFE of $\sim 10 - 30\%$ (Lada and Lada, 2003). However, Dib *et al.* (2011) show that the SFE in a given proto-stellar clump is dependent on metallicity which is primarily due to the longer time required to expel gas at lower metallicity because of weaker stellar feedback (Vink *et al.*, 2001). Furthermore, simulations show that compact clusters may have been subject to SFEs of $\sim 60 - 70\%$ (Pfalzner and Kaczmarek, 2013). The SFE is the most important parameter for setting the stellar mass of the star cluster. These quantities, along with the initial stellar density are arguably the two most important parameters in determining how massive a VMS can become (Katz *et al.*, 2015). In order for collisional runaway to ensue, it is imperative that the cluster remains bound during its formation so that the stars will not become unbound before they can collide or merge.

Similarly, understanding the stellar IMF of the primordial birth clouds has important implications for whether collisional runaway can proceed. N-body simulations of star clusters show that collisional runaway is often initiated by one of the most massive stars in the cluster because it has the largest radius (Katz *et al.*, 2015). The rate at which these stars migrate to the center is dependent on the stellar IMF (assuming that the star cluster is not initially mass segregated). Because collisional runaway often begins with the most massive star, the highest mass star in the IMF roughly sets the number of encounters needed to break the threshold of $\sim$ 250 $M_\odot$ needed to directly collapse to an IMBH (Heger *et al.*, 2003a). For example, a birth cloud which forms stars up to 150 $M_\odot$ will have a much easier time of forming a VMS compared to a birth cloud which only created 20 $M_\odot$ stars as the number of collisions required to break the 250 $M_\odot$ threshold increases from 2 to 12.

Several other parameters in addition to SFE, initial mass, initial density, and stellar IMF are needed to describe the initial state of the star cluster which forms from the primordial birth cloud. These include density profile and structure, degree of initial mass segregation, fractal degree, and binary fraction among others.

5. Star cluster evolution

For the purposes of forming an IMBH in a primordial star cluster due to collisional runaway, there are three relevant outcomes of the star cluster evolution at low metallicity which are interesting to identify. In the best case scenario, multiple stellar collisions lead to the formation of a VMS with $M > 260$ $M_\odot$ which collapses to an IMBH with minimal mass loss. A second scenario occurs when a star undergoes few collisions which creates a VMS with $260 > M > 150$ $M_\odot$ which likely results in a pair-instability supernova. Finally, in the third scenario, not enough collisions occur and the most massive star in the cluster remains below 150 $M_\odot$ and the star cluster evolves normally. The goal is to identify which parameters control the formation of a VMS and with what probability does this process occur.

5.1. *Relevant time scales*

Ignoring stellar evolution and the relevant hydrodynamics from any residual gas, the evolution of the star cluster can be described by entirely gravitational processes. The dynamical state of the star cluster is dictated by the relaxation time such that the half-mass relaxation time is

$$t_{\rm rlx} = \left(\frac{R_c^3}{GM_c}\right)^{\frac{1}{2}} \frac{N_c}{8\ln\Lambda_c}, \tag{1}$$

where R_c is the half-mass radius of the cluster, G is the gravitational constant, M_c is the cluster mass, N_c is the number of stars in the cluster, and $\ln\Lambda_c \sim 10$ is the Coulomb logarithm (Spitzer, 1987). Because self-gravitating star clusters have negative specific heat (Lynden-Bell and

Wood, 1968), they are driven to core collapse where the central density of the star cluster increases very rapidly over a finite time scale (Antonov, 1962; Spitzer and Hart, 1971). For a system of equal mass stars, the core collapse time scale scales roughly as

$$t_{\rm cc} \approx 15 t_{\rm rlx} \tag{2}$$

(Cohn, 1980). The sharp increase in density in the central regions of the cluster drastically increases the probability for a stellar collision or merger. In the case of equal mass stars, the core collapse time scale is long compared to the relaxation time scale.

The situation is categorically different for star clusters which exhibit a range of initial masses across a stellar IMF compared to a star cluster where all stars share the same mass. In this case, massive stars sink to the center of the cluster due to dynamical friction (Chandrasekhar, 1943). The inspired time scale can be computed as

$$t_{\rm df} = 3.3\frac{\bar{m}}{m} t_{\rm rlx}, \tag{3}$$

where $\bar{m}$ is the mean mass of stars in the cluster, and m is the mass of the in-spiraling star (Portegies Zwart and McMillan, 2002). The larger m is compared to $\bar{m}$, the quicker the star will in-spiral. Core collapse in a multi-mass cluster is then dictated by the time scale it takes for the most massive stars in the cluster to sink to the center. If one considers a Salpeter IMF in the range $m_{\rm min} = 0.1$ M$_\odot$ to $m_{\rm max} = 100$ M$_\odot$, then $\bar{m} = 0.39$ M$_\odot$ and $t_{\rm df} = 0.13 t_{\rm rlx}$ for the most massive star in the cluster. Empirically, Portegies Zwart and McMillan (2002) find that for realistic choices of stellar IMFs, $t_{\rm cc} \approx 0.2 t_{\rm rlx}$. The core collapse time scale of a multi-mass cluster is clearly much shorter than the relaxation time scale. This suggests, that at very early times in the cluster evolution, the densities of multi-mass clusters will grow significantly in the central regions to a point where interactions are probable.

5.2. *Collisions versus mergers*

A stellar merger or collision occurs when the radii of two stars overlap. In principle, a collision may occur randomly in the densest regions of the cluster when the orbital trajectories of two stars have an impact parameter smaller than the sum of the two radii of the stars. However, direct N-body simulations demonstrate that mergers are much more common (Portegies Zwart *et al.*, 1999). In this case, the massive stars at the center of the cluster form binaries dynamically which are hardened via binary exchanges or three-body interactions. With enough interactions, the binary orbit may lose enough energy or become sufficiently eccentric such that the two stars merge.

In Figure 1, we show both the number of binaries as a function of time as well as the mass of the VMS as a function of time in one of the fiducial star clusters simulated by Katz *et al.* (2015). The vertical dashed black lines indicate the time at which a binary merger occurred whereas the vertical dashed red lines indicate when a stellar collision occurred. In these simulations, the number of binary mergers which contributed to the growth of the VMS totaled seven while the number of collisions was only two. Furthermore, from the bottom panel of this figure, it is clear that stellar collisions provide negligible amount of mass to the growth of the VMS. Hence the growth of the VMS can be linked to the formation and hardening of binaries at the center of the cluster.

5.3. *Growth of the VMS: Analytical modeling*

Assuming all collisions in the cluster involve the same star, Portegies Zwart and McMillan (2002) provide an analytic equation for the maximum mass that a VMS can grow to such that

$$\frac{dm_r}{dt} = N_{\text{coll}} \bar{\delta m}_{\text{coll}} \tag{4}$$

where N_{coll} is the average collision/merging rate in the cluster and $\bar{\delta m}_{\text{coll}}$ is the average mass increase per collision. Since VMS grow primarily via binary mergers, the collision/merging rate is inherently linked to the formation rate of binaries at the center of the cluster.[b] Portegies Zwart and McMillan (2002) argue that for a cluster with identical mass stars,

$$N_{\text{coll}} \sim f_c n_{\text{bf}}, \tag{5}$$

where n_{bf} is the binary formation rate such that

$$n_{\text{bf}} \approx 10^{-3} \frac{N_c}{t_{\text{rlx}}}, \tag{6}$$

and f_c is the fraction of binaries that undergo a merger.

With the collision rate in hand, the only additional uncertainty is the average mass increase per collision. Following Portegies Zwart and McMillan (2002), this can be estimated by calculating the minimum mass star

[b]Here we have assumed that there are no primordial binaries.

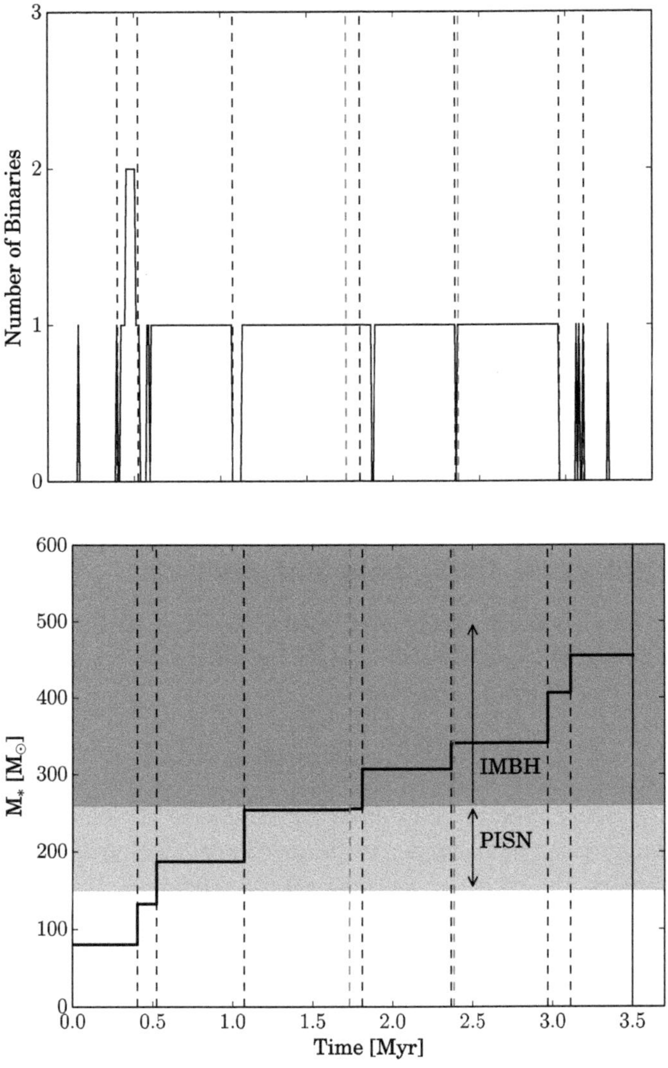

Fig. 1. (Top) Number of binaries as a function of time in a star cluster simulated by Katz *et al.* (2015). The dotted black vertical lines indicate the time at which a binary merger occurred in the cluster while the dotted red vertical lines indicate the time at which a stellar collision occurred. Every time a binary merger occurs, the number of binaries in the cluster decreases by one. (Bottom) Growth of the VMS in the same cluster as depicted in the top panel. The stellar seed mass begins at $\sim 90\ M_\odot$ and grows to more than $450\ M_\odot$ in less than 3.5 Myr. The vertical black and red dashed lines are the same as in the top panel. During a binary merger, the mass increases much more than it does via a stellar collision indicating that binary mergers are the main mechanism which dominate the growth of a VMS. Figure adopted from Katz *et al.* (2015), reproduced by permission of the Oxford University Press/on behalf of the RAS.

which can migrate to the center of the cluster at a time t due to dynamical friction and they find that

$$\delta\bar{m}_{\rm coll} \approx 4\frac{t_{\rm rlx}}{t}\bar{m}\log\Lambda_c. \qquad (7)$$

Combining Equations (4)–(7), Portegies Zwart and McMillan (2002) derive an analytical model for the mass at which a VMS grows for an individual cluster and find

$$m_r = m_{\rm seed} + 4\times10^{-3} f_c M_{c0}\ln\Lambda_c\left[\ln\left(\frac{t_{\rm dis}}{t_{\rm cc}}\right) + \frac{t_{\rm cc}}{t_{\rm dis}} - 1\right], \qquad (8)$$

where $m_{\rm seed}$ is the mass of the star that initiates the runaway collision process, M_{c0} is the initial mass of the star cluster, and $t_{\rm dis}$ is the disruption time scale of the cluster. The disruption time scale sets the limit for how massive the VMS can grow.

Depending on context, the disruption time scale may refer to any number of effects. In the presence of an external tidal field, $t_{\rm dis}$ is equal to the tidal disruption timescale of the cluster. Similarly, this could also refer to the evaporation time scale of the cluster. Although we have not yet considered how stellar evolution may affect the formation of a VMS, realistic star clusters are indeed subject to stellar evolution and it is almost certainly the main sequence lifetimes of the most massive stars in the cluster that set the disruption time scale. While supernova feedback may not destroy or unbind the cluster, it destroys the massive stars which are the ones most likely to merge, hence preventing any more significant growth of the VMS. For this reason, Devecchi and Volonteri (2009) insert the main sequence lifetime of the most massive stars into Equation (8) to calculate, using an analytical model, the mass of black hole seeds at high redshift.

5.4. *The role of stellar evolution*

Although our previous discussion neglects the effects of stellar evolution, by no means is it unimportant. As alluded to, for star clusters at high redshift, it is almost certainly the main sequence life time of a massive star which sets the disruption time scale rather than a dynamical process. Primarily, the core collapse time scale must be less than the main sequence life time of the most massive stars in the cluster or else runaway stellar collisions will not occur. Since the core collapse timescale is inversely proportional to the square root of density, this condition is almost certainly satisfied by the very high density cluster of interest at high redshift. Runaway stellar collisions simply cannot happen in a cluster where the dynamical time scales are longer than the stellar life times.

However, the effect of stellar winds should also not be underestimated. The strength of stellar winds scales with metallicity so it is expected that stars at solar metallicity will undergo significantly more mass loss than those closer to primordial composition (Vink *et al.*, 1999, 2000, 2001). Using results from direct N-body simulations, Glebbeek *et al.* (2009) calculated how massive a VMS can become if mass loss due to the collision and stellar winds are taken into account. At solar metallicity, they demonstrate that a star which may grow to over 1000 $M_\odot$ due to collisional runaway in the N-body only case may actually only become a Wolf-Rayet star of $\sim 100\ M_\odot$ at the end of its life due to the significant amount of mass loss when stellar evolution is considered. In Figure 2, we show the growth of the same VMS in simulations with and without stellar evolutionary processes from Glebbeek *et al.* (2009). While the N-body only run conserves all mass which takes part in the collision, by the end of its life time, the VMS which has formed in the run with stellar evolution has lost the majority of its mass due to stellar winds. Even with 22 mergers, collisional runaway is very unlikely to produce a massive black hole when the cluster is at solar metallicity due to the heavy mass loss from stellar winds. While stellar winds may prevent the formation of IMBHs in high metallicity clusters at low redshift, the gas at high redshift has a composition much closer to primordial. Glebbeek *et al.* (2009) show that at lower metallicity (i.e. at $Z = 0.001$ rather than $Z = 0.02$) the mass loss rates are significantly lower (in this case 13 times lower) which means that the VMS can retain much of its mass. Their simulations show that VMSs with mass $> 260\ M_\odot$ can form at this metallicity which suggests that even in the presence of stellar winds, collisional runaway is still a viable process for forming IMBHs at high redshift.

As was hinted at earlier, direct N-body simulations generally assume a sticky sphere approach such that if the radii of two stars overlap, then the stars are considered to have merged. However, smooth particle hydrodynamics simulations which have studied the mergers of stars demonstrate that not all mass which enters the collision is actually retained by the final merger remnant (Dale and Davies, 2006; Trac *et al.*, 2007; Glebbeek *et al.*, 2013). Depending on the orientation of the collision, for example, whether it is a grazing merger or head on collision, different amounts of mass can be lost from the system into the surrounding interstellar medium (Dale and Davies, 2006; Trac *et al.*, 2007). Furthermore, the amount of mass lost in

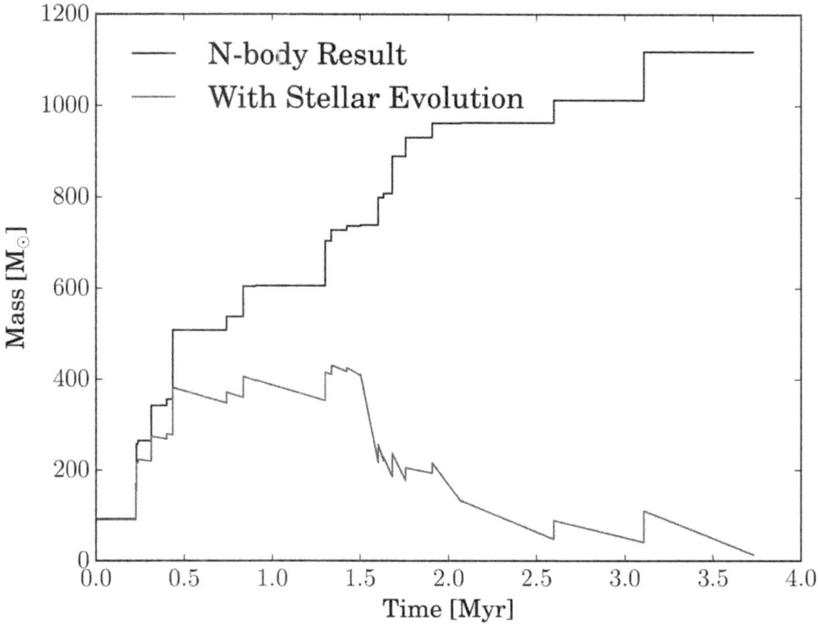

Fig. 2. Comparing the growth of a VMS due to collisional runaway with and without stellar evolution at solar metallicity. Without stellar evolution (black line), the direct N-body experiment predicts that this star will undergo 22 collisions which grow the star from an initial mass of 92.4 $M_\odot$ to a final mass of 1118.9 $M_\odot$. When stellar evolution, stellar winds and mass loss due to the collision are included (grey line), the VMS cannot maintain a large mass and ends up having a final remnant black hole mass of 13.9 $M_\odot$. The data used to make this plot was taken from Glebbeek *et al.* (2009).

the collision also depends on the types of stars which were involved in the process (Glebbeek *et al.*, 2013) For low metallicity clusters, this effect may be more important than mass loss due to stellar winds (Katz *et al.*, 2015).

In addition to the mass loss during the merger, various other physical mechanisms may affect the state of the merger remnant due to the violent process by which it formed. Kinetic energy may be transferred into the envelope of the collision remnant which can cause the cross section to swell by a factor of $\sim$ 100 for some fixed time scale (Dale and Davies, 2006). While this may enhance the probability of future collisions, if too much energy is provided to the envelope after it has been expanded, it can be completely driven off resulting in mass loss which would act against the formation of a VMS (Dale and Davies, 2006). The disruption of the star during the merger may deliver fresh hydrogen to the core (Dale and Davies,

2006; Glebbeek *et al.*, 2008, 2013). This can prolong the main sequence life time of the VMS which may allow it to grow larger due to the additional time it has to merge with other stars before undergoing supernova. While the analytical models discussed in Section 5.3 provide important insight into the systems which produce VMSs, it is clear that these may be upper limits as stellar evolution and stellar processes can have drastic effects on the formation of a VMS.

5.5. *Growth of the VMS: N-body simulations*

Direct N-body simulations of sufficiently massive and dense star clusters corroborate the idea that VMSs can form in clusters due to collisional runaway with having masses $M > 1000\ M_{\odot}$ (Portegies Zwart *et al.*, 1999; Portegies Zwart and McMillan, 2002; Portegies Zwart *et al.*, 2004). Similarly without including stellar evolution, the growth of these stars is consistent with the results from analytical models (Portegies Zwart and McMillan, 2002). Nevertheless, these systems are already deeply non-linear and with the inclusion of stellar processes and stellar evolution, numerical simulations are arguably the best way to evaluate the probability each cluster has in forming a VMS. Furthermore, with advances in computing, thousands of direct N-body simulations can be run simultaneously, accelerated by GPUs, which has opened the possibility of testing various parameters in a statistically rigorous fashion (Katz *et al.*, 2015).

Katz *et al.* (2015), using star cluster initial conditions constrained from high resolution cosmological simulations, have presented one of the most complete studies which tested the multitude of star cluster parameters which may affect the growth of a VMS. They tested the effects of varying the viral parameter, degree of initial mass segregation, the fractal degree, the initial central density, the stellar IMF, the fraction of primordial binaries, the mass of the star cluster, the initial shape of the cluster (i.e. spherical or having the same structure and rotation properties as the gas clouds in the simulation), the star formation efficiency, and the density profile of the cluster. Furthermore, all sets of parameters were simulated multiple times to determine to probability that a star cluster with a given set of parameters forms a VMS as well as how massive the VMS becomes.

For the dense star clusters simulated in Katz *et al.* (2015), core collapse occurs very rapidly compared to the main sequence life times of the most massive stars. The star clusters are relatively low mass (i.e. only a few$\times10^4\ M_{\odot}$) compared to the dense globular clusters and nuclear star clusters observed in the local Universe. In these systems they find that varying the density profile, degree of mass segregation, viral parameter and fractal degree all play minimal roles in determining how massive a VMS becomes as well as the probability of forming a VMS from the cluster. This may be due to the fact that in this type of system, only 5 – 10% of the cluster simulated underwent collisional runaway and formed a VMS with $M > 260\ \mathrm{M}_{\odot}$. Similarly, little difference was found between the spherical and non-spherical clusters. The introduction of binaries drastically increased the number of collisions/mergers in the cluster; however, the VMSs which form in these clusters do not grow significantly larger than in clusters without as the mergers may be from binaries which do not include the VMS. Three-body interactions in the dense star cluster cause the binaries to merge and therefore, a cluster initialized with binaries is much more efficient at producing stellar mergers than without. However, these mergers do not necessarily occur between the most massive stars nor at the center of the cluster. Star clusters which are initialized with 100% binaries and completely mass segregated are roughly three times less efficient at producing a VMS compared to a star cluster without primordial binaries and zero initial mass segregation. The inclusion of binaries can effectively heat the center of the cluster and delay core collapse (Heggie, 1975; Hut *et al.*, 1992). This limits the central density of the system and thus inhibits the number of mergers which can take place.

Direct N-body simulations demonstrate that the stellar IMF can have drastic differences on the mass a VMS can grow to at fixed cluster mass. Katz *et al.* (2015) find that more top heavy stellar IMFs produce VMSs which are much more massive on average. The fraction of clusters which produce a VMS is relatively consistent between the IMF but there is $\sim 50\%$ difference in the maximum mass a VMS can reach between the more top heavy Flat IMFs and the more traditional Salpeter IMF. However, most important for both the formation probability and the mass a VMS can grow to is almost certainly the initial central density of the cluster as well as the mass of the cluster. In Figure 3, we use the results from Katz *et al.* (2015) and show how the probability of forming a VMS in a given cluster as well as the mass of a VMS changes as a function of initial central density and mass. As the mass and initial central density are increased, both the probability of forming a very massive star and the maximum mass a very massive star can grow to increase steeply. In Figure 3, one can see that star clusters which have mass $M < 10^5\ \mathrm{M}_{\odot}$ can produce VMSs with $M > 1000\ \mathrm{M}_{\odot}$. This mass star cluster is small compared to those observed in the local

Universe, hence if these low mass systems at sufficiently high densities are indeed common in the early Universe, N-body experiments demonstrate that collisional runaway can be a promising mechanism for the formation of IMBHs at high redshift.

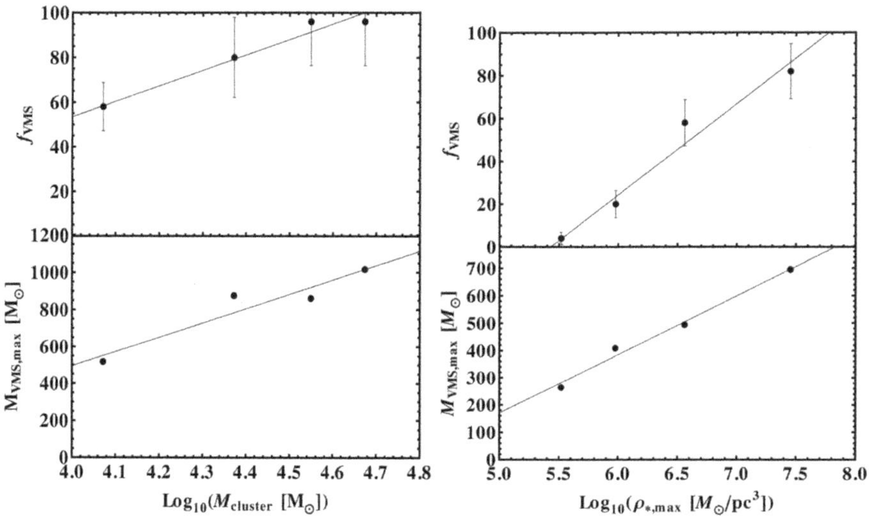

Fig. 3. (Left) Probability of forming a VMS (top) and maximum mass a VMS becomes (bottom) as a function of cluster mass. (Right) Probability of forming a VMS (top) and maximum mass a VMS becomes (bottom) as a function of initial central density. Both the probability of forming a VMS and that mass at which a VMS grows to are extremely sensitive to the cluster mass and initial central density. Data points represent results from simulations while solid lines represent linear fits to the data. All data were taken from Katz *et al.* (2015).

6. Predictions for lower redshift

Like all other models for black hole seed formation, it is crucial that the number density of black hole seeds produced by collisional runaway is greater than or equal to the number density of observed, high-redshift, supermassive black holes. For this model, calculating the number density of these seeds is far from straightforward as it requires knowledge of what fraction of star clusters produce a black hole and how massive these black holes become. Numerical simulations are the best tools for determining these values and as we have shown in the previous section, the black hole seed mass is very strongly dependent on the initial mass and central density of the cluster. Unfortunately there are still large uncertainties on the number density and properties of star clusters in the early Universe and while the cosmological hydrodynamical simulations of Katz *et al.* (2015) give insight into the plausibility of this process, determining the number of these seeds is much more difficult. Assuming that the final redshift at which collisional runaway can produce a black hole is $z_{fin} = 20$ and that the average mass of the black hole seed is 300 $\mathrm{M}_\odot$, Katz *et al.* (2015) parameterized the number density of supermassive black holes as

$$n_{\mathrm{SMBH}} = f_p f_f f_{\mathrm{edd}} n_{gal}(> M_{\mathrm{thresh}}), \tag{9}$$

where f_p is the fraction of metal enriched halos[c] at z_{fin}, f_f is the fraction of these halos that form a seed black hole, f_{edd} is the fraction of those seeds that can accrete constantly at the Eddington rate, and $n_{gal}(> M_{\mathrm{thresh}})$ is the total number density of galaxies massive enough to form a star cluster in which collisional runaway can ensue. Katz *et al.* (2015) concluded that for a threshold mass of 5×10^6 $\mathrm{M}_\odot$ needed to form a high-redshift dense star cluster, if only 1% of all star clusters formed a seed, only 1/100,000 galaxies need to be metal enriched at $z = 20$ to explain the number density of observed supermassive black holes, if all seeds accrete efficiently. There is clearly a strong tradeoff between the number of seeds which accrete efficiently and the fraction of galaxies that are metal enriched.

As described earlier, Devecchi and Volonteri (2009) use an analytical model for the formation of high redshift primordial star clusters by assuming the form in metal enriched atomic cooling halos. This model is placed into a cosmological context by assuming a redshift dependent halo mass function and metal accretion history. Knowing the mass, metallicity, and spin of each halo allows them to calculate the star cluster parameters as a function of redshift and determine which systems will undergo collisional runaway. Thus, number densities of black hole seeds can be computed as a function of redshift. In Figure 4, we show the comoving mass density of supermassive black hole seeds as computed by Devecchi and Volonteri (2009) for five different models where the metallicity-redshift relation varies, the scatter in this relation is varied, and the critical metallicity needed to form a Pop. II star cluster is also varied. The variations in these models produces a

[c]We assume that the process only occurs in metal enriched halos since simulations tend to predict that Pop. III stars form either individually or in small groups (Clark *et al.*, 2008, 2011a; Stacy *et al.*, 2010; Greif *et al.*, 2011) and thus collisional runaway is unlikely to occur.

range in black hole seed mass densities from $\sim 4\ \mathrm{M}_\odot/\mathrm{Mpc}^3 - 300\mathrm{M}_\odot/\mathrm{Mpc}^3$ by $z = 6$. The seed masses in their model are typically $\sim 1000 - 2000\ \mathrm{M}_\odot$ and thus dividing the mass density by the mean seed mass give an average number density of $\sim 2.6\mathrm{Gpc}^{-3} - 200\mathrm{Gpc}^{-3}$. Note that in all cases, the model of Devecchi and Volonteri (2009) seems to be above the observed number density of high redshift supermassive black holes at $z \sim 6-7$ which has a lower limit constraint of $1.1\mathrm{Gpc}^{-3}$ (Venemans *et al.*, 2013).

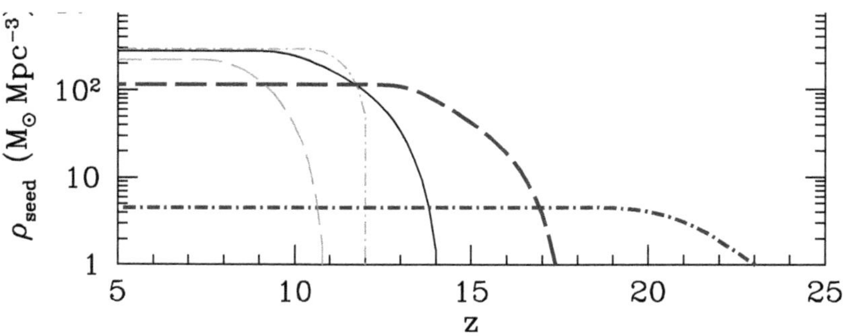

Fig. 4. Mass density of supermassive black hole seeds as a function of redshift computed from the models of Devecchi and Volonteri (2009). The solid black line is their fiducial model which assumes that the metallicity of the system scales as $10^{-0.36z}$. The dot-dashed black line has a weaker scaling of $10^{-0.18z}$. The other lines change the Pop. III star yields as well as the scatter in metallicity but the resulting mass density of black hole seeds is similar to the fiducial model at $z = 6$. In all cases, there are likely to be a sufficient number of seeds to account for the number density of supermassive black holes observed at $z = 6$. Figure adopted from Devecchi and Volonteri (2009), ©AAS. Reproduced with permission.

7. Conclusion

We have described here both the physics of high redshift star clusters that lead to runaway stellar conditions as well as the formation mechanisms for these types of objects in the early Universe. The key conditions are that the core collapse time scale of the star cluster is significantly less than the lifetimes of the main sequence stars. This is a necessary, but insufficient condition that leads to collisional runaway. The probability that a star cluster undergoes this process is highly dependent on the initial central mass and density of the system (Katz *et al.*, 2015).

Although the number density of these systems is highly uncertain, both analytical calculations and numerical simulations seem to suggest that it is plausible that a sufficient number of these IMBHs can be produced to explain the number density of $z = 6$ supermassive black holes (Katz *et al.*, 2015; Devecchi and Volonteri, 2009). The major open questions are whether or not these systems can accrete efficiently enough to grow supermassive within the first billion years. Furthermore, the properties of high redshift star clusters are still highly unconstrained. The latter is being addressed by state of the art numerical simulations that are beginning to sample multiple halos and constrain the properties of primordial star clusters (Sakurai *et al.*, 2017). The former is a particularly difficult problem numerically however, some analytical work suggests that for star clusters with strong gas inflows, certain conditions may lead to super-Eddington growth (Alexander and Natarajan, 2014). Likewise, mergers with other stellar mass black holes may allow the dominant object to accrete mass extremely efficiently (Davies *et al.*, 2011; Lupi *et al.*, 2014). Very recent studies have now also investigated collisions in Population III clusters (Reinoso *et al.*, 2018) and explored the interaction between collisions and accretion (Boekholt *et al.*, 2018). These topics certainly deserve further exploration and provide a promising channel for the formation of supermassive black hole seeds.

While we already discussed the statistical predictions of the collision-based scenario and the comparison with observations, a detailed account on the current observational status is given in Chapter 12, and predictions of other scenarios are outlined in Chapter 9. Future observational probes of the masses of the first black holes are outlined in Chapter 14. Chapter 8 next will describe the evolution of supermassive stars in more detail. Subsequent accretion physics, in particular the super-Eddington case, will be discussed in Chapters 10 and 11.

Chapter 8

Evolution and final fates of rapidly accreting supermassive stars

Takashi Hosokawa

Department of Physics, Kyoto University, Kyoto, 606-8502, Japan
hosokawa@tap.scphys.kyoto-u.ac.jp

The formation of supermassive stars (SMSs) is a possible pathway to seed supermassive black holes in the early Universe. This chapter summarizes recent theoretical efforts to understand their evolution, highlighting effects of very rapid accretion at the rates of $\dot{M}_* \gtrsim 0.1$ M$_\odot$ yr^{-1}. Stellar evolution calculations predict that such an accreting SMS has a characteristic feature, i.e., a very large radius that monotonically increases as the stellar mass increases. The radius exceeds 7000 $R_\odot$ (or 30 AU) after the star accretes the gas of more than 10^3 M$_\odot$. We show that the emergence of the "supergiant protostar" stage is a key for the formation of SMSs, because resulting radiative feedback against the accretion flow is substantially weakened during this stage. We also show that these SMSs end their lives while the accretion continues, and that their final fates vary with different accretion rates.

1. Introduction

The formation and evolution of $\gtrsim 10^5$ M$_\odot$ supermassive stars (SMSs) has been intensively studied in recent years. It has been predicted that such SMSs will leave equally massive black holes (BHs) after their deaths (Iben, 1963; Chandrasekhar, 1964). The massive BHs are favored as massive seeds to explain the quick formation of the SMBHs, which has been suggested by recent discoveries of $> 10^9$ M$_\odot$ supermassive black holes (SMBHs) at high red-shifts $z \gtrsim 6$.

Nonetheless, it is still uncertain whether the formation of the SMSs surely occurs in the early Universe. It is thought that the typical mass of the normal primordial stars is ~ 100 M$_\odot$(Hirano *et al.*, 2015, 2014; Susa *et al.*, 2014; Hosokawa *et al.*, 2011), see also discussion in Chapter 4. Hence various authors are exploring plausible processes to cause the formation of

SMSs, if any, which only rarely occurs under some non-standard conditions. One of them is the so-called direct collapse (DC) scenario (Bromm and Loeb, 2003), for which a number of studies have been conducted in the past decade, and which was described here in Chapter 5, while additional pathways were outlined in Chapter 7. We here focus mostly on one key process in the DC model, the stellar mass growth via rapid accretion. In the star formation process in general, including the DC model, a tiny protostar first appears after the gravitational collapse of a gas cloud (Omukai, 2000, 2001; Inayoshi *et al.*, 2014). Such an embryonic protostar then grows in mass via gas accretion from a surrounding envelope. For the SMS formation through the protostellar accretion, very high rates of $\gtrsim 0.1\ M_\odot\ \mathrm{yr}^{-1}$ are required because the duration of the evolution is limited by the stellar lifetime, $\sim$ Myr. The DC model provides such extremely rapid accretion, and it is considered as a possible process to yield the SMSs.

Abbreviation	Full name
SMS	supermassive star
SMBH	supermassive black hole
BH	black hole
DC	direct collapse
KH	Kelvin-Helmholtz
ZAMS	zero-age-main-sequence
GR	general relativistic

The evolution and final fates of the SMSs had been investigated by various authors (Osaki, 1966; Unno, 1971; Fricke, 1973; Fuller *et al.*, 1986), but they did not consider the realistic formation process, i.e., the rapid gas accretion to obtain its huge mass. Normally a SMS in the stable nuclear burning phase was just assumed as an initial condition, and its subsequent evolution was followed until its death. Considering the formation process with the rapid accretion, however, the timescale of the stellar mass growth is actually comparable to the stellar lifetime. Therefore, the birth and death of the SMSs cannot be separately investigated.

The stellar radiative feedback against the accretion flow is a potential obstacle for the formation of massive stars. In the normal primordial star formation, where the accretion rates are $\sim 10^{-2} - 10^{-3}\ M_\odot\ \mathrm{yr}^{-1}$, the stellar emissivity of the H-ionizing radiation rapidly rises as the star approaches the zero-age main-sequence (ZAMS) stage. The resulting UV feedback caused by the formation and expansion of an HII region easily halts the accretion toward the star (McKee and Tan, 2008; Hosokawa *et al.*, 2011). Although with some variations with different accretion rates, this occurs before the stellar mass reaches 10^3 $M_\odot$ in many cases. In the DC model, which postulates the SMS formation, the stellar bolometric luminosity should be much higher than in the normal primordial cases, exceeding 10^8 $L_\odot$ for $M_* \gtrsim 10^4$ $M_\odot$ stars. If the SMS emits a copious amount of ionizing photons, tremendous UV radiative feedback may shut off the accretion way before the formation of the SMSs. Solving the evolution of a rapidly accreting SMS is necessary to evaluate the strength of the UV feedback.

In the same vein, the final fate of the SMS, i.e., at which point the star collapses into a BH, is also affected by the mass accretion. The BH left after the death of the SMS sets the starting point for the further mass growth toward a SMBH. This is of great importance to elucidate the origin of the SMBHs.

2. Modeling of accreting protostars

Numerically modeling the evolution of accreting stars was initiated in studies on the binary evolution which causes rapid mass exchange among the stars (Kippenhahn and Meyer-Hofmeister, 1977; Neo *et al.*, 1977). The basic numerical technique was then extended and applied for studies on star formation by Steven Stahler and his collaborators (Stahler *et al.*, 1980; Stahler, 1988). They performed a series of pioneering calculations for present-day low-mass (~ 1 $M_\odot$) star formation, considering simple accretion histories with constant rates of $\dot{M}_* \sim 10^{-6\sim-5}$ $M_\odot$ yr^{-1}. Then basically the same method was extensively applied for the evolution of more massive protostars growing with higher accretion rates (Palla and Stahler, 1991; Behrend and Maeder, 2001; Hosokawa and Omukai, 2009; Haemmerlé *et al.*, 2016). In parallel, the evolution of the primordial protostars was also investigated (Stahler *et al.*, 1986; Omukai and Palla, 2003). The high accretion rates of $\dot{M}_* \sim 10^{-4\sim-2}$ $M_\odot$ yr^{-1} were commonly used in these studies on present-day and primordial star formation. Effects of the extremely rapid mass accretion with $\dot{M}_* \gtrsim 10^{-2}$ $M_\odot$ yr^{-1} have been examined recently, motivated by the developments of the DC model.

Regardless of the great differences of the considered accretion rates, the above studies deal with the same governing equations to construct stellar models,

$$\left(\frac{\partial r}{\partial M}\right)_t = \frac{1}{4\pi\rho r^2}, \tag{1}$$

$$\left(\frac{\partial P}{\partial M}\right)_t = -\frac{GM}{4\pi r^4}, \tag{2}$$

$$\left(\frac{\partial L}{\partial M}\right)_t = \epsilon - T\left(\frac{\partial s}{\partial t}\right)_M, \tag{3}$$

$$\left(\frac{\partial s}{\partial M}\right)_t = \frac{GM}{4\pi r^4}\left(\frac{\partial s}{\partial p}\right)_T\left(\frac{L}{L_s} - 1\right)C, \tag{4}$$

which are essentially the four stellar structure equations taking into account effects of the mass accretion. In the above, M is the mass coordinate, ϵ the energy production rate via nuclear fusion, s the specific entropy, L_s the radiative luminosity with adiabatic temperature gradient, C the coefficient determined by the mixing-length theory when $L > L_s$ ($C = 1$ otherwise), and the remaining quantities have ordinary meanings.

Effects of the mass accretion are incorporated in the term $T(\partial s/\partial t)_M$ in Equation (3), which is approximately discretized as

$$\left(\frac{\partial s}{\partial t}\right)_M = \left(\frac{\partial s}{\partial t}\right)_m + \dot{M}_*\left(\frac{\partial s}{\partial m}\right)_t, \tag{5}$$

where m is the inverse mass coordinate measured from the surface, $m \equiv M_* - M$. An evolutionary calculation starts with an arbitrary initial model with small mass, and advances with updating models with the mass increment via accretion $\dot{M}_*\Delta t$, where Δt is the time step.

This sort of modeling has an uncertainty in how to determine the thermal state of the accreting gas. A self-consistent treatment is possible only when assuming the spherical symmetry not only for the star but also for the surrounding accretion flow. In this case, the accretion flow directly hits the stellar surface forming an accretion shock. The jump conditions at the shock front consistently determine the thermal state of the gas which settles down into the star in a post-shock layer (Stahler *et al.*, 1980). However, the accreting gas should have some angular momentum, and it will be more realistic to consider the gas accumulating onto the star through an accretion disk. A practical way of modeling is to solve the 1D structure of the central star employing the normal photospheric outer boundary conditions, supposing that most of the stellar surface freely radiates while the accreting gas approaches the star through a geometrically thin disk. A simple ansatz often used for this case is the so-called "cold accretion" limit, where the accreting gas is assumed to have the same specific entropy as in the stellar atmosphere. Remember that this is an extreme limit; the cold accretion will be a poor approximation especially with the rapid accretion. In reality, the accreting gas should be somewhat "warmer" because it carries part of the liberated gravitational energy (i.e., the accretion energy) into the star before radiating away. For modeling this effect, we temporarily deposit the energy $L_{*,\mathrm{acc}}$ into the stellar interior,

$$L_{*,\mathrm{acc}} = \eta L_{\mathrm{acc}} \equiv \eta \left(\frac{G M_* \dot{M}_*}{R_*} \right)^{-1}, \tag{6}$$

where η is a free-parameter satisfying $0 \le \eta \le 1$. The above mentioned spherical accretion corresponds to a case with non-zero η (but $\eta \neq 1$), whose value is given by solving the flow structure across the shock. In general, even a small value of η (e.g., $\eta \sim 0.01$) is enough to change the evolution from the cold accretion limit. Fortunately, in our cases considered below, uncertainties with varying η only appear in an early stage of the evolution (see Section 3).

3. Evolution of rapidly accreting supermassive stars

3.1. *Emergence of the "Supergiant Protostar" stage*

We from here concentrate on the protostellar evolution with accretion rates $\gtrsim 10^{-2}\ M_\odot\ \mathrm{yr}^{-1}$, through which the formation of the SMSs is supposed to occur. Figure 1 summarizes the results by Hosokawa *et al.* (2012a), who have performed the stellar evolution calculations assuming the spherical accretion (see Section 2). The black curve in panel (a) represents the case with $\dot{M}_* = 10^{-3}\ M_\odot\ \mathrm{yr}^{-1}$, a relatively low accretion rate which is expected for the normal primordial star formation. It is helpful for a later reference to describe this case first. A striking feature here is the contraction stage which occurs for the mass range of $6\ \mathrm{M}_\odot \lesssim M_* \lesssim 30\ \mathrm{M}_\odot$. It corresponds to the so-called Kelvin-Helmholtz (KH) contraction, during which the stellar interior temperature rises as the star loses its energy by radiating away. The hydrogen burning starts near the center just after the stellar mass exceeds $30\ \mathrm{M}_\odot$. After that point, the stellar radius turns to increase almost following the mass-radius relation for zero-age main-sequence (ZAMS) stars. The stellar effective temperature and emissivity of ionizing photons

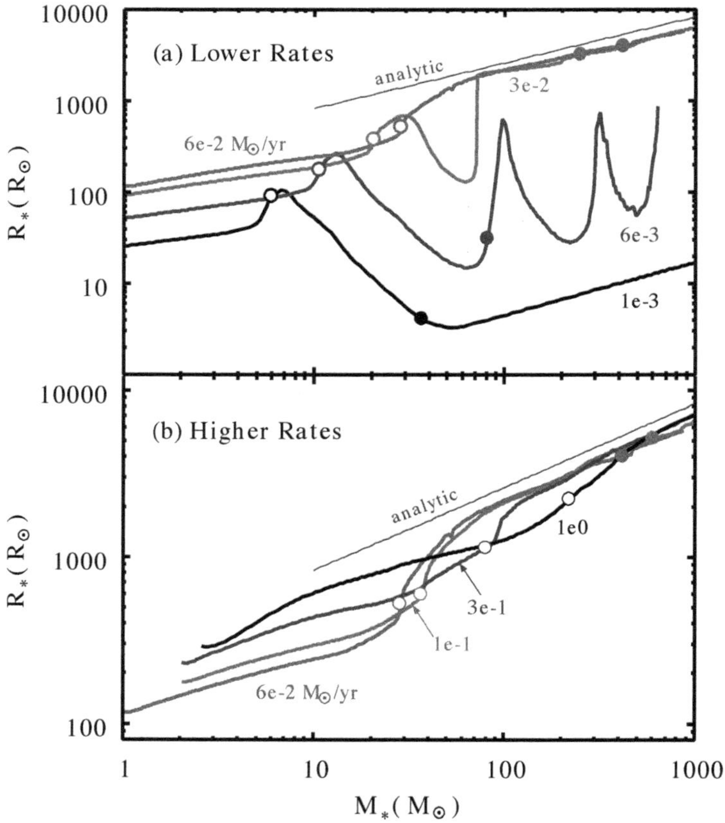

Fig. 1. Evolution of the stellar radius with the different accretion rates (Hosokawa *et al.*, 2012a). The upper and lower panels show cases with the rates lower and higher than $6 \times 10^{-2}\ M_\odot\ \mathrm{yr}^{-1}$ respectively. The accretion rate represented by each line is labeled in the figure. The thin line represents the analytic mass-radius relationship given by Equation (7). The filled circle on each line represents the epoch when the hydrogen burning starts at the center. Figure adopted from Hosokawa *et al.* (2012a), ©AAS. Reproduced with permission.

substantially increases during the KH contraction stage. The UV feedback against the accretion flows begins to operate from a late stage of such a contraction. The presence of the KH contraction toward the ZAMS stage thus marks the onset of the stellar UV feedback.

Meanwhile, the above evolution drastically changes with the more rapid accretion. Panel (a) shows that, with the higher accretion rate, the KH contraction stage only spans the narrower mass range. The contraction turns to the expansion at some point for the accretion rates $6\times 10^{-3}\ M_\odot\ \mathrm{yr}^{-1}$ and

$3\times10^{-2}\ M_\odot\ \mathrm{yr}^{-1}$. With the highest rate $6\times10^{-2}\ M_\odot\ \mathrm{yr}^{-1}$, the contraction stage never appears and the stellar radius monotonically increases with an increase of the stellar mass. The stellar radius finally reaches $\simeq 7000\ R_\odot$ (or nearly 30 AU) when the star has accreted the gas of 1000 $\mathrm{M}_\odot$ in this case. Panel (b) summarizes the evolution with even higher accretion rates, $6\times10^{-2}\ M_\odot\ \mathrm{yr}^{-1} \leq \dot{M}_* \leq 1\ M_\odot\ \mathrm{yr}^{-1}$, showing further characteristic features; the radius monotonically increases with the mass for all the cases, and moreover, the lines representing the different accretion rates converge to the same mass-radius relation. The radius at $M_* \simeq 1000\ \mathrm{M}_\odot$ is consequently $\simeq 30$ AU for all the cases, regardless of the different accretion rates.

The mass-radius relation toward which the lines finally converge is approximately given by

$$R_* \simeq 2.6\times10^3\ R_\odot \left(\frac{M_*}{100\ M_\odot}\right)^{1/2}, \tag{7}$$

which is analytically derived as follows. We begin with the general expression of the stellar luminosity

$$L_* = 4\pi R_*^2 \sigma T_{\mathrm{eff}}^4, \tag{8}$$

where σ is the Stefan-Boltzmann constant. As for very massive stars, the luminosity is very close to the Eddington value,

$$L_* \simeq L_{\mathrm{Edd}} \propto M_*. \tag{9}$$

In the current cases with the high accretion rates, a stellar surface layer is difficult to contract because it always traps part of the heat released from the contracting interior. The surface layer consequently has a high specific entropy to greatly inflate. The effective temperature accordingly drops to several $\times$ 1000 K, where H^- bound-free absorption becomes the dominant process to provide the opacity. Since the H^- opacity has a very strong temperature-dependence, the effective temperature is almost fixed at a constant value. Substituting Equation (9) and $T_{\mathrm{eff}} \simeq 5000$ K into Equation (8) provides Equation (7). The above derivation includes no dependence of the accretion rate, which explains why the different curves in Figure 1 finally converge into the same mass-radius relation.

3.2. *Evolution toward the Supermassive Stars*

Hosokawa *et al.* (2013) have further extended calculations to follow the later evolution until the stellar mass reaches $10^4 \sim 10^5\ \mathrm{M}_\odot$. As the outer

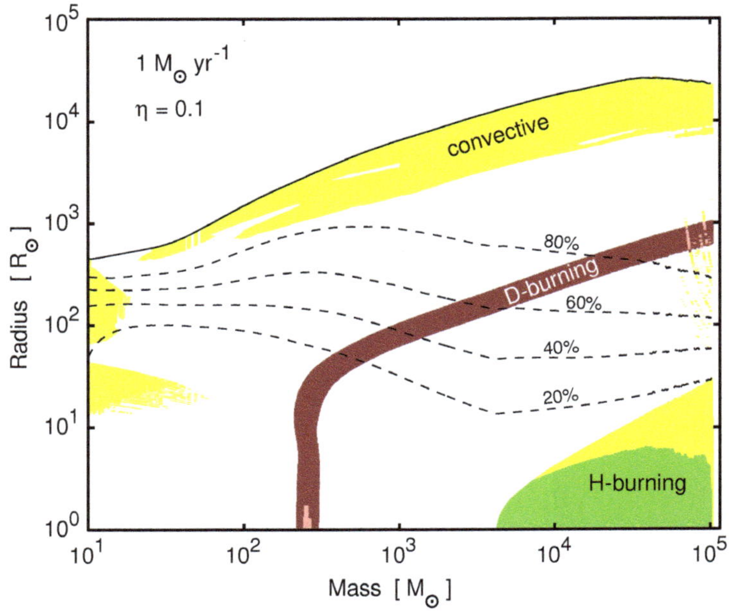

Fig. 2. Evolution of the stellar interior structure until the accreted stellar mass reaches 10^5 M$_\odot$ for the accretion rate $\dot{M}_* = 1\ M_\odot\ \mathrm{yr}^{-1}$ and energy fraction $\eta = 0.1$ (Hosokawa *et al.*, 2013). The black solid and dashed lines represent the radial positions of the stellar surface and mass coordinates of 80%, 60%, 40%, and 20% of the total stellar mass in descending order. The white and yellow areas represent radiative and convective zones without nuclear fusion, respectively. The brown stripe denotes the radiative layer where deuterium burning occurs. The green area represents a convective core where hydrogen burning occurs. The pink zones depict convective deuterium burning. Figure adopted from Hosokawa *et al.* (2013), ©AAS. Reproduced with permission.

boundary of the stellar models, the disk accretion described in Section 2 is adopted. The parameter η is set to be 0.1, i.e., 10% of the accretion energy is assumed to be deposited into the stellar interior. Figure 2 illustrates the evolution of the stellar interior structure with the constant accretion rate of 1 $M_\odot$ yr^{-1}. We confirm that Equation (7) still describes the evolution of the radius well for 100 M$_\odot$ $\lesssim M_* \lesssim$ 1000 M$_\odot$, in spite of the different prescriptions of the spherical and disk accretion. Although not presented here, effects of varying the parameter η only appear in an early evolution for $M_* \lesssim$ 100 M$_\odot$. The same evolution through the supergiant protostar appears even with $\eta = 0$, the extreme case of the cold accretion. This has been also confirmed by other authors using different numerical codes (Umeda *et al.*, 2016; Haemmerlé *et al.*, 2017b).

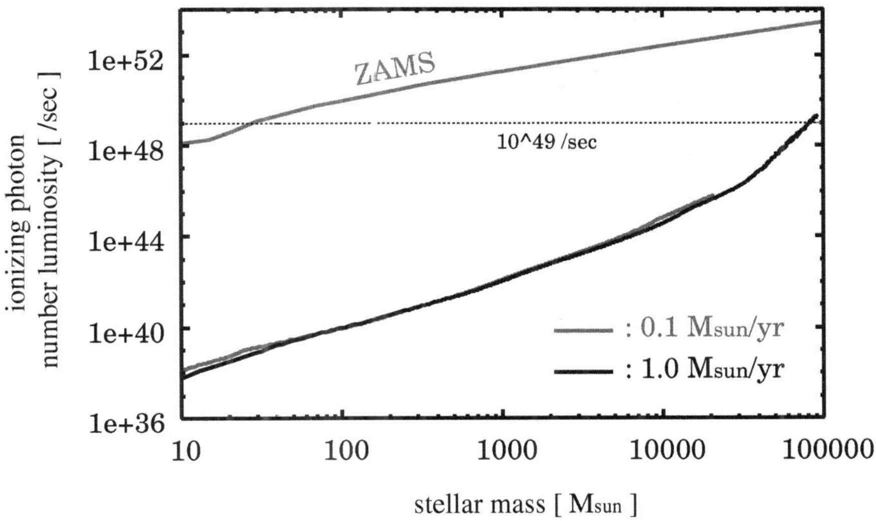

Fig. 3. Evolution of the ionizing photon number luminosity of accreting supermassive stars. The black and grey lines represent the cases with different constant accretion rates of 1 $M_\odot$ yr^{-1} and 0.1 $M_\odot$ yr^{-1}. The upper line represents the reference case of non-accreting zero-age main-sequence stars.

According to Figure 2, the monotonic increase of the stellar radius continues after the star accretes 1000 $\mathrm{M}_\odot$. The analytic formula given by Equation (7) still provides a good fit for the evolution until the stellar mass reaches $\sim 10^4$ $\mathrm{M}_\odot$. The increase of the radius slows down after that, and finally saturates at $M_* \simeq 3 \times 10^4$ $\mathrm{M}_\odot$. The star slightly contracts as the mass increases for 3×10^4 $\mathrm{M}_\odot \lesssim M_* \lesssim 10^5$ $\mathrm{M}_\odot$. The deviation from Equation (7) is caused by the decrease of the density in the stellar atmosphere, with which the H^- ions cease to dominate the opacity. Similar deviation has been also reported using different codes (Haemmerlé *et al.*, 2017b).

Figure 2 shows that a surface layer which only has a small fraction of the total mass significantly inflates to cover a large part of the stellar radial extent. The star has a bloated envelope and compact core like red giants. Most of the stellar mass is contained in the central core, which rather contracts as the stellar mass increases. The central temperature gradually rises as in the ordinary KH contraction. As a result, the nuclear burning initiates when the total stellar mass reaches $\simeq 4000$ $\mathrm{M}_\odot$. The central convective core gradually extends after that, because of the heat input by the nuclear fusion. The evolution of the radius does not follow the mass-radius relation for the ZAMS stars. The outer convective layer remains bloated as long as the rapid mass accretion continues. Such a characteristic structure of the rapidly accreting protostars (e.g., very large radius) has also been predicted by a naive analytic argument (Schleicher *et al.*, 2013). However, recall that it is the stellar atmosphere where H^- absorption dominates the opacity that regulates the evolution of the radius. Thus numerical modeling resolving the detailed structure near the surface is indispensable to accurately describe the stellar interior structure and resulting evolution.

As described above, SMSs with extremely rapid mass accretion inflate to have cool atmospheres. The resulting effective temperature is only $\simeq$ 5000 K, with which a star only emits a small amount of ionizing photons even after it accretes the gas of $\gg 1000$ $M_\odot$. Figure 3 shows that the UV emissivity of the accreting SMSs is much smaller than that of non-accreting counterparts by several orders of magnitude. The UV emissivity gradually increases as the stellar mass increases, because the bolometric luminosity rises roughly in proportion to the mass. Nonetheless, the UV emissivity of a bloating $\sim 10^5$ $M_\odot$ SMS is still comparable to that of a non-accreting star with a few $\times$ 10 $M_\odot$. The resulting UV feedback is too weak to disturb the massive accretion flow toward the accreting SMSs. The formation of SMSs via accretion is thus possible as far as such high accretion rates exceeding $\sim 10^{-2}$ $M_\odot$ yr^{-1} are maintained.

In the calculation described above, the general relativistic (GR) effects are incorporated using the post-Newton approximation, as an effective modification of the gravity constant. However, the GR effects are not strong enough yet to affect the stellar evolution. We describe the final evolutionary stage where the star dies and dynamically collapses in the next section.

4. Final fates of accreting supermassive stars

Figure 4 presents the evolution of the stellar central temperature in the final stage up to the onset of the collapse (Umeda *et al.*, 2016). The governing equations here are the same as Equations (1) – (4), except that Equation (2) is modified to consider the post-Newton approximation, and to include the acceleration term to follow the dynamical collapse. Figure 4 shows that the evolution varies with the different accretion rates. However, all the cases commonly present the characteristic stage where the central temperature is almost constant at $\log T_c \simeq 8.2$. The hydrogen burning takes place during this period. The epoch of igniting hydrogen depends on the accretion rate,

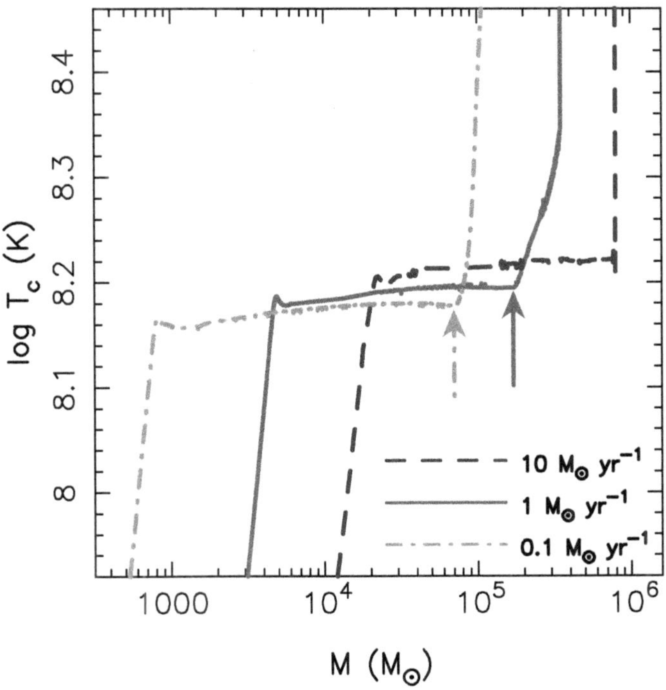

Fig. 4. The evolution of the central temperature T_c as the stellar mass increases (Umeda *et al.*, 2016). Only the values around the stable nuclear burning are shown for clarity. The different lines represent the different accretion rates, 10 $M_\odot$ yr^{-1} (dashed), 1 $M_\odot$ yr^{-1} (solid), and $\dot{M} = 0.1$ $M_\odot$ yr^{-1} (dot-dashed line). The vertical arrows indicate the points where the central hydrogen is exhausted with $\dot{M} = 0.1$ and 1 $M_\odot$ yr^{-1}. Figure adopted from Umeda *et al.* (2016), ©AAS. Reproduced with permission.

i.e., at the higher stellar mass with the higher accretion rate. After the stage of the stable hydrogen burning, the temperature sharply rises for all the cases. Although the calculations follow the evolution until the central temperature reaches $\log T_c = 9.2$, the evolution after that is too rapid for the stellar mass to increase via the accretion. We regard this epoch as the point where the final mass and fate of the star are determined.

Figure 4 shows that the final mass is higher with the higher accretion rate. Furthermore, the physical process which causes the collapse also differs with the rates. With the lowest rate 0.1 $M_\odot$ yr^{-1}, the collapse occurs at $M_* \simeq 10^5$ $M_\odot$, until which the mass growth has continued for $\sim$ Myr starting with an embryonic protostar. Its duration is almost comparable to the stellar lifetime. In fact, the helium mass fraction Y is almost zero near the center at the epoch of $\log T_c = 8.7$, meaning that the collapse occurs when the nuclear fuel has been exhausted in this case. With the higher rate 1 $M_\odot$ yr^{-1}, on the other hand, the helium fraction Y is still unity at the same epoch. The collapse is caused by GR instability at the beginning of the helium burning stage. With the extremely high rate of 10 $M_\odot$ yr^{-1}, collapse occurs with the hydrogen mass fraction $X \simeq 0.5$ and $Y = 1$. The GR instability does not cause the collapse until the stellar mass reaches $M_* \simeq 8 \times 10^5$ $\mathrm{M}_\odot$ in the course of the hydrogen-burning stage.

Since the upper mass for non-accreting SMSs is $\simeq 1.4 \times 10^5$ $\mathrm{M}_\odot$ for the initial primordial abundance (Shapiro and Teukolsky, 1983), the above mentioned variations of the final mass are caused by the rapid mass accretion. The interior structure of an accreting SMS greatly differs from that of the non-accreting counterpart with the same mass. In fact, non-accreting SMSs are normally modeled as $n = 3$ polytropic star supposing the fully-convective structure. As shown in Figure 2, however, the rapidly accreting star has a large radiative layer in its interior. The numerical results show that the more rapid mass accretion makes the stellar mass distribution less centrally condensed at a given stellar mass (Hosokawa *et al.*, 2012a). This explains why the GR instability triggers the stellar collapse at the higher stellar mass with the more rapid accretion. We conclude that the rapid mass accretion influences not only the evolution during the stellar mass growth, but also the final fate when it dies.

Similar variations of the final mass of accreting SMSs have been recently studied by other authors using different numerical codes (Woods *et al.*, 2017; Haemmerlé *et al.*, 2017b). Regardless of some technical differences expected, they all show similar overall trends; the final mass roughly logarithmically increases with increasing the accretion rate. The quantitative differences by factors of a few are to be examined in future studies.

5. Roles of the supergiant protostars

The stellar evolution calculations described above have adopted simple accretion histories with constant rates for simplicity. In realistic situations, however, the accretion histories are not steady, but variable in time. The mass accretion onto a star should occur through a massive circumstellar disk, where the mass and angular momentum transfer is caused by the gravitational torque (Greif *et al.*, 2012). The disk undergoes gravitational fragmentation in extreme cases. The fragments quickly migrate inward via

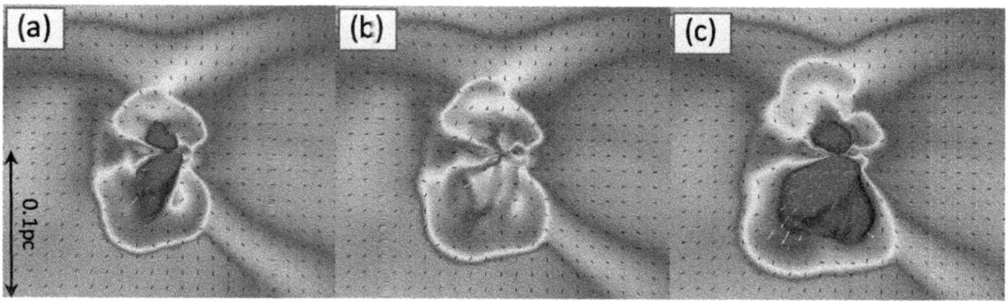

Fig. 5. (Color online) The intermittent UV feedback observed in a 3D radiation hydrodynamic simulation (case C-HR2-m0 in Hosokawa *et al.* (2016)). The panels (a)-(c) show a time sequence, where the color scale represents the gas temperature distribution around the central protostar. Note that an HII region (red) temporarily disappears in panel (b). Figure adapted from Hosokawa (2017), copyright of Società Astronomica Italiana, reproduced with permission.

the gravitational torque to eventually fall onto a star (Inayoshi and Haiman, 2014; Sakurai *et al.*, 2016). It results in an accretion burst event, during which the rate is substantially enhanced. Such a burst is normally followed by a relatively long quiescent phase, where the rate is much lower than the mean value.

The effects of such variable accretion on the evolution of the SMSs have also been studied using numerical modeling (Sakurai *et al.*, 2015). Since an accreting SMS stays in the supergiant protostar stage only with $\dot{M}_* \gtrsim 10^{-2}\ M_\odot\ \mathrm{yr}^{-1}$, the star may contract in a quiescent phase, so that its UV emissivity substantially rises. The stellar evolution calculations suggest that the contraction could occur if the duration of the quiescent phase is longer than 1000 years. Multi-dimensional hydrodynamic simulations suggest that such a long break of the accretion is unlikely to occur, as long as the star is embedded in a massive self-gravitating disk (Sakurai *et al.*, 2016).

The above is all about the formation of SMSs postulated in the DC model. In normal primordial star formation, the typical accretion rates 10^{-3}–$10^{-2}\ M_\odot\ \mathrm{yr}^{-1}$ are too low to bring an accreting star to the supergiant protostar stage. The protostar evolves through the normal KH contraction stage in this case, and the UV feedback is thought to become strong enough to halt the mass accretion. However, variable mass accretion also commonly occurs in such a case, and may modify the above picture. Hosokawa *et al.* (2016) study the interplay between the variable accretion and resulting UV feedback, performing 3D radiation hydrodynamic simulations coupled with stellar evolution calculations. Consider the KH contraction stage when a bipolar HII region is growing (Fig. 5a). Once the accretion rate exceeds $\sim 10^{-2}\ M_\odot\ \mathrm{yr}^{-1}$ with an accretion burst, a protostar rapidly inflates and its UV emissivity accordingly drops. The HII region disappears (b), and it is after a while in the quiescent phase that another HII region appears as the star contracts again (c). The cycle recurrently occurs as far as the disk accretes the gas from a surrounding envelope. The UV feedback only operates intermittently, and it does not always easily halt the mass accretion. The mass accretion continues even after the stellar mass exceeds $300\ M_\odot$ in some cases.

6. Summary and future prospects

We have described recent theoretical developments on the evolution of SMSs, highlighting effects of very rapid accretion at the rates of $\dot{M}_* \gtrsim 0.1\ M_\odot\ \mathrm{yr}^{-1}$. Stellar evolution calculations predict that such a SMS has a very characteristic feature, i.e., large radius that monotonically increases as the stellar mass increases via the accretion. The evolution of the radius is well approximated as $R_* \propto M_*^{0.5}$, independent of the accretion rate if it exceeds a critical rate. The radius exceeds $7000\ R_\odot$ (or 30 AU) after the star accretes the gas of $10^3\ \mathrm{M}_\odot$. Interestingly, the stellar effective temperature is almost fixed at a constant value $T_{\mathrm{eff}} \simeq 5000$ K during this "supergiant protostar" stage. With the very cool atmosphere, the star only emits a small amount of ionizing photons, and the resulting UV feedback is too weak to disturb the accretion flow. The mass accretion onto the star thus continues even after the stellar mass reaches $\sim 10^4 - 10^5\ \mathrm{M}_\odot$. While the mass accretion continues, the star finally collapses owing to the exhaustion of nuclear fuel, or general relativistic instability. Since the rapid accretion greatly affects the stellar interior structure, so it does also for its final fate.

The above picture has been established by numerically solving the 1D interior structure of an accreting SMS. However, there still remains additional effects that need further investigations. One of them is potential mass-loss via strong stellar winds. If the star accretes the gas through a geometrically thin disk, most of the stellar surface is evacuated from it. We normally solve the structure of such a freely radiating atmosphere assuming the hydrostatic balance. However, relaxing the assumption allows dynamical gas motion, so that very powerful stellar winds may be launched. If the resulting mass-loss rate exceeds the accretion rate, the stellar mass growth will be limited by this effect before reaching the supermassive regime. Recent theoretical studies suggest that the gas acceleration by radiation pressure and/or pulsational instability will excite some outflows, though the resulting mass-loss rates are much smaller than the accretion rates (Inayoshi *et al.*, 2013; Nakauchi *et al.*, 2017).

The angular momentum of the accreting gas may also play a critical role during the evolution of SMSs. It is still uncertain how much angular momentum should be carried into the star with the accreting gas, and how the resulting stellar evolution is modified by effects of the rotation. In general, the maximum rotation speed allowed for a star is set by the balance between the centrifugal force and gravity (Ω-limit). However, a very massive star has the lower critical speed because a large contribution of the radiation pressure effectively reduces the gravity. If the star obtains the angular momentum carried by the accreting gas, the star easily reaches such a limit (known as $\Omega\Gamma$-limit, e.g., Meynet and Maeder, 2000). The angular momentum of the star may be somehow extracted and transported outward through the disk while the star is accreting the gas (Paczynski, 1991; Popham and Narayan, 1991). Potential effects of this $\Omega\Gamma$-limit for the evolution of accreting SMSs have been controversial among recent studies (Lee and Yoon, 2016; Haemmerlé *et al.*, 2017a; Takahashi and Omukai, 2017), and are to be examined in detail in future studies.

With this chapter, we conclude the discussion on the basic black hole formation scenarios and their progenitors. Chapter 9 will summarize the statistical predictions on the first black holes. Their subsequent growth will be discussed in Chapters 10 and 11. The final chapters will present both the current observational status as well as future prospects.

Chapter 9

Statistical predictions for the first black holes

Tilman Hartwig

The University of Tokyo,
Kavli IPMU (WPI), UTIAS,
Kashiwa, Chiba 277-8583, Japan,
Tilman.Hartwig@ipmu.jp

The recent observations of supermassive black holes (SMBHs) at high redshift challenge our understanding of their formation and growth. There are different proposed pathways to form black hole (BH) seeds, such as the remnants of the first stars (Chapter 4), gas-dynamical processes (Chapter 5), direct collapse (Chapter 6), or stellar collisions in dense nuclear clusters (Chapter 7). In this chapter, we discuss the probability of forming supermassive black holes (SMBHs) via these channels and determine the expected number density of the BH seeds. We start with a brief discussion of the observational constraints on SMBHs at low and high redshift that theoretical models have to reproduce (a more detailed account is provided in Chapter 12). We further present the most popular formation channels of SMBHs, discuss under which conditions they can reproduce the observations, and compare various estimates in the literature on the expected number density of SMBHs. To account for the density of quasars at $z > 6$ requires very efficient gas accretion mechanisms or high BH seeds masses. The bottleneck to obtain sufficiently high number densities of seed BHs with masses $> 10^5$ $M_\odot$ is the interplay between radiative and chemical feedback, which constrains the conditions for primordial, isothermal gas collapse.

1. Observational constraints

Although BHs do not emit electromagnetic radiation, the accretion disk around supermassive black holes (SMBHs) converts gravitational energy into heat and the emission of the brightest of these active galactic nuclei can be seen as quasars up to high redshift. From the optically selected quasar sample in the SDSS survey we can determine a lower limit for the abundance of SMBHs with $> 10^9\,\mathrm{M}_\odot$ at $z = 6$ of about $1/\,\mathrm{cGpc}^3$ (Fan *et al.*, 2006c; Venemans *et al.*, 2013; De Rosa *et al.*, 2014). As we will see below, current models predict sufficiently many massive BH seeds to explain the number of quasars at $z > 6$. However, the direct collapse (DC) scenario, which produces massive BH seeds (see Chapter 5), cannot account for the population of less massive BHs at lower redshift under conservative parameter assumptions (Greene, 2012; Reines *et al.*, 2013). Moreover, not all halos are able to fuel gas to their center at high rates over several hundred million years to grow the BH seeds to their final masses (see Chapter 10 for the feedback limited growth rates).

The discovery of the Lyman-α emitter CR7 at $z = 6.6$ (Sobral *et al.*, 2015) might yield another constraint on the number density of direct collapse black holes (DCBHs). The strong Lyman-α and HeII emission with large equivalent widths, the steep UV slope, and the absence of metal lines in the original observation by Sobral *et al.* (2015) suggests a massive population of metal-free stars with a total stellar mass of $\sim 10^7\,\mathrm{M}_\odot$. However, several studies demonstrate that the observational signature is more likely to emerge from an accreting DCBH in the center of CR7 (Pallottini *et al.*, 2015; Agarwal *et al.*, 2016b; Hartwig *et al.*, 2016a; Smidt *et al.*, 2016; Smith *et al.*, 2016; Pacucci *et al.*, 2017c) than by a population of metal-free stars, but see Visbal *et al.* (2016); Bowler *et al.* (2017). The expected number density of CR7-like sources is of the order $10^{-6} - 10^{-7}\,\mathrm{Mpc}^{-3}$ (Pallottini *et al.*, 2015; Hartwig *et al.*, 2016a; Visbal *et al.*, 2016), which yields an additional observational constraint on the abundance of the first BHs. However, see also the recent observation that challenges the interpretation of CR7 as a DCBH (Matthee *et al.*, 2017; Sobral *et al.*, 2017; Bowler *et al.*, 2017).

Abbreviation	Full name
SMBH	supermassive black hole
BH	black hole
DC	direct collapse
DCBH	direct collapse black hole
SN	supernova
LW	Lyman-Werner
PDF	probability distribution function

In the next section we will see that stellar mass BHs, as the remnants of the first stars (see Chapter 4), are sufficiently abundant and it is not their number density that poses a problem, but rather the necessary mass accretion over more than six orders of magnitude and the associated feedback effects that pose the biggest challenge (see Chapter 10). We will hence mostly focus on DCBHs, which are predicted to be born with a higher seed mass, but which form only under very peculiar conditions and it is worth to investigate the probability of these environmental conditions of DC to assess the likelihood and the halo occupation fraction for DCBHs.

2. Probability for stellar mass seed black holes

Stellar mass seed BHs of $100 - 1000\,\mathrm{M}_{\odot}$ can either form as remnants of the first metal-free stars (see Chapter 4) or as a consequence of runaway stellar collisions in metal-poor nuclear clusters (see Chapter 7). Devecchi *et al.* (2012) estimate the seed density for these two scenarios with a semi-analytical model of structure formation. Their model confirms recent observations that all halos with masses above $10^{11}\,\mathrm{M}_{\odot}$ host a BH and predicts a BH occupation fraction of 10% for halos less massive than $10^{9}\,\mathrm{M}_{\odot}$. Habouzit *et al.* (2016b) demonstrate that the number density of Pop III remnant BHs is $\gtrsim 10^{-2}\,\mathrm{cMpc}^{-3}$ at $z < 10$, significantly higher than the number density of DCBH seeds in their numerical simulations. We compare the predicted number densities of stellar mass BHs in Fig. 1; see also Fig 4, Chapter 7. The number density of BH seeds formed by runaway stellar collisions in nuclear clusters in slightly below the number density of Pop III remnant BHs, but both are significantly above the expected number density of DCBHs (see Section 3). However, since BH formation via runaway collisions in dense stellar clusters can also occur in slightly metal-enriched gas, this formation channel dominates the mass density of BH seeds at $z < 10$ (Devecchi *et al.*, 2012). In their model, BH form via runaway stellar collision only in the narrow metallicity range $10^{-3.5} < Z/Z_{\odot} < 10^{-3}$, but recent simulations show a wider range of conditions for this BH formation channel (Sakurai *et al.*, 2017; Reinoso *et al.*, 2018; Boekholt *et al.*, 2018).

Hence, the bottleneck to form SMBHs at high redshift from stellar mass BHs is not the abundance of their seeds but rather the required high accretion rate over a sufficiently long time and the associated feedback (see Chapter 10). In the following section we will hence focus on DCBHs and quantify the probability to form such a massive BH seed.

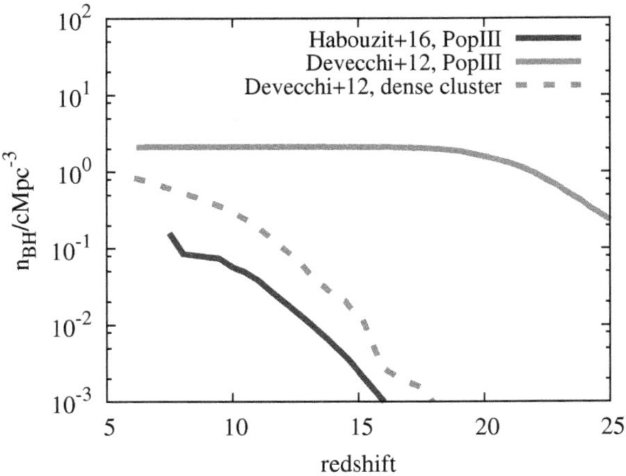

Fig. 1. Comparison of the comoving number density of stellar mass BHs as a function of redshift based on models by Devecchi *et al.* (2012) and Habouzit *et al.* (2016b). To convert from the given mass density in Habouzit *et al.* (2016b) to number density we assume an average mass of $100\,M_\odot$ for Pop III remnant BHs and of $700\,M_\odot$ for BHs formed by runaway collisions in dense stellar clusters. The number of Pop III seed black holes in Habouzit *et al.* (2016b) is a lower limit, because their simulation does not resolve minihalos as potential formation sites of these BHs. The number density of DCBHs is always below $10^{-5}\,\mathrm{cMpc}^{-3}$ (Habouzit *et al.*, 2016b), but see Chapter 5 for a more detailed discussion.

3. Probability for the direct collapse scenario

The idea of forming a massive BH as the result of the monolithic collapse of a proto-galactic gas cloud was first proposed by Rees (1984). Since then, various flavors of this formation scenario have emerged with different processes triggering the collapse of the cloud: galaxy mergers (Mayer *et al.*, 2010), gas instabilities (Begelman *et al.*, 2006; Choi *et al.*, 2013), or contraction due to radiative cooling (Bromm and Loeb, 2003).

The main requirement for these scenarios is a high mass infall rate so that the accretion timescale of the infalling gas is shorter than the Kelvin-Helmholtz timescale of the protostar (Hosokawa *et al.*, 2013; Schleicher *et al.*, 2013; Sakurai *et al.*, 2016). Under this condition, the stellar radius monotonically increases with mass and enough gas can be accreted before the star reaches the main sequence of hydrogen burning. In this section, we mostly focus on the DC scenario as a massive BH seed forming as the result of the thermodynamically triggered collapse of a pristine gas cloud with high mass infall rates and suppressed fragmentation. We will also comment on alternatives and explore the formation, such as the formation of massive seed BHs in metal-poor gas or under the influence of dust cooling.

Pristine halos that cool mainly by atomic hydrogen are expected to have high mass infall rates, because they isothermally collapse at $T \approx 10^4\,\mathrm{K}$ and the inefficient atomic cooling prevents fragmentation of the gas and allows for high mass infall rates ($\dot{M}_{\mathrm{in}} \propto T^{3/2}$). A photodissociating flux from a nearby star-forming galaxy is required to keep the gas atomic. Phrased differently, a pair of halos has to be well synchronized in space and time to trigger DC (Visbal *et al.* 2014; Regan *et al.*, 2017): they have to be close enough to generate a sufficiently high photodissociating flux, but they have to be sufficiently far separated to prevent metal enrichment and tidal stripping (Chon *et al.*, 2016; Regan *et al.*, 2017). Regarding the temporal synchronization, there is only a short window after the initial starburst in which the LW flux is high enough, but the metal-enriched, supernova (SN)-driven winds have not yet reached the neighboring halo (Dijkstra *et al.*, 2014; Habouzit *et al.*, 2016b; Hartwig *et al.*, 2016a; Agarwal *et al.*, 2016b).

3.1. *Atomic cooling halos*

The cooling function of hydrogen rises steeply around $\sim 8000\,\mathrm{K}$ (see Fig. 3 in Chapter 3) and we follow the convention to define halos with $T_{\mathrm{vir}} \gtrsim 10^4\,\mathrm{K}$ as atomic cooling halos. The virial temperature of a halo of mass M at redshift z is given by

$$T_{\mathrm{vir}} = 2 \times 10^4 \left(\frac{M}{10^8\,\mathrm{M}_\odot}\right)^{2/3} \left(\frac{1+z}{10}\right) \mathrm{K} \tag{1}$$

for atomic gas of primordial composition (Loeb, 2010) and the critical mass for an atomic cooling halo as a function of redshift is hence given by

$$M_{\mathrm{ac}} = 3.5 \times 10^7\,\mathrm{M}_\odot \left(\frac{1+z}{10}\right)^{-3/2}. \tag{2}$$

The number density of atomic cooling halos can be calculated via Press-Schechter theory (Press and Schechter, 1974) as outlined in chapter 2. The latter is based on the assumption of a Gaussian random fluctuation field, employing the results of linear theory to predict when the evolution becomes non-linear. The approach adopted here includes the corrections by Sheth *et al.* (2001), yielding:

$$n_{\mathrm{ac}} = \int_{M_{\mathrm{ac}}}^{\infty} \sqrt{\frac{2}{\pi}} \frac{\rho_m}{M} \frac{-\mathrm{d}(\ln\sigma)}{\mathrm{d}M} \nu_c \exp(-\nu_c^2/2)\,\mathrm{d}M, \tag{3}$$

with $\nu_c = \delta_{\rm crit}(z)/\sigma(M)$, where $\delta_{\rm crit}(z)$ is the critical overdensity for collapse, $\sigma(M)$ is the variance of the density power spectrum at mass M, and ρ_m is the mean matter density of the Universe. However, this constraint is not very restrictive because any host galaxy of a SMBH once passed this mass threshold of atomic cooling and was hence a potential candidate for DC. In the next section we will discuss more constraining criteria such as the photodissociating radiation and the absence of metals.

3.2. *Pristine gas*

In order to prevent fragmentation and maintain a high temperature and hence mass infall rate of the collapsing gas cloud, the halo has to be pristine (but see Omukai *et al.* (2008); Latif *et al.* (2016a); Agarwal *et al.* (2017c) for alternative scenarios). Metals are a much more efficient coolant than hydrogen and the higher cooling rate makes gas more susceptible to fragmentation and prevents DC via ordinary star formation. Halos can either be enriched internally by Pop III stars or externally by SN-driven metal-enriched winds from nearby star-forming halos and external enrichment seems to be more efficient in suppressing the formation of DCBHs (Dijkstra *et al.*, 2014). See Fig. 2 which illustrates how the fraction of pristine gas and halos evolves with redshift.

At $z = 4$ only $\lesssim 10\%$ of the cosmological volume is enriched with metals, see also Johnson *et al.* (2013a). However, this volume filling factor also accounts for voids without star formation. When we consider atomic cooling halos as the halos of interest for DC, Latif *et al.* (2016a) find that more than half of them are still pristine until $z \approx 9$ (see Fig. 1 of Chapter 5). These candidate halos have not been externally enriched; also, star formation in these halos was suppressed by e.g. a photodissociating background (Machacek *et al.*, 2001). Considering this criterion for DC independently may not seem very constraining, because every second target halo is still pristine at the redshift of interest. However, we discuss in the next section that DC generally requires a nearby star-forming galaxy, which drastically reduces the probability of not being polluted by metals.

3.3. *Photodissociating radiation*

A photodissociating LW flux is another crucial ingredient for DC. The concept of a critical flux threshold, $J_{\rm crit}$, above which a DCBH can form is well established in the literature, but there is no consensus on its value and most likely it is a distribution of critical values depending on e.g. the

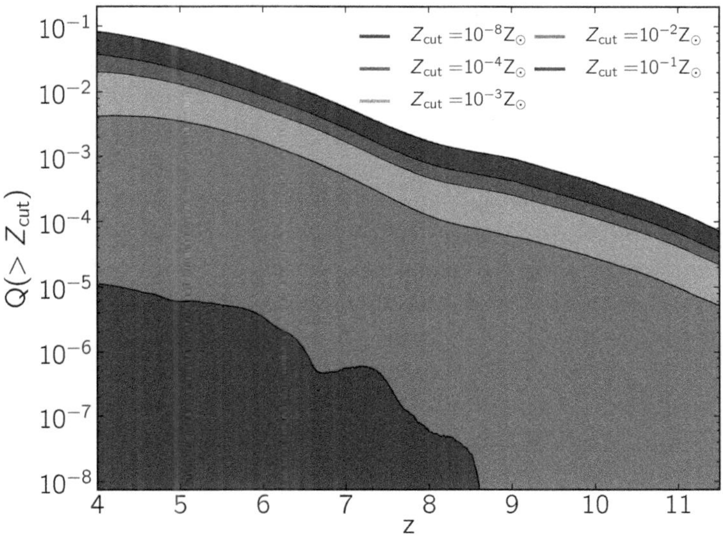

Fig. 2. Left: Redshift evolution of the metal volume filling factor for different values of the metallicity cut, Z_{cut}. Figure adopted from Pallottini *et al.* (2014), reproduced by permission of Oxford University Press/on behalf of the RAS.

spectral shape of the stellar population (Shang *et al.*, 2010; Sugimura *et al.*, 2014), the anisotropy of the illuminating flux (Regan *et al.*, 2014, 2016b), the treatment of H_2 self-shielding (Wolcott–Green *et al.*, 2011; Hartwig *et al.*, 2015c), the details of the chemistry model (Glover, 2015a,b), and the star formation history (Agarwal *et al.*, 2016a). We refer here also to the detailed discussions on this topic in Chapter 3 on chemistry, Chapter 5 on the determination of J_{crit} from dynamical simulations and Chapter 6 describing the dependence on the radiation spectrum. Although there is no universal value for J_{crit}, it is convenient to quantify the importance of different effects by this value. Once we have chosen a threshold value for DC, we can quantify the probability to find a pristine atomic cooling halo that is exposed to a LW flux above this threshold. In this section we will review and summarize different models to address this question and to derive the number density of DCBH formation sites. We use the standard notation for the LW flux and express its intensity in units of $J_{21} = 10^{-21} \mathrm{erg\,s^{-1}\,cm^{-2}\,sr^{-1}\,Hz^{-1}}$.

The build-up of the cosmological LW background is self-regulated (Johnson *et al.*, 2008), because a high flux suppresses star formation and hence the production of further LW photons. The mean value in the redshift range $z = 10 - 20$ is on the order of a few times of $10\,J_{21}$ (Dijkstra *et al.*, 2008; Ahn *et al.*, 2009; Holzbauer and Furlanetto, 2012; McQuinn, 2012; Agarwal *et al.*, 2012; Dijkstra *et al.*, 2014; Chon *et al.*, 2016), but possible formation sites for DC are hidden in the high-end tail of the $J_{\rm LW}$ distribution.

Ahn *et al.* (2009) present the first self-consistent radiative transfer calculations of the inhomogeneous LW background. By using high-resolution N-body simulations and an equivalent gray opacity for the radiative transfer they show that the LW background is inhomogeneous and correlates with the clustering of large-scale structure. In contrast to ionizing photons in the high redshift Universe, LW photons can travel cosmological distances due to the small pre-reionization H_2 abundance of $\sim 10^{-6}$ and hence an optical depth of $\tau_{\rm LW} < 1$ (Shapiro *et al.*, 1994). However, due to the expansion of the universe they get redshifted out of the LW bands and can no longer contribute to H_2 photodissociation once their energy is below $\sim 11.5\,$eV (see also Haiman *et al.* (2000) for a discussion of radiative transfer of LW at high redshift on cosmological scales).

Ahn *et al.* (2009) define the "LW horizon" as the maximum comoving distance from a source that a H_2 dissociating photon can reach, and determine this radius to be $\sim 100\,\alpha\,$cMpc, with

$$\alpha = \left(\frac{h}{0.7}\right)^{-1}\left(\frac{\Omega_m}{0.27}\right)^{-1/2}\left(\frac{1+z_s}{21}\right)^{-1/2}, \tag{4}$$

where z_s is the redshift of the source. Within this LW horizon the cosmic expansion reduces the effective LW flux by

$$f_{\rm mod} = \begin{cases} 1.7\exp[-(r_{\rm cMpc}/116.29\alpha)^{0.68}] - 0.7 & \text{for} \quad r_{\rm cMpc}/\alpha \leq 100 \\ 0 \quad \text{for} \quad r_{\rm cMpc}/\alpha > 100, \end{cases} \tag{5}$$

where $r_{\rm cMpc}$ is the distance from the source in comoving Mpc.

Due to the long mean free path of LW photons, many sources contribute to the local flux and Dijkstra *et al.* (2008) find that $> 99\%$ of all halos are exposed to a LW flux within a factor of two of the mean value. They calculate the LW background based on the clustering of halos and account for the non-linear Eulerian bias of the two-point correlation function (Iliev *et al.*, 2003), which fits the correlation function derived in N-body simulations. Accounting for this non-linear bias is crucial since we are especially interested in close pairs of halos, which are the most promising candidates for DC (Dijkstra *et al.*, 2008; Visbal *et al.*, 2014; Regan *et al.*, 2017). The differential probability to find a halo of mass M at redshift z in the distance r to an atomic cooling halo is given by

$$\frac{\mathrm{d}^2 P_1(M,r,z)}{\mathrm{d}M\,\mathrm{d}r} = 4\pi r^2 (1+z)^3 [1+\xi(M_{\mathrm{ac}},M,r,z)]\frac{\mathrm{d}n}{\mathrm{d}M}, \tag{6}$$

where $\xi(M_{\mathrm{ac}},M,r,z)$ is the two-point correlation function, which yields the excess probability to find two halos of masses M_{ac} and M at distance r, and $\mathrm{d}n/\mathrm{d}M$ denotes the Press-Schechter mass function (Press and Schechter, 1974; Sheth *et al.*, 2001). Dijkstra *et al.* (2008) model the LW luminosity of each halo by randomly sampling from a log-normal distribution of UV luminosities per unit star formation rate, $P_2/\mathrm{d}\log(L_{\mathrm{LW}})$. The probability distribution function (PDF) of a given atomic cooling halo to be exposed to a LW flux J_{LW} is hence given by

$$\frac{\mathrm{d}P}{\mathrm{d}\log(J_{\mathrm{LW}})} = \int_{M_{\mathrm{min}}}^{\infty} \mathrm{d}M \int_{r_{\mathrm{min}}}^{\infty} \mathrm{d}r \frac{\mathrm{d}^2 P_1}{\mathrm{d}M\,\mathrm{d}r}\frac{P_2}{\mathrm{d}\log(L_{\mathrm{LW}})}. \tag{7}$$

Dijkstra *et al.* (2008, 2014); Inayoshi and Tanaka (2015) assume $M_{\mathrm{min}} = M_{\mathrm{ac}}$, but the choice of the minimal radius r_{min} is debated, because its value is crucial: small values will populate the high-end tail of the PDF because the flux scales with the inverse square of the distance, but a too-small value is physically prohibited from preventing external enrichment of the atomic cooling by the star forming halo and from preventing tidal stripping due to dynamical effects (Chon *et al.*, 2016; Regan *et al.*, 2017).

We refer to Dijkstra *et al.* (2008) for a discussion of how the choice of r_{min} (and other model parameters) affects the PDF of J_{LW}. The high-end tail of the PDF is dominated by close pairs of halos, and the illuminating flux, as seen by the atomic cooling halo, should hence be highly anisotropic (Regan *et al.*, 2014, 2016a). Moreover, the relative velocity of this close pair of halos is relevant for the escape fraction of LW photons out of the star-forming galaxy, because photons might be shifted into the LW bands in the far-field (Schauer *et al.*, 2015; Schauer *et al.*, 2017a).

In Dijkstra *et al.* (2014) they improve their previous model by explicitly accounting for pollution by metal-enriched winds from the star forming galaxy. They approximate the radius of the metal enriched region around a star-forming halo by

$$r_{\mathrm{metals}} = 0.3\,\mathrm{kpc}\left(\frac{M_*}{10^5\,\mathrm{M}_\odot}\right)^{1/5}\left(\frac{n}{1\,\mathrm{cm}^{-3}}\right)^{-1/5}\left(\frac{t}{\mathrm{Myr}}\right)^{2/5}, \tag{8}$$

where M_* is the stellar mass and n the density of the gas into which the SN-driven metal bubble expands. Based on this metal-polluted radius,

Dijkstra *et al.* (2014) correct the PDF of $J_{\rm LW}$ (Eq. (7)) by multiplying it with $\Theta(r - r_{\rm metals})$ to exclude atomic cooling halos as potential candidates for DC if they are too close and already polluted. This approach was also used in Habouzit *et al.* (2016a); Hartwig *et al.* (2016a) to estimate the likelihood of finding DCBH candidates.

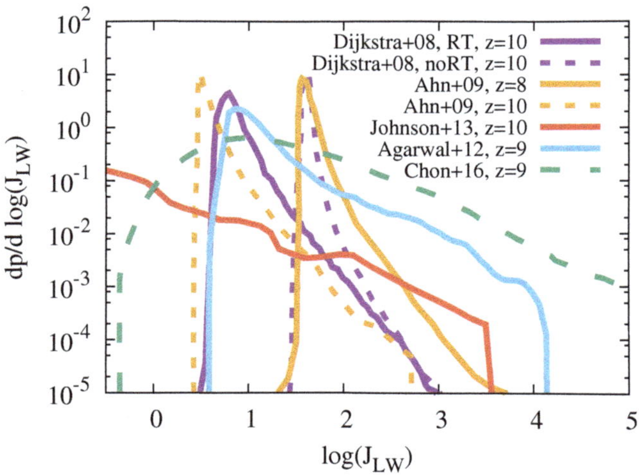

Fig. 3. Comparison of the PDF of the LW flux. The data are based on Dijkstra *et al.* (2008); Ahn *et al.* (2009); Johnson *et al.* (2013a); Agarwal *et al.* (2012); Chon *et al.* (2016). Note the slightly different redshifts and how the peak of this distribution shifts towards higher fluxes with decreasing redshift due to the increasing cosmic star formation rate. All studies determine the LW flux based on the clustering of halos. Chon *et al.* (2016); Agarwal *et al.* (2012) derive the distribution for pristine halos, Ahn *et al.* (2009) show the PDF for all grid cells inside their computational domain, Johnson *et al.* (2013a) for primordial gas with $n \geq 1\,{\rm cm}^{-3}$, and Dijkstra *et al.* (2008) provide the flux as seen by atomic cooling halos of mass $M = 4 \times 10^7\,{\rm M}_\odot$. In their fiducial model (noRT) they assume the IGM to be transparent to LW photons and in their second model with radiative transfer they account for intergalactic H_2, which reduces the contribution from far away sources and hence the mean and minimum value of the PDF. However, the distribution at high $J_{\rm LW}$ is not affected by intergalactic absorption, because it is dominated by close pairs of halos.

A comparison of the literature on the PDF of $J_{\rm LW}$ can be seen in Fig. 4. All studies broadly agree on the shape and mean value, which increases with decreasing redshift and all studies predict potential DCBH hosts, which are exposed to a flux above $J_{\rm crit} \approx 10^2 - 10^3$. The final density of DCBHs depends crucially on the high-end slope, which differs amongst the authors. While the recent simulation-based results of Agarwal *et al.* (2012); Chon *et al.* (2016); Johnson *et al.* (2013a) agree on the high-end slope, the models by Dijkstra *et al.* (2008); Ahn *et al.* (2009) yield a steeper slope. Inayoshi and Tanaka (2015), who follow the approach by Dijkstra *et al.* (2008, 2014), find an even steeper slope of $\mathrm{d}p/\mathrm{d}\log(J_{\mathrm{LW}}) \propto J_{\mathrm{LW}}^{-5}$ in the range $10^3 \lesssim J_{\mathrm{LW}} \lesssim 10^4$. They attribute this discrepancy to the choice of r_{min}.

The sharp cut-off at low J_{LW} is due to the very long mean free path of LW photons and hence the mean, median, and minimum value of this distribution are very close. The less sharp cut-off at low J_{LW} in Johnson *et al.* (2013a); Chon *et al.* (2016) is caused by the smaller box size and the hence smaller number of sources contributing to the local flux. In the other studies, the box sizes are larger or they implicitly account for the LW background radiation by the cosmic star formation history.

Another crucial question is for how long an atomic cooling halo has to be exposed to a photodissociating flux above J_{crit} for it to collapse isothermally. The collapse time of the gas core of a typical atomic cooling halo is $t_{\mathrm{coll}} \approx 10\,\mathrm{Myr}$ at $z = 10$ (Visbal *et al.*, 2014) with $t_{\mathrm{coll}} \propto (1+z)^{-3/2}$. Other authors require the conditions for DC to be fulfilled over at least the freefall time, which is of the order 100 Myr at the redshifts of interest and yields more conservative estimates (Dijkstra *et al.*, 2014). Habouzit *et al.* (2016b) analyze the impact of the choice of the collapse time on the probability to find DCBH candidates and find a significant difference: At $z = 7.3$ they find 17 (3) DC formation sites in their computational volume for an assumed collapse time of 10 Myr (150 Myr).

Chon *et al.* (2016) demonstrate that the requirement of having a flux of at least J_{crit} at the time of virialization of the atomic cooling halos is too strict for DC. They reduce the LW intensity artificially to 2% of its original value at virialization and still find the halo to collapse isothermally. They argue that the local flux increases over time due to the approaching halos by an order of magnitude and that lower intensities than previously thought are sufficient to induce DC, also see Fernandez *et al.* (2014).

3.4. *Tidal and ram pressure stripping*

The conditions for DC are generally met in close halo pairs, where one halo provides the photodissociating radiation and the other one is still pristine and about to pass the mass threshold of atomic hydrogen cooling. However, dynamical effects such as tidal forces and ram pressure stripping can no longer be neglected for the thermal evolution of the gas collapse (Chon *et al.*, 2016). Tidal stripping occurs in close pairs of galaxies, where the tidal force exerted by the more massive galaxy results in an effective transfer of gas and stars from the lower mass to the more massive galaxy. Ram pressure stripping can expel gas out of the gravitational potential of a galaxy that moves with a sufficiently high relative velocity with respect to the ambient medium.

To study these dynamical effects of the environment, Chon *et al.* (2016) select DC candidate halos in a semi-analytical model and perform 3D hydrodynamical follow-up simulations of these pristine atomic cooling halos that are illuminated by a sufficiently high LW flux. Out of 42 selected halos they find only two successful candidates and show that tidal forces and ram pressure stripping prevent the gas collapse in the other cases.

They also point out that while minor mergers tend to disrupt the gas distribution and prevent collapse, major mergers rather help in triggering the collapse due to the induced gravitational instability at the halo center, caused by the asymmetric mass distribution on large scales, also see Mayer *et al.* (2010, 2015). In summary, they find that only 5% of the atomic cooling halos that are pristine and illuminated with $J_{\rm LW} \geq J_{\rm crit}$ eventually collapse to form a DCBH.

3.5. *Density of direct collapse seed black holes*

In this section, we compare different estimates for the probability to form DCBHs, their expected number density, and the resulting occupation fractions. As we have seen before, the PDF to find pristine atomic cooling halos exposed to a certain LW flux (Eq. (7)) can be calculated analytically (Dijkstra *et al.*, 2008, 2014; Inayoshi and Tanaka, 2015), by post-processing cosmological N-body simulations (Ahn *et al.*, 2009; Agarwal *et al.*, 2012), or by performing cosmological simulations (Johnson *et al.*, 2013a; Agarwal *et al.*, 2014; Habouzit *et al.*, 2016a; Chon *et al.*, 2016). The number density of DCBHs can then be approximated by

$$n_{\rm DCBH}(z) = \int_{M_{\rm ac}}^{\infty} {\rm d}M \frac{{\rm d}n}{{\rm d}M} P_{\rm DCBH}(\geq J_{\rm crit}, z). \tag{9}$$

However, this approach assumes that all pristine atomic cooling halos that are exposed to $> J_{\rm crit}$ collapse to a DCBH, but the analytical and semi-analytical models do not account for hydrodynamical effects such as tidal or ram pressure stripping. Chon *et al.* (2016) show that these effects can reduce the number density of DCBHs by more than one order of magnitude.

A comparison of the number density of DCBH as a function of redshift by different authors can be seen in Fig. 3.5.

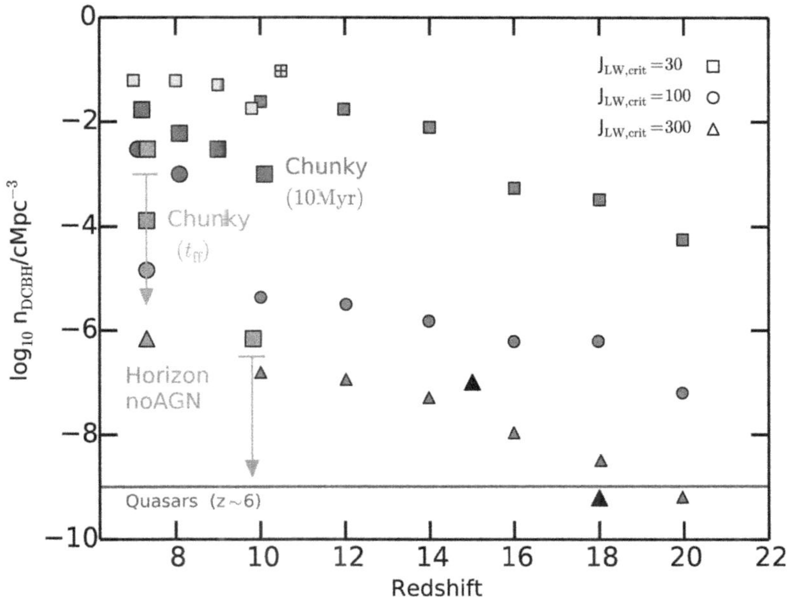

Fig. 4. (Color online) Comoving number density of DCBH formation sites, as a function of redshift. Symbol shapes represent different radiation intensity thresholds. Squares: $J_{\rm crit} = 30$, circles: $J_{\rm crit} = 100$, triangles: $J_{\rm crit} = 300$. The horizontal solid blue line shows the comoving number density of quasars at $z \sim 6$. The light gray crossed square at $z = 10.5$ is from the hydrodynamical simulation by Agarwal *et al.* (2014), the light gray squares in the range $z = 7-10$ are from Agarwal *et al.* (2012), dark gray squares and black triangles are the results of Dijkstra *et al.* (2014) and Valiante *et al.* (2016), respectively. The orange square and purple symbols show the number density for Habouzit *et al.* (2016a) for different assumptions on the collapse time. The cyan squares, circle and triangle represent the large-scale cosmological simulation Horizon-noAGN (Dubois et al. 2014b). Adapted from Habouzit *et al.* (2016a), reproduced by permission of Oxford University Press/on behalf of the RAS.

The number density increases with decreasing redshift and with decreasing $J_{\rm crit}$, but even for the same value of $J_{\rm crit}$ the absolute values of the number density differ by several orders of magnitude. All studies agree that chemical enrichment from previous star formation in progenitor halos or by metal-enriched winds from nearby star-forming halos play a crucial role in determining the final number density of DCBHs. Also the implementation of SN feedback strongly affects the possibility to form a DCBH (Habouzit *et al.*, 2016a). Hydrodynamical simulations suffer from the limited volume which allows in some cases only for upper limits on the DCBH density (Agarwal *et al.*, 2014; Habouzit *et al.*, 2016a). Other simulations find values of $2.5 \times 10^{-4}\,\mathrm{cMpc}^{-3}$ for $J_{\mathrm{crit}} = 100$ (Chon *et al.*, 2016) and $(0.1-5) \times 10^{-6}\,\mathrm{cMpc}^{-3}$ for $J_{\mathrm{crit}} = 300$ (Habouzit *et al.*, 2016b). Using an analytical estimate for the clustering of halos, Yue *et al.* (2014) find that the DCBH mass density rises from $\sim 5\,\mathrm{M}_\odot\,\mathrm{Mpc}^{-3}$ at $z \sim 30$ to the peak value $\sim 5 \times 10^5\,\mathrm{M}_\odot\,\mathrm{Mpc}^{-3}$ at $z \sim 14$ in their fiducial model. However, the abundance of accreting DCBHs decreases after $z \sim 14$, because potential formation sites get either polluted or photoevaporated. We refer the interested reader to Habouzit *et al.* (2016a) for a more detailed quantitative comparison of the different semi-analytical models.

Bellovary *et al.* (2011) use cosmological simulations to derive the BH occupation fraction and they show that halos with $M > 3 \times 10^9\,\mathrm{M}_\odot$ host massive BHs regardless of the efficiency of seed formation. Habouzit *et al.* (2016a) find, based on post-processing state-of-the-art cosmological simulations (Dubois *et al.*, 2014a), that $\sim 30\%$ of halos more massive than $10^{11}\,\mathrm{M}_\odot$ at $z = 6$ have at least one progenitor that fulfils the conditions for DC, assuming a critical value of $J_{\mathrm{crit}} = 30$. However, this occupation fraction is a strong function of the critical flux and falls below $1 - 10\%$ for $J_{\mathrm{crit}} > 100$. In Habouzit *et al.* (2016b) they confirm an occupation of $\sim 10\%$ for Pop III remnant BHs and an occupation fraction of the order $10^{-6} - 10^{-5}$ for DCBHs, also see Tanaka and Haiman (2009).

Although there are still large theoretical uncertainties in the number density of DCBH formation sites, caused by e.g. the model of metal enrichment, the value of J_{crit}, or the necessary duration of the photodissociating feedback, all models are able to yield more than 1 DCBH per Gpc^3 by $z = 6$, which is the lower limit set by current observations. The most optimistic scenarios are even able to explain the majority of BHs in local galaxies via the DC scenario. The remaining bottleneck is to grow these seed BHs to SMBHs by circumventing the self-regulating accreting feedback as we will see in the next chapter.

In this chapter, we presented the rather consensual picture on the statistical predictions on BH formation in the early Universe. However, we note that most presented results are subject of ongoing debate and reality might lie beyond what we currently assume to be realistic. We have presented DC as a promising pathway to form massive BH seeds, but other formation channels than the isothermal collapse may also be relevant, or different scenarios, such as runaway collisions in the first stellar clusters could be more efficient than we have assumed (Katz *et al.*, 2015; Sakurai *et al.*, 2017). Some predictions based on the latter scenario were therefore presented in Chapter 7. Within the framework of the DC, the conservative conditions that the gas for the BH formation via gas-dynamical processes has to be pristine might be too strict and BH seed formation may also be possible in gas that is slightly enriched with metals or dust (Omukai *et al.*, 2008; Mayer *et al.*, 2015; Latif *et al.*, 2016a; Agarwal *et al.*, 2017c), see also discussion in chapter 5. All these caveats affect the predicted formation efficiency and number density of BH seeds in the early Universe.

The statistical predictions on the abundances of seed black holes are of course only the first step. In chapter 10 next, we will describe their further growth in the presence of feedback, while scenarios for super-Eddington accretion are outlined in chapter 11. The subsequent chapters describe both the current observational status as well as future prospects.

Chapter 10

Growth and feedback from the first black holes

John H. Wise

Center for Relativistic Astrophysics, School of Physics, Georgia Institute of Technology, 837 State Street, Atlanta, GA 30332, USA

jwise@physics.gatech.edu

Regardless of their initial seed mass, any active galactic nuclei observed at redshifts $z \geq 6$ must have grown by several orders of magnitude from their seeds. In this chapter, we will discuss the physical processes and latest research on black hole growth and associated feedback after seed formation. Fueling is initially slowed down by radiative feedback from the black hole itself and supernova explosions from nearby stars. Its growth however accelerates once the host galaxy grows past a critical mass.

1. Observational motivation

Since the beginning of the century, there have been approximately 40 supermassive black holes (SMBHs) discovered above a redshift of six when the universe was barely one billion years old (Fan *et al.*, 2003; Willott *et al.*, 2010; Bañados *et al.*, 2014). The vast majority of these distant quasars exist in the bright-end of the active galactic nuclei (AGN) luminosity function (Kashikawa *et al.*, 2015). The most distant observed SMBH has a redshift of $z = 7.54$ and a mass of $\sim 8 \times 10^8\ M_\odot$ (Bañados *et al.*, 2018b) while the most massive SMBH observed in the early Universe has a mass of $\sim 1.2 \times 10^{10}\ M_\odot$ at $z = 6.30$ (Wu *et al.*, 2015). These rare luminous objects are only the tip of the iceberg of the high-redshift SMBH population, only occurring about once per comoving Gpc^3 (Fan *et al.*, 2006a), with the remaining population having more moderate growth histories that exist in more normal galaxies like our Milky Way (Gültekin *et al.*, 2009).

In addition to inspecting growth histories of individual SMBHs, cumulative statistics and quantities, such as the total BH mass density and

AGN/galaxy luminosity functions, place further constraints on BH growth in the early Universe. Because only the brightest AGN are detected at $z \geq 6$, the Soltan argument (Sołtan, 1982) can estimate the BH mass density from the AGN luminosity function. It argues that accreted mass density into SMBHs, which is derived from the integrated quasar light and some radiative efficiency factor, must be less than the total BH mass density, i.e. $\rho_{\rm acc} \leq \rho_{\rm BH}$. Another important statistical measure is the present-day M-σ relation that suggests that there is some connection between galaxy and BH growth (Silk and Rees, 1998; Ferrarese and Merritt, 2000; Gebhardt *et al.*, 2000; Gültekin *et al.*, 2009). Its behavior at high-redshift is difficult to observe, but in theoretical models, it provides a window in how the M-σ relation is established as both the host galaxy and central BH grow over billions of years (Robertson *et al.*, 2006; Volonteri and Natarajan, 2009). By combining the M-σ relation and galaxy luminosity (or mass) function, the total BH mass density in SMBHs with $M_{\rm BH} > 10^6\ M_\odot$ rises from $\sim 10^{-3}\ M_\odot$ per comoving Mpc^{-3} at $z = 7$ to the present-day value of $5 \times 10^5\ M_\odot$ per comoving Mpc^{-3} (Shankar *et al.*, 2004; Somerville *et al.*, 2008; Willott *et al.*, 2010; Schulze and Wisotzki, 2011; Volonteri *et al.*, 2017). A more detailed account on the current status of observations is given in Chapter 12.

These individual and statistical observational SMBH measures provide a basis for the theoretical models of early BH growth to match at $z \sim 6$. This chapter will explore the initial buildup of the first black holes and its impact on the first galaxies, following its behavior prior to galaxy formation to the end of reionization when the farthest quasars have been observed. First, we will estimate the growth rates from abundance matching in Section 2. Then in Section 3, we will review the important physical processes that govern the gas inflow and associated feedback. Section 4 covers the bulk of this chapter and focuses on the environment and growth shortly after the formation of the first seed black holes (see Chapters 5 and 7 for their formation scenarios). Then we follow their journey into larger halos in Section 5, becoming central massive black holes in the first generations of galaxies. We close this topic with a discussion of outstanding questions and the possible paths forward to illuminate these mysteries in Section 6.

2. Host halo progenitors and timescale estimates

To begin, we can use the comoving number density of the brightest $z \sim 6$ AGN to estimate the typical seed formation time in the latest Planck

Abbreviation	Full name
BH	black hole
SMBH	supermassive black hole
DCBH	direct collapse black hole
AGN	active galactic nuclei
IGM	intergalactic medium
ISM	interstellar medium

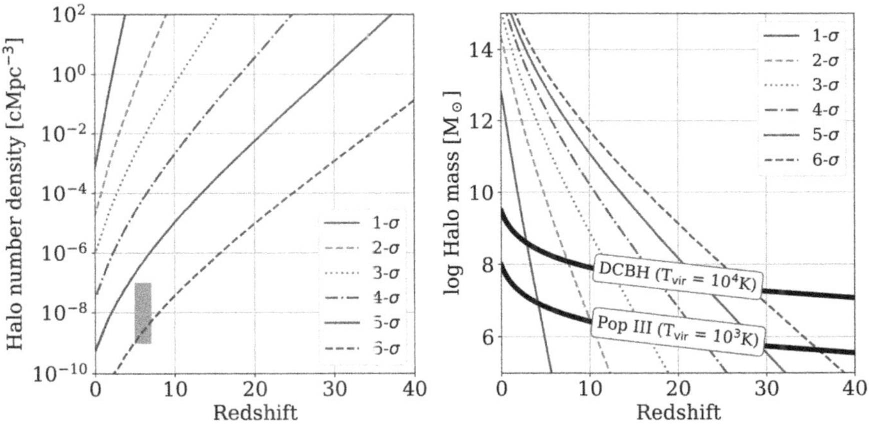

Fig. 1. *Left panel:* Each line shows the comoving number density of dark matter halos for a particular rarity, i.e. n-σ. The shaded box shows an estimate of the SMBH number density at $z \sim 6$, considering a duty cycle between 1–100%. Simple abundance matching of $z \sim 6$ AGN suggests that their host halos are very rare density fluctuations (5–6 σ) in large scale structure. *Right panel:* Dark matter halo masses for particular σ values compared with critical halo masses for Population III (metal-free) star formation and direct-collapse black hole (DCBH) formation. The 5- and 6-σ tracks can be used to estimate the $z \sim 6$ progenitor halo masses at earlier redshifts, suggesting that they hosted their first star around $z \sim 35$ (80 million years after the Big Bang) and started galaxy formation around $z \sim 25$ (130 million years after the Big Bang).

cosmology (Planck Collaboration *et al.*, 2016). Using an ellipsoidal variant of the Press-Schechter formalism (Press and Schechter, 1974; Sheth *et al.*, 2001), we calculate the number density of dark matter halos, shown in the left panel of Figure 1. Here n-σ denotes the rarity of each halo in a Gaussian density field, and the grey shaded area shows the estimated number density, assuming that some fraction $f = 0.01 - 1$ of SMBHs are currently in an active phase (Luo *et al.*, 2011). At $z = 6$, this simple abundance matching exercise suggests that the brightest AGN are hosted in very rare and massive halos (5-σ to 6-σ). These halos have masses around $10^{13}\ M_\odot$ at this redshift, shown in the right panel of Figure 1. We can then trace back the most massive progenitor mass at the same σ value, where it must have formed its first star at redshifts $z \simeq 30-35$ and could efficiently cool through hydrogen atomic line cooling at redshifts $z \simeq 25-30$. At these redshifts, the universe is around 100 million years old. Thus, the seed black holes only have 800 million years to grow by 4–7 orders of magnitude in order to transform into the observed SMBHs at redshift 6. Aiding this difficult ascent, their host halos will grow by similar factors.

3. Relevant physical processes and methods

The growing BH will have to compete with star formation and feedback processes for any accreted gas. Through a myriad of mergers and smooth IGM (intergalactic medium) accretion, the halo will have an ample gas supply, but the big question is what fraction makes the long journey to the SMBH event horizon, affected by many physical processes along the way.

3.1. *Accretion*

Any gas inside of the innermost stable orbit at three Schwarzschild radii ($R_{\rm sch} = 2GM_{\rm BH}/c^2$) will be accreted by the BH. However, it first must overcome the "angular momentum barrier" to exist at such a small radius. The gas inside pre-galactic halos is mainly turbulent from virialization (Wise and Abel, 2007a; Greif *et al.*, 2008) and has some coherent rotational motions that are usually around 30–50% of the Keplerian value $\sqrt{GM/R}$ (Regan and Haehnelt, 2009). Being a turbulent medium, there always exists some gas with very low specific angular momentum (Wise *et al.*, 2008) that can migrate toward the central regions without being rotationally supported (Lodato and Natarajan, 2006; Lodato, 2007).

Once the gas is within the central regions of the halo, the next barrier is becoming gravitationally bound to the BH, which was first described by Bondi and Hoyle (1944) and Bondi (1952) in the context of point masses and a spherically symmetric system. For gas with an adiabatic equation of state ($\gamma = 5/3$), this transition occurs at the Bondi radius, $r_{\rm B} \equiv GM_{\rm BH}/c_{\rm s}^2$, where $c_{\rm s}$ is the sound speed. The gas entering this sphere is assumed to accrete onto the BH at the Bondi-Hoyle rate, $\dot{M}_{\rm BH} = 4\pi G^2 M_{\rm BH}^2 \rho/(c_{\rm s}^2 + v_{\rm rel}^2)^{3/2}$, where ρ is the gas density at the Bondi radius and $v_{\rm vel}$ is the gas

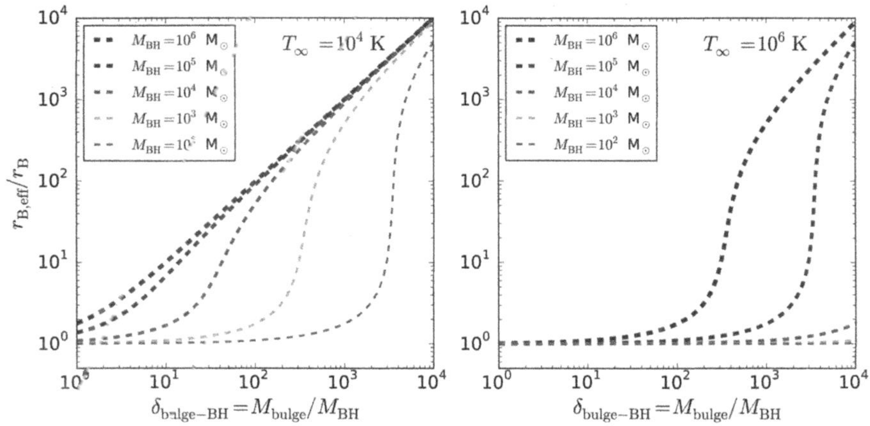

Fig. 2. The effective Bondi radius in units of the Bondi radius as a function of bulge-to-BH mass ratio for BH masses ranging from 10^2 to 10^6 $M_\odot$ that reside in gas reservoir with $T = 10^4$ K (left) and 10^6 K (right). Adopted from Park *et al.* (2016), ©AAS. Reproduced with permission.

velocity relative to the BH. Park *et al.* (2016) modified this prescription to include an external gravitational potential. The most likely sources for this potential are a stellar bulge or a cusped dark matter halo. They found that the effective Bondi radius, shown in Figure 2, dramatically increases when the total bulge mass exceeds 10^6 $M_\odot$ that would boost BH fueling.

Once the gas flows into the Bondi radius, it is assumed that some fraction will deposit onto the accretion disk, in which it will lose further angular momentum through viscous forces (Power *et al.*, 2011) and magnetic instabilities (Balbus and Hawley, 1991). It is a common assumption that the accretion rate onto the BH is equal to the Bondi-Hoyle rate or some scaled version (Schaye *et al.*, 2010), limited by the Eddington rate, $\dot{M}_{\rm Edd} = 4\pi G M_{\rm BH} m_{\rm p}/\epsilon c \sigma_{\rm T}$, where $m_{\rm p}$ and $\sigma_{\rm T}$ are the proton mass and Thomson scattering cross-section. Here ϵ is the radiative efficiency factor that dictates what mass-energy fraction of the accreted material is converted into radiation. If the BH grows at the Eddington limit, the e-folding time is $4.5 \times 10^7 (\epsilon/0.1)^{-1}$ yr. Applying this growth rate to the early universe, the largest SMBH at redshift 6 only has 15–18 e-folding times, corresponding to a mass ratio of $\log(M_{\rm BH}/M_0) = 6.5 - 7.8$, to grow to 10^9 $M_\odot$, where M_0 is the seed BH mass. In the following subsection, we will see that the associated feedback can hamper SMBHs from growing at this limit.

On the other hand, the Eddington limit is derived in spherical symmetry. Super-Eddington accretion rates can occur when

(1) the medium is not spherically symmetric or is porous, such as a disk (Abramowicz *et al.*, 1988; Novak *et al.*, 2012) or (supersonic) turbulence (Van Borm and Spaans, 2013; Van Borm *et al.*, 2014), or
(2) the radiation can be trapped with the gas flow, is advected with the fluid flow, and does not impart all of its momentum to the surrounding medium (Begelman, 1979; Jiang *et al.*, 2014).

In these cases, BH accretion rates can easily exceed the Eddington rate up to a factor of $\sim$200 for short periods at a radiative efficiency of $\sim$5% (Jiang *et al.*, 2014). These bursts somewhat alleviate the constraints on the growth history of the most massive SMBHs at $z \gtrsim 6$ (Volonteri and Rees, 2005).

3.2. *Feedback*

A fraction of the accreted gas mass-energy will inevitably be returned to the surrounding environment (Fabian, 2012). What fraction is unclear, but given the tremendous growth rates necessary to build the observed SMBHs, any fraction may have a substantial effect on the subsequent fueling. These feedback processes can be broadly categorized into radiative and mechanical feedback, both of which can launch winds from AGN that are ubiquitous in intrinsic quasar absorption spectra (Hamann *et al.*, 2008; Grier *et al.*, 2015). They will transition between a radiative mode (quasar mode) when $\dot{M}_{\rm BH} \lesssim 0.02\, \dot{M}_{\rm Edd}$ and a mechanical mode (radio mode) at higher accretion rates (Croton *et al.*, 2006; McNamara and Nulsen, 2007).

Many different astrophysical processes, such as blackbody radiation and inverse Compton scattering in the disk and its corona respectively, can be categorized into radiative feedback with the luminosity $L = \epsilon c^2 \dot{M}_{\rm BH}$, which is appropriate when radiation is decoupled from the gas. However in a dense optically-thick environment, such as the disk or jet, the radiation is trapped as photons scatter within the medium, causing $L \propto \ln \dot{M}_{\rm BH}$, that limits the amount of feedback (Begelman, 1979). This radiation will photo-heat the surrounding gas, creating a H II region, and will impart its momentum $|p| = E/c$ onto the baryons, i.e. radiation pressure. Heating the gas will reduce its susceptibility to accretion, correspondingly reducing the Bondi-Hoyle accretion rate (Alvarez *et al.*, 2009; Park and Ricotti, 2012; Jeon *et al.*, 2012). In addition, the radiation pressure will slow down or, in some cases, reverse the gas inflow toward the BH (Cowie *et al.*, 1978; Pacucci *et al.*, 2015; Inayoshi *et al.*, 2016).

Observations have shown that a significant portion of energy is released through bipolar relativistic jets (Blandford and Znajek, 1977; Bridle and Perley, 1984; Begelman *et al.*, 1984; Pounds *et al.*, 2003; Peterson and Fabian, 2006), also known as mechanical feedback, that entrains some fraction of mass along its journey through the host galaxy. These jets can inflate cavities in the largest of AGN in galaxy clusters (Fabian *et al.*, 2002; McNamara *et al.*, 2005). Contrary to radiative feedback, the kinetic energy of the jet will be lost as it exits the SMBH gravitational potential well. As the jet propagates through the immediate region (McKinney, 2006; Sadowski *et al.*, 2013) and the interstellar medium (ISM), it will shock-heat the gas, eventually launching an outflow from the host halo/galaxy (Sijacki *et al.*, 2007; Dubois *et al.*, 2012; Bieri *et al.*, 2017).

3.3. *Methods*

Capturing the growth and feedback of SMBHs in a cosmological setting is a difficult task primarily originating from the enormous dynamic spatial and temporal range of the problem. To follow the full growth history, it is essential to calculate the host halo merger tree, as the mergers bring in new gas supply for further accretion and other SMBHs that will merge after some time. Furthermore, following the host galaxy evolution is critical because there will be a feedback loop between the two (Silk and Rees, 1998; Hopkins *et al.*, 2006; Somerville and Davé, 2015). There are two major players in modeling galaxies and BHs over their lifetimes—semi-analytical models and numerical simulations—both of which can include the physics previously described directly or indirectly. In the process, the physical properties of the SMBH and galaxy are converted into key observables (Vogelsberger *et al.*, 2014; Schaye *et al.*, 2015) and compared and tuned against current observations, which allows for predicting observations with the next-generation observatories (Barrow *et al.*, 2017). Typical observables include luminosity functions, event rates for gravitational wave detections, mass functions, number counts on the sky as a function of redshift, and their spectra.

Semi-analytical models. use a halo merger tree as a basis for galaxy and BH formation and evolution (Kauffmann and White, 1993; Somerville *et al.*, 2008; Benson, 2010). It can be calculated either from extended Press-Schetcher (Bond *et al.*, 1991; Lacey and Cole, 1993), perturbative Lagrangian techniques (Monaco *et al.*, 2002; Munari *et al.*, 2017), or

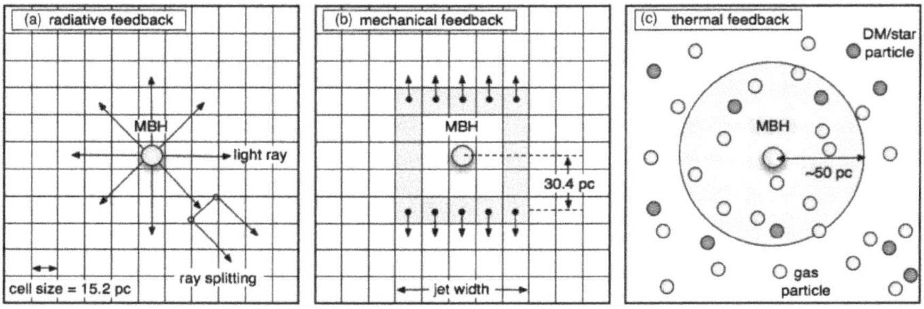

Fig. 3. Schematic views of numerical implementations of BH feedback. (a) Radiative feedback through ray tracing, (b) Mechanical jet feedback through momentum injection in "supercells", and (c) thermal feedback that assumes some thermal conversion efficiency between radiative/mechanical feedback and thermal energy. Adopted from Kim *et al.* (2011), ©AAS. Reproduced with permission.

extracted from an N-body simulation (Guo and White, 2014). Physically motivated models are included into each halo and followed through the merger tree. These recipes control processes such as gas cooling, star formation, chemical enrichment, BH accretion, stellar and gas radial distributions, and any associated feedback mechanisms. Specific to the BH, different accretion models (e.g. Bondi, Eddington-limited, super-Eddington), feedback models (e.g. radiative, jets, mass-loading, efficiency factors), accretion disk properties (e.g. slim disk), and what triggers bursts of high accretion (e.g. mergers) can be altered to inspect the resulting AGN and galaxy evolutionary track to understand how each process governs their properties and of their importance. The most serious disadvantage of semi-analytical models is the lack of spatial information, such as clustering, but they are computationally inexpensive and can be used to explore vast regions of parameter space. Furthermore, they can be utilized in Markov Chain Monte Carlo methods to place constraints on the importance of each physical process (Lu *et al.*, 2011).

Numerical simulations. self-consistently model the large-scale structure, explicitly following the buildup of many halos from Gaussian density perturbations. Their usual repertoire includes hydrodynamics, non-equilibrium gas cooling (Smith *et al.*, 2017), and formation and feedback models for stars and BHs (Vogelsberger *et al.*, 2013), with the most sophisticated simulations now including magnetic fields (Pakmor and Springel, 2013), cosmic ray feedback (Pfrommer *et al.*, 2017), and radiative transfer (Xu *et al.*, 2016b; Pawlik *et al.*, 2017). They can be solved using an adaptive Cartesian grid, smooth particle hydrodynamics, a moving mesh, or meshless methods. Figure 3 shows an example of different methods of injecting (radiative, mechanical, and thermal) feedback energy from an accreting SMBH (Kim *et al.*, 2011). The biggest physics limitation is the limited spatial and mass resolution, necessitating "sub-grid" models of star/BH formation and feedback and any other type of physics occuring below this limit. Particular attention must be paid toward energy balance, where any feedback should not be irradiated away before being converted into the kinetic gas energy through the hydrodynamical solver (Katz, 1992; Stinson *et al.*, 2006; Wise *et al.*, 2012a). In the process, these models must be tuned to match observed properties of galaxies and their BHs, in which subsequently they can make predictions for future observations and determine the relative importance of each process, not unlike semi-analytical models. A more logistical downside is the computational expense of these calculations, sometimes taking several months while using thousands of compute cores. Nevertheless by directly simulating the system, these calculations can capture the complex and non-linear interplay between astrophysical processes, especially in the context of feedback loops.

4. Initial growth of the first black holes

The initial growth of the first BHs is highly dependent on its stellar progenitor and how it affected the adjacent environment, which can be encapsulated by the Bondi-Hoyle rate (Section 3.1). Afterward cosmological effects, such as mergers and associated gas mass accretion rates, play a significant role in the evolution of the seed BH and its host galaxy. As the galaxy grows, both star formation and SMBH fueling will become self-regulated through their feedback processes (Milosavljević *et al.*, 2009; Park and Ricotti, 2012). In this section, we will review the gaseous properties surrounding the seed BH from a metal-free (Population III; Pop. III) star, a dense stellar cluster, and a direct-collapse BH.

4.1. *Population III stellar remnant seeds*

Observations of metal-poor stars and simulations of Pop. III formation (see also Chapter 4) have suggested that they have a top-heavy initial mass function (Aoki *et al.*, 2014; Susa *et al.*, 2014; Hirano *et al.*, 2015), resulting in a large production of BH remnants. Furthermore being massive,

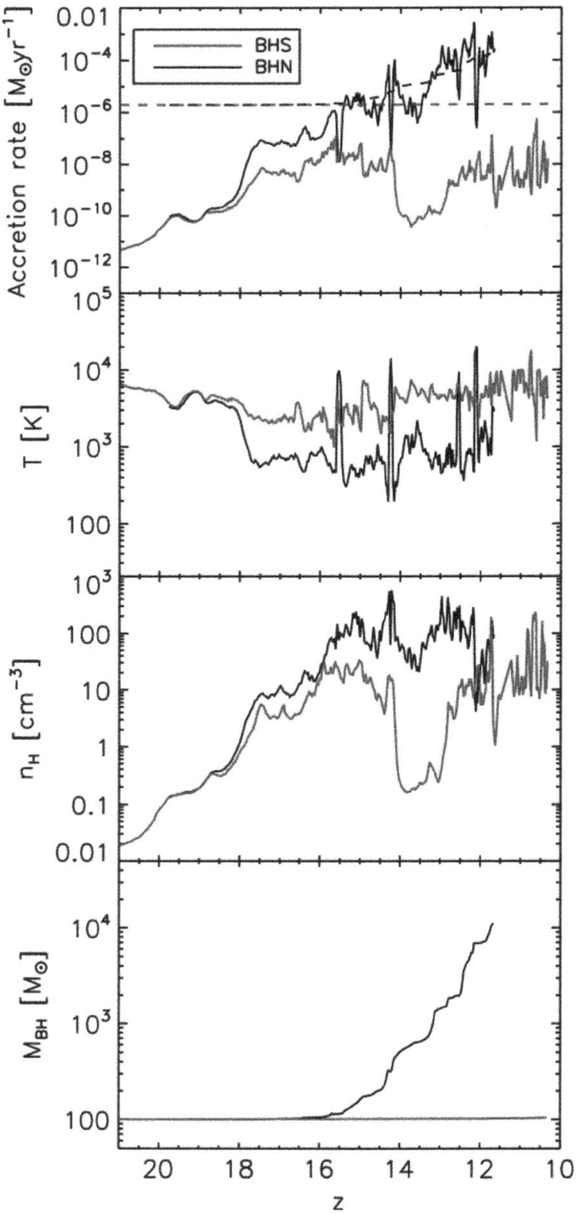

Fig. 4. Evolution of a Pop. III seed BH. *Top to bottom:* BH accretion rate, temperature and gas number density of the immediate vicinity, and BH mass with (grey) and without (black) radiative feedback. The dashed line in the top panel marks the Eddington limit. Adopted from Jeon *et al.* (2012), ©AAS. Reproduced with permission.

they are prodigious producers of ionizing radiation, creating H II regions on the kpc-scale. Most of the gas in the host (mini-)halo with masses $\sim 10^6\ M_\odot$ is swept up by the $\sim 30\ \mathrm{km\,s^{-1}}$ shock associated with the ionization front, leaving behind a warm ($\sim 10^4$ K) and diffuse ($\sim 1\ \mathrm{cm^{-3}}$) medium (Kitayama *et al.*, 2004; Whalen *et al.*, 2004; Alvarez *et al.*, 2006; Abel *et al.*, 2007). The escape velocities of these halos are only a few km $\mathrm{s^{-1}}$, so the halo is evacuated of nearly all of its gas after the shock leaves the halo (Muratov *et al.*, 2013; Jeon *et al.*, 2014b). This medium is not conducive for BH growth, and it must wait until there is a gas-rich merger that resupplies the dense gas reservoir.

Simulations following the growth of these seed BHs have been restricted to halos with a quiescent assembly history, only increasing by a factor of 10–100 to the atomic cooling limit of $\sim 10^8\ M_\odot$ over $\sim$200 million years. However during this phase, X-ray radiative feedback greatly suppresses any further fueling, and the seed does not grow more than 1% during this time at a typical rate of $\sim 0.01\dot{M}_{\mathrm{Edd}}$ (Alvarez *et al.*, 2009; Jeon *et al.*, 2012). Figure 4 compares the accretion rate, BH mass, and temperature and density of the surrounding gas of a simulation with and without radiative feedback [Jeon *et al.* (2012)]. Accretion radiation allows the gas to maintain $T \sim 10^4$ K and lowers its density through thermal pressure forces. If feedback is neglected, the BH grows by a factor of 100 at the Eddington rate when the halo undergoes a series of mergers at $z \simeq 14$ and approaches the atomic cooling limit at $z \simeq 12$, which is driven by a denser and cooler gas reservoir. Jeon *et al.* (2014a) then considered Pop. III high mass X-ray binaries. After consuming the envelope of the secondary star, they have accretion rates up to eight orders of magnitude lower than the Eddington limit when hosted by minihalos, similar to isolated Pop. III stars.

Pop. III star formation relies on H_2 cooling in minihalos, which can be dissociated by a Lyman-Werner background (Dekel and Rees, 1987; Machacek *et al.*, 2001). In this case, they will first form in larger minihalos with masses in the range $10^6 - 10^8\ M_\odot$ (Machacek *et al.*, 2001; Wise and Abel, 2007b; O'Shea and Norman, 2008). Although there is more gas available, the star formation efficiencies are not much higher than lower mass minihalos, only totaling up to a few thousand solar masses (Johnson, 2010; Xu *et al.*, 2016a). Their total UV luminosity might not be sufficient to blow out the gas from the host halo. This affected gas is then more likely to fall back into the halo center, aiding in early fueling of the BH seeds. Some fraction of these stars will explode in supernova, perhaps triggering prompt star formation (Whalen *et al.*, 2008), and the gas reservoir would be consumed by further star formation and disrupted by their supernovae. However, there has been little numerical focus on BH growth and feedback in these more massive minihalos.

4.2. *Intermediate mass black hole seeds*

Stellar collisions and dynamical instabilities can cause dense stellar clusters to form seed BHs with masses $\sim 10^3\ M_\odot$ (Rees, 1984; Devecchi and Volonteri, 2009; Davies *et al.*, 2011; Alexander and Natarajan, 2014, see also Chapter 7). Regardless of the formation mechanism, they all have massive stellar precursors that will photo-heat the immediate environment. During the formation phase, dynamical effects between massive stars or BHs ($\sim 10\ M_\odot$) will drive dense gas inflows toward the center, fueling the seed BHs at super-Eddington rates (Davies *et al.*, 2011; Alexander and Natarajan, 2014). After this phase, further gas supply may be disrupted by stellar (radiative and supernova) and BH feedback. If they are hosted in a minihalo, they might have a difficult time growing without external gas supplies from mergers, similar to the Pop. III seed BHs. However, being embedded in a stellar cluster with $M_\star \simeq 10^4\ M_\odot$ (Katz *et al.*, 2015), the deeper potential well could mitigate any fueling issues (Park *et al.*, 2016). If the host halo has a virial temperature $T_{\rm vir} \gtrsim 10^4$ K (i.e. an atomic cooler), then the seed BH should avoid these initial growing pains if it is centrally located in the halo. However, most of the attention has been on the formation of such intermediate-mass seed BHs, not the ensuing growth.

4.3. *Direct collapse black hole seeds*

Models of supermassive star (SMS) formation suggest that they form in metal-free atomic cooling halos that have little H_2 content, limiting their cooling to $\gtrsim 8000$ K (Omukai, 2001; Oh and Haiman, 2002, see also Chapter 5). In addition, they must be fueled by rapid infalling gas flows, providing over $0.1\ M_\odot\ {\rm yr}^{-1}$ onto the SMS. They can grow up to a mass of $10^6\ M_\odot$ with $\sim$10% collapsing into a direct-collapse black hole (DCBH) (Begelman *et al.*, 2006, 2008; Johnson *et al.*, 2013b), resulting in massive seed BHs in the range $M_{\rm BH} \simeq 10^4 - 10^5\ M_\odot$.

During the SMS main sequence, the stellar envelope is supported by radiation pressure with surface temperatures around 5000 K (Begelman, 2012; Hosokawa *et al.*, 2012a, 2013; Schleicher *et al.*, 2013; Sakurai *et al.*, 2015, see also Chapter 8). Thus their ionizing photon luminosities are extremely low $Q \sim 10^{44} - 10^{46}$ ph s^{-1} and will have little effect on the surrounding

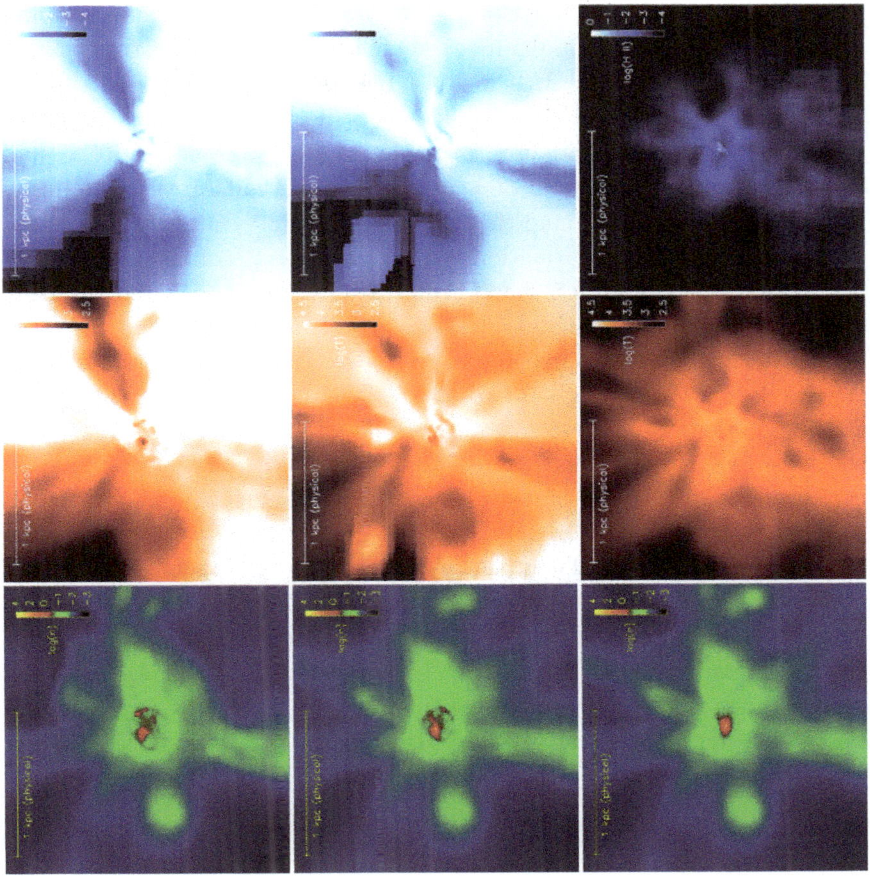

Fig. 5. Radiative feedback from massive BH seeds with $M_{\rm BH} = 5 \times 10^4\ M_\odot$ (top), $2.5 \times 10^4\ M_\odot$ (middle), and $10^4\ M_\odot$ (bottom). Projections of number density, temperature, and H II fraction three million years after BH formation in a field of view of 2 proper kpc. The more massive BH seed create larger H II regions along with disrupting the nearby dense gas. Adopted from Johnson (2010), reproduced by permission of Oxford University Press/on behalf of the RAS

medium that is rapidly falling toward it. Also, a relativistic jet might be launched from the collapsing SMS powering a gamma-ray burst, disrupting any leftover accreting material [Matsumoto *et al.* (2015)]. In addition, some SMSs may have exploded in extremely energetic ($\sim 10^{53} - 10^{55}$ erg) supernovae (Montero *et al.*, 2012; Chen *et al.*, 2014), totally disrupting the gaseous structures in the host halo (Whalen *et al.*, 2013b).

Once the DCBH has fully formed, its feedback through radiation and jets should have a substantial impact on the host halo. Because their masses are between 10^4 and 10^5 $M_{\odot}$, most of their radiation will lie in the UV region with a minor component in the X-rays (Johnson *et al.*, 2011; Pacucci *et al.*, 2016). The UV radiation photo-heats the gas to $10^4 - 10^5$ K and drives down its density to 10–300 cm^{-3} after three million years (Johnson *et al.*, 2011). They also found that the DCBH accretes near the Eddington limit immediately after formation but decreases to 1% of the limit after a few million years. Figure 5 shows projections of the density, temperature, and ionized fraction of the host halo and its surroundings. This radiation propagates through the surrounding metal-free medium and prompts Pop. III star formation through enhanced H_2 formation (Aykutalp and Wise, 2017). A fraction of these metal-free stars enriches the inner regions of the halo and partially disrupts the inflowing gas. But the now metal-enriched gas will have a higher opacity in the X-rays. These X-rays will photo-heat the gas to $T \simeq 10^6 - 10^7$ K, driving outflows in the process (Aykutalp *et al.*, 2014). The massive BH seed eventually settles into a self-regulated accretion pattern, having a duty cycle 5–50%, depending on the gas supply, only growing by 10–20% within 100 million years. Similar to the less massive seed BHs, extended super-Eddington growth periods are probably dependent on cosmological mergers, bringing a fresh gas supply that could smother the central BH, creating an opacity-thick medium in and around the accretion disk that favors super-Eddington accretion (Wyithe and Loeb, 2012; Pacucci *et al.*, 2015; Inayoshi *et al.*, 2016; Pezzulli *et al.*, 2016; Sakurai *et al.*, 2016).

5. Incorporation into larger halos and galaxies

The observed $z \sim 6$ AGN are located in rare halos that have extremely rapid accretion histories, forming their first luminous objects at $z \sim 30$ and growing by a factor of $\sim 10^6$ within 800 million years (see Figure 1). They will be bombarded with infalling halos and gas, fueling intense star formation and central BH growth. In the process, they will have several major mergers and hundreds, if not thousands, of minor mergers, all containing the first generations of galaxies and most likely a population of seed BHs. On the other hand, more typical galaxies, like the Milky Way, will grow at a more leisurely pace, whose progenitor halos can only support star formation after $z \sim 10$. But it is not a problem if their initial growth is stunted because they have several more billion years to grow into the observed AGN population at lower redshifts (Miyaji *et al.*, 2015). Nevertheless, theories of SMBH growth during the early universe should capture both the typical and extreme cases.

At early times, a halo/galaxy is not guaranteed to have a central BH because there is some probability that either a Pop. III star does not leave behind a BH remnant in the mass ranges $M_\star < 20\ M_\odot$ and $M_\star \simeq 140 - 260\ M_\odot$ (Heger *et al.*, 2003a; Heger and Woosley, 2010; Takahashi *et al.*, 2014) or a dense stellar cluster does not experience a runaway collapse. When the most massive progenitor does not host a central BH, the early galaxy will initially have a swarm of tens of stellar-mass BHs from Pop. III stars as their halos merge into the primary galaxy, moving at a velocity dispersion similar to the halo circular velocity (Xu *et al.*, 2013). Habouzit *et al.* (2017) found that only 20% of galaxies with stellar masses $M_\star \sim 10^6\ M_\odot$ have central BHs in their simulations, and all galaxies with $M_\star > 10^8\ M_\odot$ have a central object. For the remainder of this section, we will focus on early galaxies with central BHs.

Halo mergers can take a substantial fraction (tens of millions of years) of the Hubble time at $z > 10$. During the merger, the BHs must sink into the potential well center through dynamical friction (Chandrasekhar, 1943), which takes even more time than the halo merger (Tremmel *et al.*, 2017). For DCBH seeds, this is especially important because they are theorized to mainly form in satellite halos that are falling into an newly born galaxy. They are thought to form ~ 1 kpc from the galaxy (Dijkstra *et al.*, 2008; Agarwal *et al.*, 2014; Chon *et al.*, 2016; Regan *et al.*, 2017) and then infall and migrate toward the galactic center, where conditions for further accretion are more favorable.

The tight time constraints for the $z \sim 6$ AGN suggest that SMBHs grow mainly through accretion rather than mergers (Volonteri and Rees, 2005). BH mergers also have the complication of the merged remnant receiving a kick with $v \sim 300\ \mathrm{km\,s^{-1}}$, depending on the BH mass ratios and spins, which is greater than the vast majority of the first generation of galaxies. These kicked compact objects escape the galaxy and take several billion years to return to a galaxy (Micic *et al.*, 2007; Whalen and Fryer, 2012).

High-redshift galaxies tend to have high specific star formation rates (sSFR $\equiv \dot{M}_\star/M_\star \simeq 1 - 50\ \mathrm{Gyr^{-1}}$), doubling their stellar masses in as little as 20 million years and rarely having any organized rotation (Wise *et al.*, 2012b; Ma *et al.*, 2017). However, there is a lack of low- and moderate-luminosity AGN at $z \gtrsim 6$ (Willott *et al.*, 2011; Weigel *et al.*, 2015). Can the central BH keep pace with this furious growth rate? Volonteri *et al.* (2017) used empirical scaling relations between low-mass galaxies and their

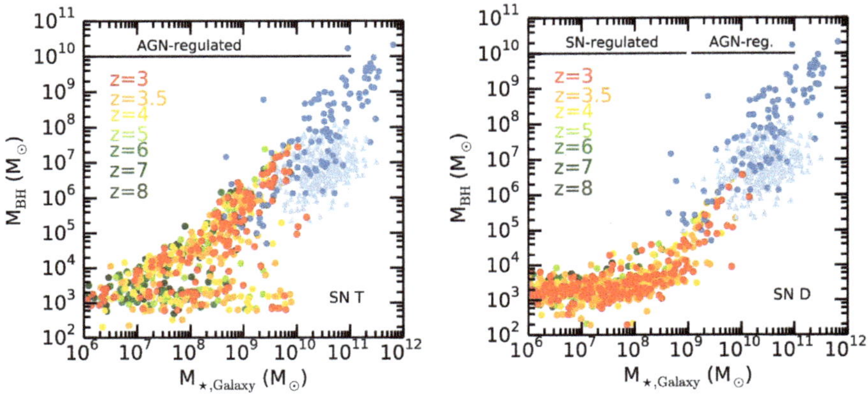

Fig. 6. Black hole mass as a function of stellar mass for a thermal supernova feedback (left) and delayed cooling supernova feedback (right) with the latter better matching observations. The blue points show the observed present-day relation (Reines and Volonteri, 2015). In galaxies with $M_\star \leq 10^9\ M_\odot$, supernova feedback launches winds from the central regions, removing potential gas for black hole accretion. Adopted from Habouzit *et al.* (2017), reproduced by permission of Oxford University Press/on behalf of the RAS.

central BHs to determine that such objects are more likely to have lower Eddington ratios ($\dot{M}_{\rm BH}/\dot{M}_{\rm Edd}$) than the observed luminous quasars. This difference suggests that either moderate-luminosity AGN are intrinsically X-ray weak (Luo *et al.*, 2014) or that they are heavily obscured by dense and dusty gas (Pezzulli *et al.*, 2017), which could be especially true when gas inflow rates are high. Using high-redshift galaxy formation simulations that captured the formation of Pop. III remnant BH seeds, Habouzit *et al.* (2017) showed that SN feedback drove outflows that suppress BH accretion in low-mass ($M_\star < 10^9\ M_\odot$) galaxies, only growing by a factor of 10–100 up to $10^4\ M_\odot$ by $z = 3$, as depicted in Figure 6. However above this mass scale, the central BHs begin to grow faster at $\dot{M}_{\rm BH} \gtrsim 0.01 \dot{M}_{\rm Edd}$ for a substantial fraction of time.

Although the BH seeds and host galaxies do not lie on the M-σ relation, they should approach the observed relation at the present-day. Volonteri and Natarajan (2009) explored how it evolves from high redshift to $z = 1$ when high-redshift progenitors were populated with "light seeds" and "heavy seeds". The former have masses $\sim 100\ M_\odot$ corresponding to Pop. III remnants, and the latter exist in a mass range $10^2 - 10^6\ M_\odot$ that were derived from the Lodato and Natarajan (2006) DCBH formation model. The light seeds have a characteristic plume of ungrown seeds existing in galaxies with a velocity dispersion $\sigma = 50 - 100\,\mathrm{km\,s^{-1}}$ that then transition to the observed M-σ relation. The heavy seeding model exhibits a BH mass floor at $10^5\ M_\odot$ in galaxies below this same velocity dispersion (Agarwal *et al.*, 2013; Natarajan *et al.*, 2017).

It is common practice for numerical simulations with a focus toward present-day galaxies and BHs to seed the smallest resolved halos ($M_{\rm vir} \sim 10^9 - 10^{10}\ M_\odot$) with a central BH with $M_{\rm BH} \sim 10^5\ M_\odot$ that obeys the present-day M-σ relation (Li *et al.*, 2007; Di Matteo *et al.*, 2012; Dubois *et al.*, 2014a; Taylor and Kobayashi, 2014; Vogelsberger *et al.*, 2014; Schaye *et al.*, 2015; Feng *et al.*, 2014; Sijacki *et al.*, 2015; Smidt *et al.*, 2017; Tenneti *et al.*, 2018). These seeding models assume that central BHs grow efficiently in halos below their mass resolution, however as previously discussed, they can be heavily affected by their own feedback and depletion of dense gas through supernova winds, if they host a central BH at all (Aykutalp *et al.*, 2014; Habouzit *et al.*, 2017). Nevertheless, these large-scale galaxy simulations are accurate for galaxies similar in mass to the Milky Way and larger, but they may be lacking accuracy at high redshifts and low masses and cannot make predictions about such objects. Because it is currently computationally prohibitive to directly follow the smallest star forming minihalos all the way to the most massive SMBHs, a good compromise in these large-scale simulations is to seed the smallest galaxies stochastically according to the stellar-BH mass ratios found in high-redshift, small-scale simulations or analytical models (Bellovary *et al.*, 2011; Habouzit *et al.*, 2017) as a way forward before clever algorithms and more powerful supercomputers can model the entire formation sequence.

6. Summary and future directions

Although there are still many open questions about the yet-to-be observed universe when the first generations of stars, galaxies, and black holes were abundant, we have a good understanding about the general progression of early galaxy formation and the growth of their (if any) central BHs. One major uncertainty is the initial mass function of seed BHs (see also discussion in Chapter 9), which then affects their subsequent growth and feedback. However it is clear that all of the progenitors have massive stars in common, suggesting that the initial growth phase will be slow in the warm and diffuse medium left behind. Substantial growth will be delayed until gas-rich halo mergers provide additional fuel.

Better resolution and larger dynamic range are always an easy answer for future work, but we should continue to place emphasis on the importance of accurate sub-grid BH models (Negri and Volonteri, 2017) because, barring a breakthrough, a cosmological simulation that resolves and evolves the accretion disks and the associated radiation over the first billion years is not possible in at least the next decade. Bondi-Hoyle accretion is idealized, whereas reality is everything but ideal. One could imagine having a low-resolution accelerated calculation existing inside the simulation, using the properties from the cosmological simulation as boundary conditions. For example, better models could include (1) radiation transport at many energies, (2) sub-grid models for a multi-phase accretion from the ISM to an accretion disk to the BH / jet (Power *et al.*, 2011), and (3) better estimating angular momentum transport and BH spin that would respectively influence the jet direction and radiative efficiency. Including these effects in sub-grid models could inform us when super-Eddington accretion occurs, if at all, in BHs evolving from their seeds to the monsters we observe at $z \sim 6$.

Future observatories, such as JWST, WFIRST, ATHENA, and thirty-meter class telescopes, will enlighten us with the spectral properties of a multitude of $z > 6$ galaxies and their AGN, which is the subject of the following chapters. Mock observations of these objects (Barrow *et al.*, 2017; Natarajan *et al.*, 2017; Volonteri *et al.*, 2017) are of critical importance to constrain BH growth scenarios and their relationship to host galaxies, using these upcoming spectacular photometric and spectral datasets of the first galaxies and black holes.

In Chapter 11, the discussion on the growth of SMBHs will be complemented by specific scenarios that allow super-Eddington accretion and thus a more rapid growth. An account of the current observational status is given in Chapter 12. Predictions for gravitational wave observatories are outlined in Chapter 13, and future observational possibilities in Chapter 14.

Acknowledgments

John H. Wise is supported by National Science Foundation grants AST-1333360 and AST-1614333, NASA grant NNX17AG23G, and Hubble theory grants HST-AR-13895 and HST-AR-14326. This research has made use of NASA's Astrophysics Data System Bibliographic Services.

Chapter 11

Super-Eddington accretion; flow regimes and conditions in high-z galaxies

Lucio Mayer

Center for Theoretical Astrophysics and Cosmology, Institute for Computational Science, University of Zürich, Winterthurestrasse 190, 8057, Zürich, Switzerland

lmayer@physik.uzh.ch

We review and discuss theoretical studies addressing the possibility of gas accretion onto black holes occurring at rates exceeding the Eddington limit. Our focus is on the applications to the growth of black hole seeds at high redshift. We first present the general notion of Super-Eddington accretion, and then summarize the different models and numerical simulations developed to study such regime. We consider optically thick flows in accretion disks as well as in spherically symmetric envelopes, and devote particular attention to the widely adopted model based on the SLIM disk solution. While attractive for its simplicity, the SLIM disk solution is challenged by the latest generation of three-dimensional radiation (magneto)-hydrodynamical simulations, in which radiative losses can be an order of magnitude higher, and the mechanisms of radiation transport is more complex than straight advection as it takes place in a complex turbulent regime. We then discuss the gas supply rate to the sub-pc scale accretion disk or envelope from larger scales, revisiting gas inflow rates in protogalaxies under various conditions. We conclude that in the dense gaseous nuclei of high-z galaxies the conditions necessary for the onset of Super-Eddington accretion regimes, such as a high optical depth and high gas supply rates from large scales, should be naturally met. Feedback from the growing BH seed should not alter significantly such conditions according to the results of radiation magneto-hydrodynamical simulations of super-critical flows in accretion disks. Furthermore, based on the required nuclear gas inflow rates and the tendency of stellar feedback to remove efficiently gas in low mass halos, we argue that super-critical accretion will be more easily achieved in relatively sizable halos, with virial masses $M_{vir} > 10^{10} M_{\odot}$, which become more common at $z < 15$.

1. Introduction

In this chapter we will cover and summarize work carried out to model Super-Eddington flows on massive black hole seeds. These are flows in which the accretion rate onto a massive seed black hole occurs at a rate higher than predicted by the Eddington limit. We refer to Chapter 10 for the basic concepts of accretion onto black holes and the general understanding of massive black hole growth on cosmic timescales. Here we recall that for a black hole of mass M_{BH}, the Eddington limit on the mass accretion rate $\dot{M}_{Edd}$ is set by imposing perfect balance between the gravitational force pulling matter towards the black hole and radiation pressure of the emerging photons exerting a force in the direction opposite to the accretion flow. It is obtained under the assumption of spherical symmetry and is given by $\dot{M}_{Edd} = 4\pi G M_{BH} m_p / \epsilon c \sigma_T$, where G is the gravitational constant, m_p is the proton mass, ϵ is the radiative efficiency of the accretion process, and σ_T is the Thomson scattering cross section for the emerging photons by which radiation pressure is communicated to matter. We also recall that the characteristic accretion timescale expressed as a function of the Eddington luminosity $L_{Edd} = \epsilon \dot{M}_{Edd} c^2$ reads:

$$t_{\rm acc} = \left(\frac{\epsilon}{1-\epsilon}\right)\left(\frac{L_{Edd}}{L}\right) t_E = (4.3 \times 10^7 {\rm yr}) \left(\frac{L_{Edd}}{L}\right), \tag{1}$$

where $t_E = M_{BH} c^2 / L_{Edd}$ is the Eddington timescale, and the last equality assumes $\mu_e = 1.15$ (valid for primordial gas) and a radiative efficiency of $\epsilon = 0.1$.

As mentioned in Chapter 10, unless the formation of massive black hole seeds by e.g. direct collapse is invoked, Super-Eddington accretion phases are required in order to grow light seeds from Pop. III stars, with typical masses $\sim 100 M_\odot$ or lower, to the gargantuan masses of bright QSOs, $10^9 - 10^{10} M_\odot$ (Mortlock *et al.*, 2011; Bañados *et al.*, 2018b) by $z = 6$. It is thus of crucial interest to understand if Super-Eddington accretion can occur under realistic conditions.

For astrophysically relevant models of Super-Eddington accretion there are at least two key aspects that warrant consideration. One is, of course, under which conditions an accretion disk or, more generically, a gaseous envelope, can supply gas to a black hole seed at a rate higher than Eddington, and whether this can be achieved in a nearly steady-state regime

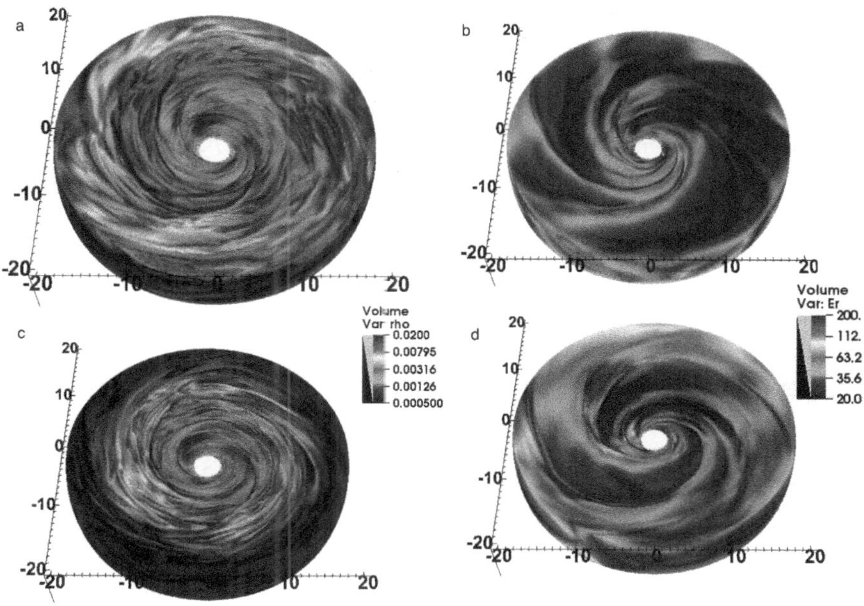

Fig. 1. (Color online) Density (left panels) and radiative flux (right panels) snapshots from the Jiang *et al.* (2017) 3D MHD simulations carried out with the VET method for radiative transfer. The highly turbulent nature of the flow, the generation of large scale spirals and the fact that the radiative flux has marked local variations are all features evident from the color-coded maps (see text). Figure provided via private communication.

or only episodically (Pezzulli *et al.*, 2016). The other aspect concerns understanding the conditions under which the accretion disk/envelope can be fed at a rate well above Eddington via gas inflows initiated at much larger scales. Tackling both requires studying and modeling the gas flow along with the various physical processes governing its behavior across a prohibitively wide range of scales, from several kiloparsecs to centiparsecs. Such complexity is conceptually and computationally challenging, and is absent in the case of stellar mass black holes for which only the scale of the accretion disk is relevant as external gas flows are negligible (unless the black hole is part of a binary system, in which case other considerations may apply). The problem of studying gas inflows at galactic scales is already by itself a multi-scale process since its natural boundary conditions are determined by gas dynamics in the circumgalactic medium. This is especially true when a galaxy is still in the early assembly phase and thus is hosted in a dark matter halo vigorously fed by cold filaments originating from the cosmic web (the so-called "cold accretion" mode, see e.g. Dekel and Birnboim (2006); Keres *et al.* (2009). The timescales of gas inflows may vary from scale to scale as different phenomena drive them, e.g. nuclear bars and spiral density waves in galactic nuclei, tidal interactions, mergers and bar-instabilities on galactic scales, cold flows and feedback-driven galactic fountains on larger scales. Ultimately, the duty cycle of gas feeding to the black hole seed will be determined by the concerted action of all these processes, hence a characteristic timescale is hard to define. Because of that the simplest approach will have to assume a boundary value for $\dot{M}$, constant or time-dependent, and explore whether or not Super-Eddington regimes can persist. This is indeed the approach followed by numerical simulations that model directly an accretion in a disk/envelope surrounding the black hole.

While the possibility of super-critical accretion onto black holes has been considered since a long time (Begelman, 1979; Abramowicz *et al.*, 1988), it is only recently that hydrodynamical simulations have become capable of studying its feasibility in realistic astrophysical environments, including in the specific case targeted by this chapter, namely massive black holes and black hole seeds. Pioneering numerical work over a decade ago has been carried out by Ohsuga *et al.* (2005), who identified very clearly two key factors leading to super-critical accretion; photon trapping and the anisotropy of the radiation field. These two aspects play a different role in the various super-critical simulations of accretion disks or accreting spherical envelopes that were developed later, as we will discuss in this chapter.

Following the rationale of the book we will focus on host galaxies at very high redshift, between $z = 20$ and $z = 6 - 7$, namely up to as late as the epoch of appearance of the first QSOs (as early as $z = 7.5$ according to Bañados *et al.* (2018b). We recall that the first motivation for studying Super-Eddington accretion at such early epochs is to find out if light BH seeds formed by the collapse of primordial metal-free Pop. III stars could grow rapidly and become as massive as required to explain the population of bright high-z QSOs at $z > 6$ discovered in the last decade (see Chapter 12). It is indeed the only plausible alternative route to postulating a direct collapse scenario, a topic covered in other chapters of this book. As we will see, though, simulations are revealing that galaxy assembly is highly dynamical and complex at such high redshifts, and has a wide range of flow conditions. Therefore, the nature and strength of gas inflow rates at scales above the accretion disk/envelope depend on galaxy mass as well as on the nature of gas dynamics in the host galaxy. In other words, galaxies at such high redshifts are by no means simple. Observational evidence on the nature of gas flows and stellar components of galaxies is still lacking, so this is still mostly theorists' playground. Yet the advent of ALMA, JWST, WFIRST, EUCLID and large ground based telescopes such as E-ELT will change the landscape soon, not only by providing enough sensitivity to detect emission from the early stages of massive BH seed growth, but also allowing to distinguish, in principle, between different scenarios for their initial birth and subsequent growth, including eventually the regime of super-critical growth (see Chapter 14).

The chapter is organized as follows. In Section 2, we begin by discussing theoretical scenarios for the realization of Super-Eddington accretion, starting from relatively idealized models such as the popular SLIM disk model, and we will then move on to discuss the results of complex three-dimensional simulations that include radiation, the effect of the magnetic and, in some cases, also solve the set of equations in the relativistic regime. A major distinction, in this context should be done between accretion occurring in a rotationally supported configuration, namely an accretion disk, and accretion occurring in a spherically symmetric envelope. For the latter reason two separate subsections are devoted to the two different flow regimes. Section 3 concerns the conditions of the large scale gas inflows in proto-galaxies and galaxies at high-z, namely focusing on how we can inform the boundary condition for the accretion flow onto a black hole seed in a galactic/proto-galactic nucleus. Finally, Section 4 describes the first attempts made to combine models of super-critical accretion such as those discussed in Section 2 with the galactic astrophysics framework covered in Section 3. hydrodynamical simulations of galactic/protogalactic nuclei. We conclude with Section 5, which summarizes the whole chapter.

2. Super-Eddington flows: direct models of the accretion process

This section is devoted to discuss theoretical models and simulations carried out at scales smaller than the Bondi radius of the black hole to study directly the accretion process, eventually down to the radius of the last stable orbit. The works summarized here represent the backbone of our

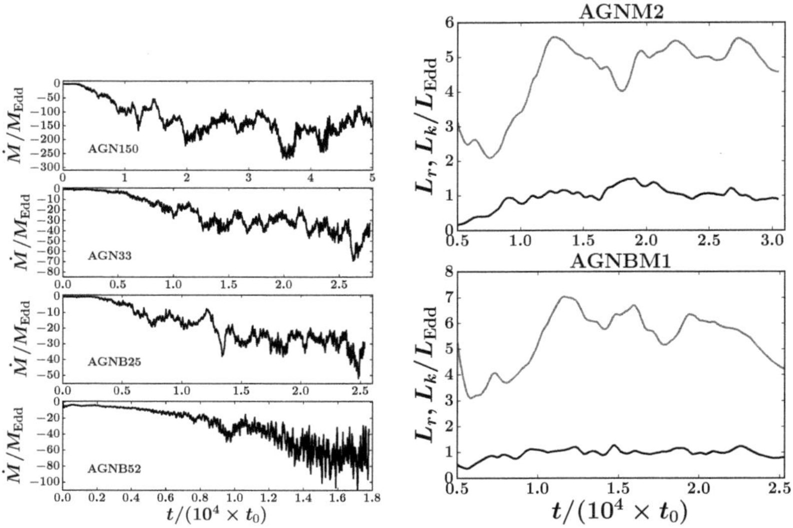

Fig. 2. In the left panel we show the history of the black hole accretion rate at 10 Schwarzschild radii for four representative simulations of accreting SMBHs in Jiang *et al.* (2017), in units of the accretion rate corresponding to the Eddington limit. In the right panel the radiative and kinetic luminosity in units of the Eddington luminosity, time-averaged and plotted as a function of distance, are shown for two of such runs. Time is shown in dimensional units, the same in each run. Figure provided via private communication.

current understanding of Super-Eddington accretion. There is, however, no direct information on nuclear and galactic-scale gas dynamics as these models, due to their computational complexity, cannot cover a wide range of scales. We have divided the section into two main parts, one focusing on flows in accretion disks and one on flows in extended envelopes that are not rotationally supported.

2.1. *Super-Eddington flows in accretion disks: from the SLIM disk models to 3D MHD simulations*

For accretion flows occurring in rotating disk-like configurations the SLIM disk model (Begelman, 1979; Abramowicz *et al.*, 1988) has a well-established framework yielding well understood super-critical flow solutions. Analogously to Shakura-Sunyaev "alpha" disks, in SLIM disks dissipation is described by a viscous stress tensor that is proportional to pressure. Slim disk solutions are obtained by solving vertically-averaged radial momentum equations. Later Sadowski (2011) improved the model, introducing a vertical structure by solving the fully relativistic, axisymmetric hydrodynamical equations in a Kerr metric augmented with hydrostatic equilibrium, a vertical energy profile and energy transport differential equations. In this steady-state model accretion can occur in a Super-Eddington regime if the medium is optically thick, so that radiation is advected with the flow rather than escaping. In order for the latter condition to be satisfied the photon diffusion timescale has to be much longer than the radial advection timescale, which is governed by viscous transport. By construction this requires a very high optical depth, and indeed steady-state models such as those in Sadowski (2011) have an optical depth $\tau > 10^3$ beyond ISCO (the innermost stable circular orbit around a Kerr black hole). Note that, because of the high optical depth, the vertical energy transport equations can be solved using the diffusion approximation (as in Sadowski, 2011). The timescale constraints in the SLIM disk model imply that no outflows are possible because, for that to happen, radiation pressure has to overcome bulk thermal and turbulent pressure. This result is at variance with what will happen in a standard radiative accretion disk in which, as the inflow rate rises above the Eddington limit, a radiation pressure-driven outflow would occur. As we will see later, realistic 3D MHD simulations suggest that neither picture is correct, namely while advection is important and partial photon trapping can occur in a dense, heavily mass loaded accretion disk, outflows can also occur. The extent by which outflows are important, and how luminous above Eddington can the disk be, is dependent on the nature of radiative transport. The jury is still out on which numerical approach to radiative transport is capturing the energy flow via radiation most realistically. Therefore, this is an area of research in which new insight is strongly tied to advances in simulation methods.

A straightforward way to characterize the SLIM disk model is to state that accretion is highly radiatively inefficient compared to the standard efficiency in Shakura-Sunyaev accretion disks around a Kerr black hole. This is because the photon diffusion timescale becomes much longer than the other characteristic timescales in the disk, namely the sound crossing timescale and the viscous timescales associated with the mechanism of mass transport. Indeed SLIM disk solutions imply radiative efficiencies as low as $< 0.1\%$ (Sadowski *et al.*, 2013; Sadowski and Narayan, 2016) as opposed to $\sim 40\%$ in a standard viscous disk around a spinning BH. As a result, $\dot{M}$ can be 1-2 orders of magnitude above M_{EDD} while the emitted luminosity remains close to the Eddington limit. This has important consequences on the energetic feedback onto the surrounding interstellar medium, as discussed below (see Section 4). A low radiative efficiency also

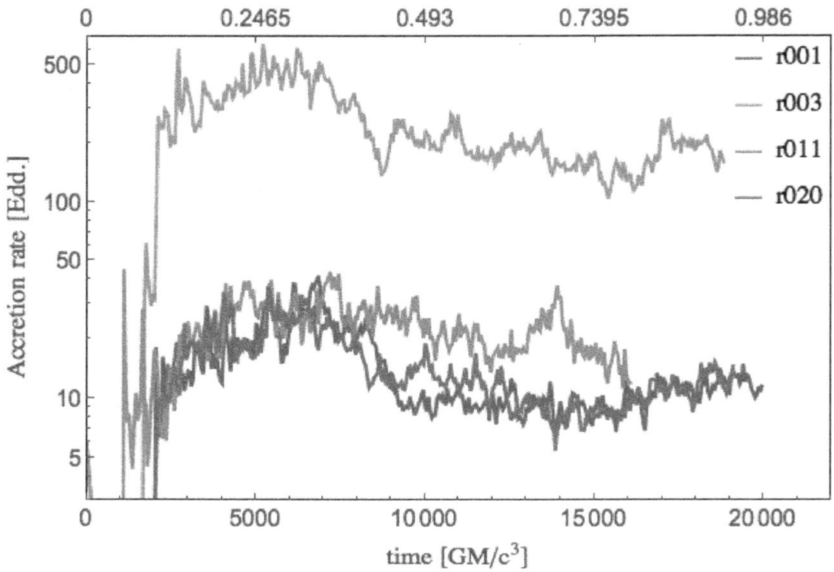

Fig. 3. Accretion rate histories in units of the Eddington accretion rate for four of the 3D relativistic runs of Sadowski and Narayan (2016) with stellar mass black holes (see text). Figure adopted from Sadowski and Narayan (2016), reproduced by permission of Oxford University Press/on behalf of the RAS.

characterizes the advection dominated flow (ADAF) model, except that in that the medium is optically and geometrically thin, which can only occur with very low accretion rates (Sub-Eddington). The latter is the opposite situation of what is needed in the SLIM disk model, which requires a medium that is optically thick, which yields long diffusion timescales, and thus relatively hot. An attractive feature of the SLIM disk model is that a simple fitting formula for the radiative efficiency, weakly dependent on the spin of the Kerr black hole, can be found (see Sadowski, 2009; Madau *et al.*, 2014). The latter was readily employed in large-scale MBH accretion models in place of the standard viscous disk radiative efficiency, as we will discuss in Section 4.

While 2D radiation-hydro simulations of accretion disks or spherically symmetric envelopes have been available since the early 2000s (see Ohsuga *et al.* (2005) and section 1), it is only recently that simulations that include the third dimension as well as MHD to study the Super-Eddington regime have appeared in the literature (see example in Fig. 1, from Jiang *et al.* (2017). Currently there are just a handful of 3D super-critical accretion

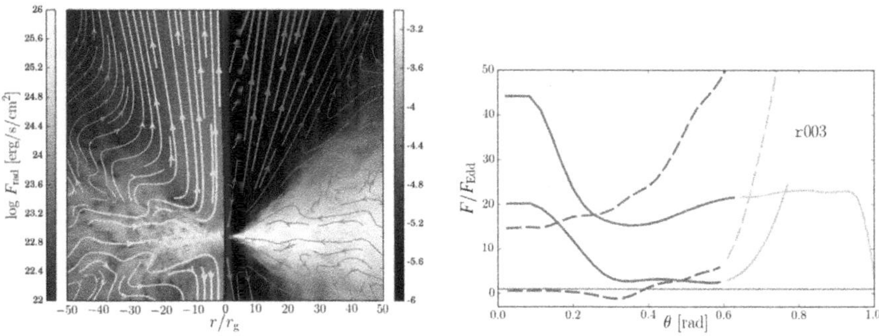

Fig. 4. On the left a slice through the midplane of the accretion disk at the end of one of the simulations of Sadowski and Narayan (2016) shows the radiative flux (left) and the density map (right) The powerful winds leaking out radiation via advection are evident, as is the photon trapping through inward advection in the high density midplane. On the right we show an example of the radiative flux (solid lines) and kinetic flux (dashed lines) in units of the Eddington flux from one of the runs, using blue for quantities measured near the edge of the disks and red for quantities measured near the inner annulus of the computational domain. From the plot it is evident that the kinetic luminosity can supersede the radiative luminosity, in contrast with Jiang *et al.* (2017) (see text).

disk simulations. They are carried out by different research groups employing different numerical techniques and approaches, each with its own advantages and disadvantages. Overall they can achieve $\dot{M}$ in the range several tens to a few hundred times $\dot{M}_{Edd}$ (Figures 2 and 3). A large, supercritical $\dot{M}$ is imposed as a boundary inflow condition at the beginning of the simulation (Sadowski and Narayan, 2016; Jiang *et al.*, 2017). Modern MHD simulations are carried out at high resolution with low-diffusivity numerical schemes that can capture accurately the magneto-rotational instability (MRI). The latter is found to be the main angular momentum and mass transport mechanism in the accretion disk, in agreement with the findings for the more conventional regime of accretion in presence of low rates of mass loading. MRI is thus the main source of viscous transport, although evidence is mounting that in dense optically thick disks other transport mechanisms are also at play.

One group of simulations, primarily by Sadowski and collaborators, comprises *fully relativistic* 3D MHD calculations with a finite difference code combined with a relatively simple treatment of the radiation part, either using flux-limited diffusion (FLD, e.g. Sadowski *et al.*, 2013), or, more recently, the M1 approximation (Sadowski and Narayan, 2016). These simulations so far have focused on accretion onto stellar mass black holes or

SMBHs, although there is one single attempt to study the accretion around a 1000 $M_\odot$ MBH in Sadowski and Narayan (2016) which yielded qualitatively similar results.

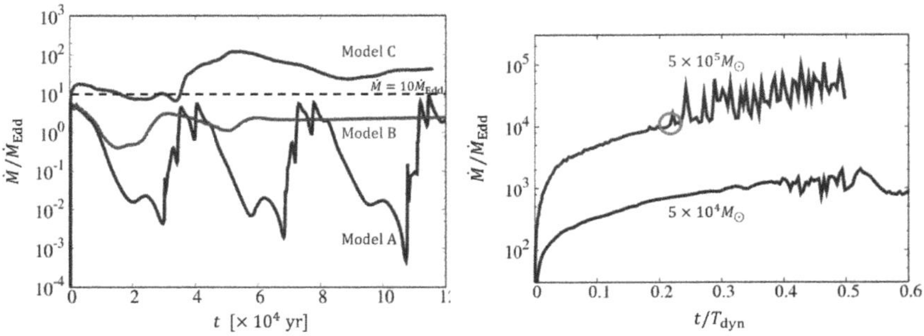

Fig. 5. Examples of Hyper-Eddington accreting light (left) and massive BH seeds (right) from Takeo *et al.* (2018). The mass of the BH is $10^3 M_\odot$ on the left, and the different models differ for how the radiation field geometry is imposed (isotropic for Model A and increasingly aniostropic for Model B and C). Super-critical accretion sets in for anisotropic radiation propagation. For seeds with masses $> 10^5 M_\odot$, shown on the right, the ionization region can recombine rapidly and collapse, leading to the highest accretion rates (after the time marked with the circle). Figure adopted from Takeo *et al.* (2018), reproduced by permission of Oxford University Press/on behalf of the RAS.

A second type of simulations (Jiang et al., 2014, Jiang *et al.*, 2017) comprises *non-relativistic* 3D MHD calculations with finite volume hydro methods (ATHENA) using a highly sophisticated radiative transport schemes such as the variable Eddington tensor (VET) scheme coupled with solving the steady-state radiative transfer equation at each timestep. With the latter approach the very inner region of the disk, including the generation of a relativistic jet, cannot be properly modeled, but radiation transport is more accurate and realistic as it takes into account the actual flow configurations. This can have significant effects on the nature of energy and momentum transport through the disk as well as on the net energy loss. Very recently the latter simulation method has been applied to studying accretion onto a SMBH of mass $10^8 M_\odot$ (Jiang *et al.*, 2017).

Perhaps not surprisingly given the differences in the numerical methods, the results of simulations carried out by different groups are quite different. Both the radiative efficiency and the relative proportion between the radiative and mechanical energy output generated by accretion differ significantly. In particular, the simulations with the VET method produce higher radiative efficiencies, at least one order of magnitude above those of Sadowski and Narayan (2016). The latter result is still mostly based on studying accretion onto stellar mass black holes. Jiang *et al.* (2017) have suggested that the cause of the difference might be the more pronounced anisotropy of the radiative flux captured with the VET method as opposed to the M1 method. The M1 approximation is indeed still a moment-based approach to the radiative transfer equation, which means it still integrates out the information on directionality of photon propagation to some extent. A higher anisotropy in the propagation pattern allows radiation to percolate more efficiently through a clumpy, turbulent medium with large inhomogeneities. MRI, as expected, gives rise to a turbulent flow regime, which naturally creates a highly inhomogeneous flow (Figure 1).

When compared to the standard SLIM disk model, the simulations of Jiang *et al.* (2017), which are the only ones to focus exclusively on accretion onto supermassive black holes, are those that show the most striking differences. Photon transport occurs via advection, not diffusion as in the SLIM disk model, and disks evolve in a regime in which the advection timescale is shorter than the sound crossing timescale. Advection is primarily generated by MRI turbulence. Advection of photons along with the gas flow is also the main mechanism of radiative emission in the simulations of Sadowski and collaborators. Despite higher luminosities, because accretion and emitted radiation are both highly anisotropic, high mass accretion rates exceeding 100 times the Eddington rate do occur (Figure 2).

In addition, Jiang *et al.* (2017) show that radiation can leak out also via other forms of turbulent transport, in which, while mass is not advected, radiation can be advected due to effects such as magnetic buoyancy. *In this case radiation transport is not through diffusion nor through advection, rather it occurs in a third way which is of course not captured by simple models.* On the contrary, Sadowski and Narayan (2016) found turbulent transport of radiation to be negligible compared to advection in their simulations. The reason of the latter discrepant conclusion is still unclear. While Sadowski and collaborators find a higher kinetic luminosity, especially in the outer parts of the disks (Figure 4), the total luminosity in their most recent three-dimensional simulations (Sadowski and Narayan, 2016), hence the sum of kinetic and radiative luminosity, is found to be only a factor of 2 lower than that expected in the standard viscous thin disk accretion theory for both zero and high spin BHs, being, respectively, 3% and $9\%\dot{M}c^2$. In Jiang *et al.* (2017) the kinetic energy luminosity is always

15% – 30% of the radiative luminosity, and increases with higher accretion rates.

Finally, while there is discrepancy in the predicted radiative luminosity, both groups find radiative efficiencies higher than predicted by the SLIM disk model in the regime of very high accretion rates, corresponding to at least 10 times the accretion rate at the Eddington limit. For lower accretion rates the radiative efficiencies of Sadowski and Narayan (2016) agree with the predictions of the SLIM disk model, while those of Jiang *et al.* (2017) are in the range 5–7%, hence are significantly higher. In Jiang *et al.* (2017), the radiative efficiency first increases with increasing accretion rate, then, as the accretion rate becomes higher than $1000 M_{\odot}/\mathrm{yr}$, it starts to decrease to $\sim 1\%$, a finding that is further supported by additional numerical simulations being analyzed as we write (Y. Jiang, private communication). As a reference, Sadowski and Narayan (2016) find typically a radiative efficiency of 0.006 and 0.01 – 0.02 for, respectively, zero spin and near maximally rotating black holes. This is about an order of magnitude lower than the standard radiative efficiency in a thin viscous disk.

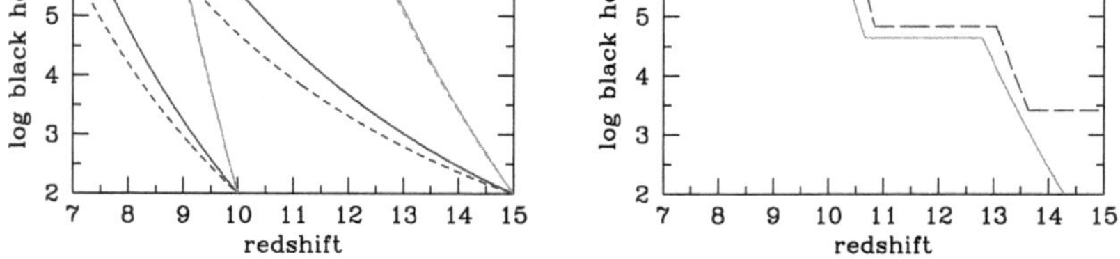

Fig. 6. (Color online) The left panel shows the different growth rate of an individual light BH seed, with mass of $100 M_{\odot}$, for models adopting, respectively, SLIM disk accretion (orange, for an accretion rate constantly at 3 times the Eddington rate) and a standard viscous accretion at the Eddington limit (blue). Solid and dashed lines are for, respectively, a zero spin and a highly rotating black hole (see labels in panel). The right panel shows only super-critical growth curves for the same BH mass but for intermittent accretion with a duty cycle of 0.5 (solid line, for 3 times the Eddington rate) and 0.2 (dashed line, for 4 times the Eddington rate). Redrawn with permission from Madau et al. (2014).

2.2. *Super-Eddington flows in spherically symmetric envelopes*

In this section we will discuss spherically symmetric accretion flows, without dynamically important angular momentum. This is the configuration in which the conventional Eddington limit can be applied more naturally, as this assumes no directionality in the gas flow or in the radiation field. Nevertheless, the Super-Eddington accretion regime is possible under certain conditions. Radiation hydrodynamics simulations have been used since more than a decade to show that, if the BH is embedded in an optically thick medium, efficient photon trapping can reduce the radiative force enough to allow super-critical accretion from an envelope (Ohsuga *et al.*, 2005). As in the case of accretion disks, though, the problem has always been how to maintain a sustained super-critical regime. The problem often documented in the literature is that, as soon as radiative feedback from the accreting hole is accounted for, the emerging hot ionizing flux stifles gas accretion from larger scales effectively (Milosavljević *et al.*, 2009; Alvarez *et al.*, 2009; Park and Ricotti, 2011; Park and Ricotti, 2012; Park and Ricotti, 2013). Therefore, the time-dependent infall rate at the edge of the envelope, a boundary condition that in principle could be set based on gas inflow rates found at larger scales (e.g. in protogalaxy simulations), and the anisotropy of the radiation field, are both important factors. Inayoshi and Haiman (2016) have developed a self-consistent model in which radiative heating and ionization is accounted for in 1D radiation-hydro simulations. They found a steady-state solution corresponding to Super-Eddington accretion at the remarkably high rate of more than $5000 L_{edd}/c^2$, which they dub as *Hyper-Eddington accretion* (Figure 5). The configuration is that of a radiation-dominated core where photon trapping occurs surrounded by an accreting optically-thin envelope. The core thus shields the envelope from the effect of radiation pressure. The BH thus needs to be embedded in a dense gas cloud satisfying the following condition:

$$M_{\rm BH} \geq 10^4 M_\odot \left(\frac{n_\infty}{10^5 cm^2}\right)^{-1} \left(\frac{T_\infty}{10^4 K}\right)^{3/2}, \tag{2}$$

where n_∞ and T_∞ are the density and temperature of the ambient gas, respectively. The condition above corresponds to the Bondi radius being larger than the size of the ionizing region (the Stromgren sphere). If this is satisfied accretion occurs essentially at the Bondi rate. We recall that the Bondi accretion rate, which is derived under the assumption of spherical symmetry and an infinitely large radius at which gas begins to acquire a net inward radial motion, is expressed as:

$$\dot{M}_B \equiv \pi e^{3/2} \rho_\infty \frac{G^2 M_{BH}^2}{c_\infty^3}, \tag{3}$$

where an adiabatic index $\gamma = 1$ was assumed. The Bondi rate can be much larger than the Eddington rate given sufficiently high density and temperature, but also significantly smaller, as in the case of mini-halos discussed in Chapter 10. As the temperature in galactic nuclei should not be much different from the mean temperature of the galactic ISM, namely in the range $10^3 - 10^4$ K (molecular gas would be colder but it should be dissociated efficiently by the warm ionizing flux resulting from accretion onto the BH seed), the ambient gas density n_∞ becomes the most important variable. For example, for light BH seeds formed by Pop. III remnants, with masses around $100 M_\odot$, the corresponding density must be $> 10^7$ atoms/cm^2 for the above condition to be satisfied. Qualitatively, the condition of a dense, highly optically thick flow yielding very long photon diffusion timescales reminds of the conditions in the SLIM disk model, except that here there is no role of advection. Furthermore, the seemingly high gas densities required to initiate sustained Super-Eddington accretion are commonly found in the center of protogalaxies or circumnuclear regions of high-z gas rich galaxies in hydrodynamical simulations (see e.g. next two sections and Lupi *et al.*, 2016).

Takeo *et al.* (2018) have improved such models further by studying the effect of the anisotropy of the radiation field emitted around the black hole, to mimic the effect of an unresolved accretion disk, for which one expects radiation to be primarily emitted around the rotation axis of the disk (Ohsuga *et al.*, 2005; Jiang *et al.*, 2014; Sadowski and Narayan, 2016). The starting point is still a spherically symmetric gas cloud, and a large-scale gas infall rate is given as a boundary condition. They used a 1D multi-frequency radiative transfer code coupled with a 2D hydrodynamical simulation and a non-equilibrium primordial chemistry network. Self-gravity of the gas cloud, assumed to be initially of uniform density, was neglected, and a finite difference approach to solve the hydro equations was employed. The cloud is initially static. They consider a range of BH seed masses ($10^3 - 5 \times 10^5 M_\odot$) along with different geometries for the radiation field. The radiation field geometry was chosen as an *ansatz* in the initial conditions by projecting the radiative flux vector over a fraction of the solid angle. The underlying assumption is that the accretion flow will be organized in a disk or torus, or, more generically, would have an anisotropic pattern, thus leaving funnels of low density gas along which radiative losses would preferentially occur. Furthermore, only the radial component of the radiation force was considered. Because of anisotropic emission, matter was found to accrete preferentially in the plane perpendicular to the emission direction. The main difference with the calculations for isotropic radiation fields is that, for the same envelope density, lower mass BH seeds, down to $10^3 M_\odot$, can enter a sustained super-critical accretion (see left panel of Figure 5). This is interesting since this is in the mass range of Pop. III remnants. Another novel result found by such calculations is that, as the mass of the BH seed increases, a transition is observed in the qualitative evolution of the gaseous envelope. Indeed above a BH mass of $5 \times 10^5 M_\odot$ the ionized gas around the hole collapses due to super-sonic non-radial motions of neutral gas and radiative recombination. The neutral gas thus absorbs the momentum carried by photons and is accelerated outward in funnels along the poles triggering warm outflows with $T \sim 8000$ K. The latter finding could have interesting observational implications that may tell the model apart from others that have been proposed for Super-Eddington accretion.

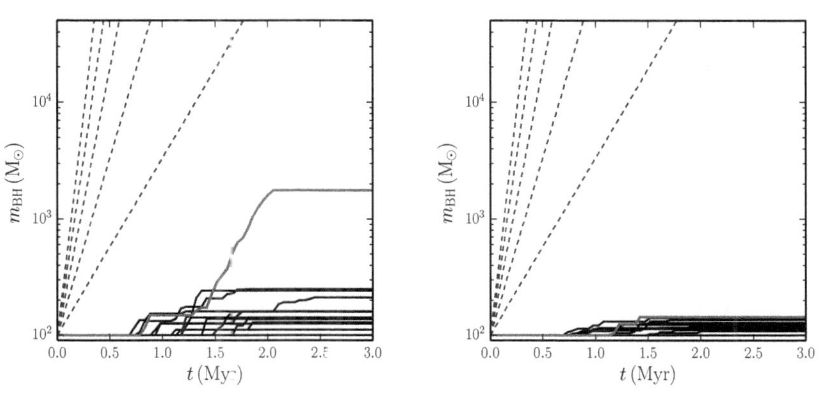

Fig. 7. Mass growth of light black hole seeds in a simulation using the SLIM disk model for accretion (left panel) and in one using the standard 0.1 radiative efficiency for a viscous disk (right panel), from Lupi *et al.* (2016). The grey lines correspond to the most massive BHs at the end of the runs, the black lines correspond to all other BHs in the CND, while the grey dashed lines depict reference accretion histories at fixed Eddington ratios of 500, 400, 300, 200, and 100, respectively. The marked difference in growth rate and absolute growth between standard and SLIM disk-like radiative efficiency is evident. Figure adopted from Lupi *et al.* (2016), reproduced by permission of Oxford University Press/on behalf of the RAS.

3. Large scale flow conditions in primordial galaxies: Super-Eddington gas supply rates?

In order to sustain Super-Eddington accretion a necessary condition is that the accretion disk/envelope itself is fed at its edge from large scales with gas impinging at very high rates, as assumed indeed in the Jiang *et al.* (2017) and Sadowski and Narayan (2016) simulations. We can ask what the current evidence of high inflow rates is at scales just above the accretion disk. We can also ask if such flow has a high angular momentum or not, and thus whether or not accretion will occur indeed through a disk-like interface.

First of all let us set the spatial scale of interest. While that specifically will depend on the mass of the MBH under consideration, we can proceed conservatively and decide that any result concerning the flow properties below pc scales, possibly below 0.1 pc, would be relevant as it would provide the natural boundary condition for smaller scale models/simulations starting at the edge of the accretion disk. Indeed the accretion disk around an MBH of a few thousand solar mass or larger, hence ranging from the mass of a Pop. III remnant to the mass of a full-fledged MBH, will have a physical radius in the range 10^{-3} and 10^{-2} pc roughly. The most interesting case is that of an MBH weighing the few hundred to a few thousand solar mass, which would be the expected mass for the remnant of a Pop. III star. The latter scenario has been considered in recent works modeling accretion from a spherically symmetric envelope as in Inayoshi et al. (2017).

In the remainder we will summarize the gas inflow rates in the range of kiloparsecs to 0.1 pc found in simulations of protogalaxies and high-z galaxies. First we cover the results concerning metal-free protogalaxies, namely galactic-scale objects forming in mini-halos with a virial mass $< 10^{10} M_\odot$, most typically in the range $10^8 - 10^9 M_\odot$, at $z \sim 15 - 20$. These are the typical target of the simulations designed to study the possibility of massive BH seed formation by direct gas collapse in metal-free halos, which were thoroughly described in Chapter 5 of this book. In this case inflow rates are in the range $0.1 - 1 M_\odot$/yr (Latif et al. 2016) at scales of 0.1 pc. If we assume that a light BH seed would be present in such a protogalaxy, such inflow would be well above the Eddington limit for BHs of a hundred to several thousand solar masses. Indeed the accretion rate at the Eddington limit, $\dot{M}_{Edd}$ can be conveniently expressed as:

$$\dot{M}_{Edd} = (4\pi G/\epsilon\kappa c) M_{BH} = 2.2 \times 10^{-8} (M_{BH}/M_\odot) M_\odot yr^1 \quad (4)$$

from which it is clear that the Eddington limit for a BH with mass of $1000 M_\odot$ corresponds to $2.2 \times 10^{-5} M_\odot/\mathrm{yr}$. In the equation above we have assumed an opacity $\kappa = 0.4\ \mathrm{cm}^2/\mathrm{g}$ and $\epsilon = 0.1$, as in standard Eddington-limited accretion flows. However, Super-Eddington accretion models discussed in the last section predict rates that are in the range of a few tens to several thousand M_{Edd} for both super-critical flows in accretion disk and in extended envelopes (see Section 2). For a BH of $1000 M_\odot$ these numbers would imply an inflow rate of $10^{-3} - 0.1 M_\odot \mathrm{yr}{-1}$ to feed the nuclear region at the corresponding rate, which are still well within the range of the pc-scale inflow rates in the simulations of metal-free protogalaxies.

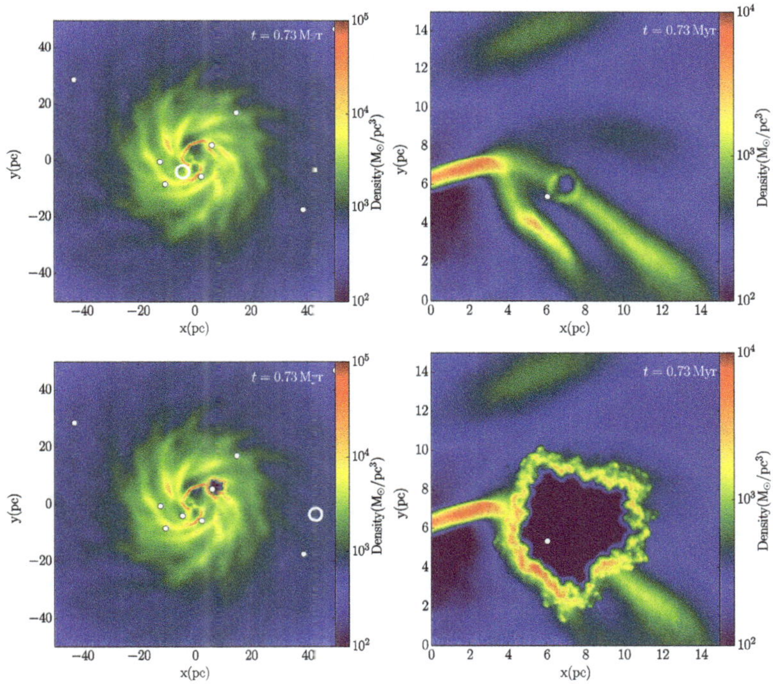

Fig. 8. Density maps of the circumnuclear disk (CND) at large (left) and small scales (right) in simulations carried out with a SLIM disk sub-grid model (top) and with a standard 0.1 radiative efficiency as expected in conventional viscous disk accretion (bottom), from Lupi *et al.* (2016). These are the same runs analyzed in Figure 7. The maps highlight the different effect of BH (radiative) feedback in the two cases. Indeed in the SLIM disk case, due to much lower radiative efficiency, the bubble created by feedback around the BH is much smaller and weaker (top-right panel), promoting faster sustained growth relative to the standard accretion case (bottom-right panel). Figure adopted from Lupi *et al.* (2016), reproduced by permission of Oxford University Press/on behalf of the RAS.

The mass inflow rate of gas with negligible pressure support pulled in by the gravitational action of the halo, and neglecting the effect of angular momentum, is $\dot{M} \sim \alpha V_{inf}^3/G$, where V_{inf} is the infall velocity and α is the coefficient of viscous drag relenting the inflow in an eventual disk-like envelope (Mayer and Bonoli, 2018). To first order inflow rates should thus scale as αV_{vir}, where we have taken the characteristic infall velocity to be of the order of the virial velocity scale of the halo. Note $\alpha \sim 0.01 - 0.1$ if self-gravitating instabilities are the main transport mechanism and they are treated as a local effective viscosity acting on the flow (Lodato and Rice, 2004). By assuming $\alpha = 0.01 - 0.1$ one recovers inflow rates in the range $0.1 - 1 M_\odot$/yr, comparable to those in the protogalaxy simulations (Latif *et al.*, 2013a, 2015), which have halo masses $\sim 10^9 M_\odot$. Based on the scaling $\dot{M} \propto M_{vir} \sim {V_{vir}}^3$ one expects much higher rates in more massive halos, 100-1000 times larger indeed in $10^{10} - 10^{11} M_\odot$ halos, hence in the range 100-1000 $M_\odot$/yr. Equation (4) indicates that an inflow rate of 1000 $M_\odot$/yr would be super-critical even for a $10^8 M_\odot$ BH, suggesting that in these larger galaxies super-critical accretion is possible throughout all stages of black hole growth, across many decades in mass. Halos with virial masses in the range $10^{11} - 10^{12} M_\odot$ appear below $z = 12$, hence in an epoch in which gas in such biased halos will be already metal-enriched, rendering radiative cooling is more efficient, which also can boost gas inflow further (Mayer *et al.*, 2015). Both observations, which are just beginning to probe this phase (Grazian *et al.*, 2015; Vito *et al.*, 2018) and simulations (Feng *et al.*, 2014, 2016), agree on the point of fast metal enrichment in such biased high density peaks at high-z.

The MassiveBlack (Di Matteo *et al.*, 2012), MassiveBlackII (Khandai *et al.*, 2015), and BlueTides (Feng *et al.*, 2016) simulations have studied gas infall in these massive high-z galaxies with the goal of assessing the growth history of putative BH seeds at their centers. All these simulations modeled large cosmological volumes in order to be able to identify highly biased density fluctuations that should correspond to the hosts of high-z QSOs (Li *et al.*, 2006). Given the large volumes adopted they limited their analysis to scales of kiloparsec order, corresponding to their resolution limit. Due to the same resolution limit they assumed initial BH seeds with already large masses, $> 10^5 M_\odot$. Therefore, they mainly assessed the rates of gas infall from the cosmic web onto the nascent galaxies in very massive halos. Di Matteo *et al.* (2012) were the first to show that the feeding occurs mainly in the form of cold gas flows ($T \sim 10^4$ K) accreting directly from the cosmic web, although a circumnuclear disk, perhaps unstable and clumpy, could be present at smaller unresolved scales (Gabor and Bournaud, 2013). Infall rates of order $100 M_\odot$/yr sustained on Gyr timescales were observed in these simulations, even in the presence of AGN feedback (starting from an initial BH seed with mass $\sim 10^5 M_\odot$ implanted in each halo). While in the simulations the sub-resolution accretion onto BH seeds was capped at the Eddington limit (except in the BlueTides suite, in which it was allowed to reach 3 times the Eddington value, see Feng *et al.* (2016)) these prominent inflows would be highly Super-Eddington if they could continue unimpeded all the way to the edge of the accretion disk/envelope. Whether or not they can really continue unimpeded to smaller scales will depend on the gas dynamics inside the galaxy, hence on the nature of the galactic potential and of the density distribution.

Only very recently it has become possible to tackle directly this question using the zoom-in technique by increasing many-fold the mass and spatial resolution in a selected object within a large cosmological volume. The MassiveBlackHR simulation (Capelo, Mayer, Di Matteo & Feng, in prep.) achieves this in *Halo 3* of the precursor MassiveBlack cosmological volume. The selected halo has clustering properties consistent with those of bright $z \sim 6$ QSOs (Di Matteo *et al.*, 2012). The simulation, which employed the unprecedented resolution of 500 million fluid elements (SPH particles in this scale) reaches a spatial resolution of a few tens of parsecs in the gas phase. Preliminary analysis shows that a prominent gas disk a few kpc in size forms at the center of the main dark matter halo at $z \sim 8$, following a merger between two massive proto-galaxies. The disk, with regular grand-design spiral structure and a central bar-like distortion, is reminiscent of low-z disks in galaxies, has a mass comparable to the disk of our own Milky Way but it is a size factor of 10 more compact, has a much higher gas content relative to the stellar content (50%) and thus a much higher density. This simulation shows that not only do high gas inflow rates ($> 100 M_\odot$/yr) persist down to sub-kpc scales, but they occasionally increase further, peaking at $1300 M_\odot$/yr, mostly in coincidence with mergers and flybies. Vigorous bars and spiral density waves do aid angular momentum transport via non-axisymmetric torques at sub-kpc scales, confirming the findings for much smaller simulated protogalaxies at $z > 10$ (Chapter 10). If such inflows continue to pc scales at similar rates they would, of course, be highly Super-Eddington accretion.

We have seen that massive halos with $M > 10^{10} M_\odot$ are in principle capable of sustaining super-critical accretion over a range of masses, from light Pop. III seeds to black holes with masses $\sim 10^8 M_\odot$. There is, though, another important consideration that goes in the direction of favoring halos with masses above $10^{10} M_\odot$ as natural environments for super-critical accretion; this is the effect of stellar and SN feedback as a function of halo mass. At moderate redshifts, $z < 5$, it is now fairly established that dwarf galaxies with a virial mass $M_{vir} < 10^{10} M_\odot$ cannot preserve large amounts of dense gas in their nuclei because of the strong effect of winds/outflows driven by SNe (Governato *et al.*, 2010; Shen *et al.*, 2014; Tollet *et al.*, 2016; Habouzit *et al.*, 2017), see also Chapter 10. While no specific studies have been carried out to explore the detailed conditions of nuclear gas flows in realistic high-z dwarf galaxy hosts, one expects a qualitatively similar behavior. Moreover, feedback-driven outflows should be widespread in low mass galaxies and become even stronger at high redshift reflecting the notion that specific star formation rates increase at a fixed galaxy stellar (as indeed verified in the simulations of Fiacconi *et al.* (2017b)). As for mini-halos at even higher-z, episodic accretion of the central BH would contribute further to heating and expeling the gas (see Chapter 10). Therefore the consistently large gas inflows necessary for sustained Super-Eddington accretion should be more prevalent in such galaxies hosted in massive halos since these are more resilient to the effect of feedback-driven outflows. We also note that, typically, in simulations of high-z metal-free protogalaxies (see Chapter 5) *normal* star formation beyond the stage of Pop. III stars, and thus its resulting feedback is not taken into account, hence the effect of feedback is underestimated. It follows that halos with $M_{vir} > 10^{10} M_\odot$ are the natural candidates to support sustained super-critical accretion onto their central BHs.

Episodic Super-Eddington accretion might occur for a wider range of halo masses, especially in the presence of strong tidal forces promoting temporarily gas inflows during mergers (major and minor), or in strong tidal interaction events, all particularly frequent at high-z. Episodic intense accretion might still be capable of growing some BHs to large sizes even starting from light seeds, especially in highly biased regions such as those that will later host bright QSOs since these are by construction growing their mass faster (Pezzulli *et al.*, 2016, 2017). Pezzulli *et al.* (2017) found only a 50% decrease in the overall population of MBHs that grow efficiently via super-critical accretion at $z > 7-8$ in their semi-analytical model when the effect of stellar and SNe feedback are taken into account. However it is not known how much bigger such effects would be in halos with $M_{vir} < 10^{10} M_\odot$ since the latter work focuses on halos that reach a mass as large as $10^{13} M_\odot$ at $z =\sim$. In any case, in their semi-analytical model mergers play a major role in revitalizing inflows even in low mass halos, an aspect that should be tested with detailed high-resolution cosmological hydrodynamical simulations.

4. Models of Super-Eddington growth of massive BH seeds in high redshift galaxies

In principle the growth of BH seeds in the early Universe should be studied using multi-scale simulations that capture the accretion disk flow as well as the larger scale feeding of gas from the surrounding interstellar medium to the disk. However this is not computationally feasible at the moment as the range of densities and timescales is prohibitive. The difficulty is exacerbated further by the need of treating radiative processes at all scales, and eventually MHD and relativistic effects at the scale of the accretion disk. This is a daunting task with current hydro codes even on large parallel supercomputers as the multiple timescales introduce load balancing bottlenecks that prevent eficient scaling on large node counts. Already at the scale of the accretion disk itself the inclusion of advanced radiative transfer schemes in global 3D simulations, such as those of Jiang *et al.* (2017) described in Section 2, results in a computational challenge. Therefore connecting such simulations to the protogalaxy simulations starting from cosmological initial conditions, the topic of the previous section, requires some modeling interface to simplify the computation. This interface can be realized in the form of a sub-grid model of disk accretion applied to the BH seeds that evolve in simulations of the protogalactic environment. Similar sub-grid approaches are routinely adopted to study the growth of MBHs at lower redshift in cosmological simulations, which describe mass growth as well as radiative and momentum feedback from the accreting MBH onto the surrounding ISM (the so-called AGN feedback, see Chapter 10).

The main difference is that in conventional cosmological/galaxy-scale simulations accretion rates are capped at the Eddington limit (Di Matteo *et al.*, 2005; Hopkins and Quataert, 2011), whereas in this Chapter we want to discuss how the gas flow behaves when one drops such constraint. The SLIM disk model lends itself to be cast in the form of a sub-grid model given its relative simplicity, in that its main features can be encapsulated in the resulting radiative efficiency (see Section 2). This indeed has been done in Lupi *et al.* (2016), who have studied the growth of light BH seeds in the nuclear regions of protogalactic disks using hydro simulations by allowing accretion to become super-critical, as well as in Pezzulli *et al.* (2016, 2017), who have modeled super-critical accretion in a semi-analytical model of galaxy and black hole formation. In a preceding work by Madau *et al.* (2014) the same SLIM disk model has been used in a semi-analytical code to show that even very light seed BHs originating from Pop. III stars can grow to billion solar masses by $z \sim 6 - 7$. In such work the seed BH was considered to be isolated and with an arbitrary large gas reservoir to feed onto. This was shown to be the case for both steady and intermittent super-critical accretion. The accretion rate was only a few times above Eddington (Figure 6), which is a very conservative assumption in light of the results of the small-scale simulations described in Section 2. Madau et al (2014) proposed the following spin-dependent fitting equation to the numerical results of Sadowski (2009):

$$L/L_E = A(a)\left[\frac{0.985}{\dot{m}_E/\dot{m} + B(a)} + \frac{0.015}{\dot{m}_E/\dot{m} + C(a)}\right], \tag{5}$$

where the functions A, B, and C scale with the spin of the black hole as

$$A(a) = (0.9663 - 0.9292a)^{-0.5639}, \tag{6}$$

$$B(a) = (4.627 - 4.445a)^{-0.5524}, \tag{7}$$

$$C(a) = (827.3 - 718.1a)^{-0.7060}. \tag{8}$$

Because photon-trapping occurs in Super-Eddington accretion the emitted luminosity is not linearly proportional to the accretion rate anymore, Madau *et al.* (2014) also propose a convenient way to recast the accretion timescale equation (see above equation (1)):

$$t_{\rm acc} = \frac{t_E}{16(1-\epsilon)}\left(\frac{\dot{m}_E}{\dot{m}}\right)(8.4 \times 10^6 {\rm yr})\left(\frac{3\dot{m}_E}{\dot{m}}\right), \tag{9}$$

where the last inequality holds for moderate Super-Eddington rates independent of the value of the black hole spin (Madau *et al.*, 2014). While low radiative efficiency of the SLIM disk model is accretion-rate dependent, the equation shown above shows that it is very mildly dependent on the spin, at stark variance with what occurs for the standard viscous disk model for which spin dependence is important. The next key question is whether or not the conditions for Super-Eddington accretion, namely sustained high gas accretion rates, are physically possible in protogalaxies. Lupi *et al.* (2016) indeed addressed this aspect using idealized simulations that only model the circumnuclear region of dense gaseous protogalaxies (the circumnuclear disk, hereafter CND). This approach allows to reach very high resolution and thus study the nature of the gas flow at scales even as small as 0.02 pc. Note, though, that this is still an order of magnitude larger than the sphere of influence of a stellar black hole coming from the collapse of even the largest Pop. III stars (on the order of a few thousand solar masses). Controlled experiments with isolated systems allow the parameter space to be surveyed extensively at high resolution, which would require prohibitive computational costs with cosmological simulations. Lupi *et al.* (2016) performed their simulations using two different numerical hydrodynamics techniques, the Lagrangian finite mass meshless method (MFM) in the GIZMO code (Hopkins, 2013) and the adaptive mesh refinement code RAMSES (Teyssier, 2002). Star formation and feedback from Supernovae were included, adopting the so-called blast wave model by Stinson *et al.* (2006) in both codes, in which cooling of the gas heated by SN ejecta is temporarily suspended to mimic unresolved turbulence and momentum deposition. Black holes were treated as massive collisionless particles that accrete at a rate determined by their radiative efficiency. Initially black holes with a range of masses were distributed at random locations inside the CND.

Generally speaking there are two ways by which a low radiative efficiency can affect BH growth. First of all, for a given accretion rate BHs grow faster with lower radiative efficiency because less mass is converted into radiant energy (by a factor proportional to $1 - \epsilon$). The second way is that BH feedback is reduced with a lower ϵ, once again because the radiant energy emerging from the accretion process is lowered, which implies a larger accretion rate can be sustained. The mechanical component of feedback manifests itself as momentum-driven winds and/or outflows. Its strength does not have a natural scaling behavior with radiative efficiency, but still has to decrease because of the first effect. Feedback onto the surrounding ISM by the accreting black hole was included in Lupi *et al.* (2016) by adopting the thermal coupling model in Booth and Schaye (2009), hence neglecting momentum feedback. Lupi *et al.* (2016) then compared the results obtained with standard viscous disk radiative efficiency ($\epsilon = 0.1$) against those with a reduced efficiency based on the SLIM disk model, and found striking differences concerning the growth rate of the BH seeds. Furthermore, black holes were also allowed to merge, which introduced another channel for growth. The resulting scenario is thus more complex compared to the toy model presented in Madau *et al.* (2014), in which only growth by gas accretion of a single, isolated black hole seed was considered.

Furthermore, as the CNDs were dense and massive self-gravity played an important role, this caused fragmentation into clumps which changed the nature of the gas flow (see Figure 8). These very dense clouds were weakly affected by feedback owing to their high density (much higher than present-day Giant Molecular Clouds (GMCs)), and even more so in the case of the SLIM disk accretion regime. This clumpy flow favored a sustained accretion rate, quite irrespective of hydro method and resolution.

Despite the intrinsic complexity, Lupi *et al.* (2016) found a recurrent outcome in their simulations, namely that one black hole, which had experienced faster growth since the beginning, was overtaking all the others by capturing competitively most of the available gas in the surroundings as well as merging with the other black holes. This recalls the behavior in competitive accretion in star formation (Bonnell *et al.*, 2001). Growth was typically terminated when all the gas was consumed, predominantly into star formation rather than accretion (the mean star formation rate in a sphere of 1 pc was $0.1M_{\odot}$/yr as opposed to $10^{-3}M_{\odot}$/yr for the black hole accretion rate). We remark in the simulations there was no gas infall onto the CND from exterior regions that could prolong sustained growth of the black hole seeds. By the end of the simulations, which extended for 3 Myr, when the gas in the CND was depleted, the mass of the major seed BH reached values $10^4 - 10^5 M_{\odot}$ in the SLIM disk case, while it barely reached $10^3 M_{\odot}$ when using standard disk accretion efficiency (see e.g. Figure 7). Remarkably, the accretion rates in the SLIM disk case were in the range 100-200 times the Eddington mass accretion rate, which is consistent with the results of the accretion-scale simulations described in Section 2. The most massive BH will inevitably migrate to the center via dynamical friction on a few orbital times, explaining why MBHs are found at the center of galaxies.

Finally, it is important to remark that these CND simulations, having a resolution of 0.1 pc, do not resolve the scale of the accretion disk, hence it is not yet possible to connect the gas flow properties at the numerical accretion radius to the gas flow in the MHD disk simulations of Super-Eddington accretion discussed in Section 2. Nevertheless, one can measure the angular momentum of the gas that accretes on the BH sink particles. It was found that this is very low, as most of the gas accretes in the dense clumpy phase, namely in a regime in which the gas flow in the CND by nature highly asymmetric and chaotic. As a result in the regime of these simulations accretion, while not spherical and isotropic, ends up having a rate comparable to the Bondi rate for the temperature and density of the local medium

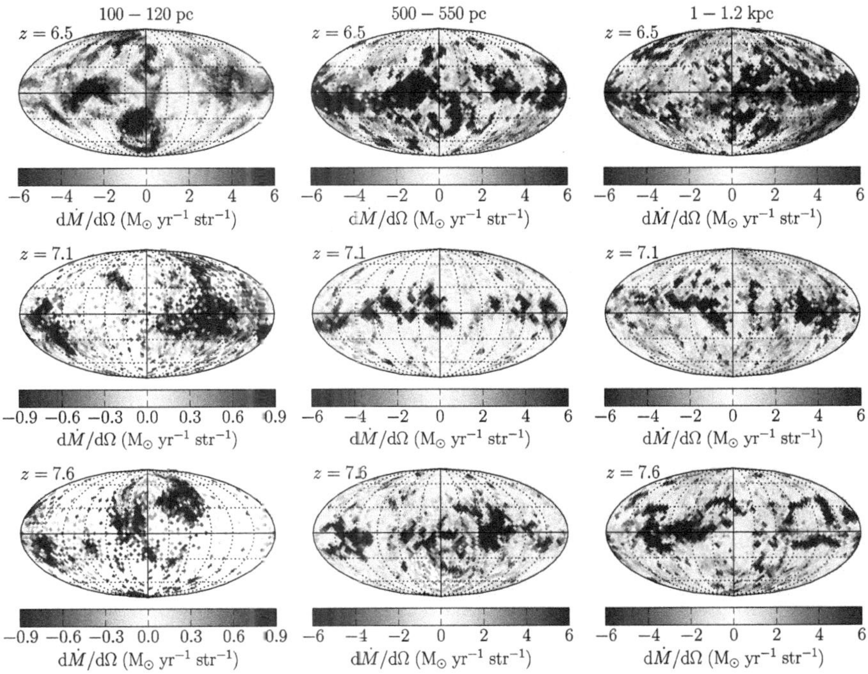

Fig. 9. Mollweide projections of gas density through the center showing the inflow and outflow to/from the nucleus at different scales (left to right) and different redshift (from bottom to top redshift decreases) for the "PONOS" cosmological galaxy formation simulation of Fiacconi *et al.* (2017a). The gas inflow pattern is anisotropic at all scales, down to the nuclear region. The hydrodynamical resolution of the simulation reached a few pc but the gravitational softening was set to 40 pc. Figure adopted from Fiacconi *et al.* (2017a), reproduced by permission of Oxford University Press/on behalf of the RAS.

around the BHs. Whether this is realistic or not has to be ascertained with cosmological simulations, which provide realistic boundary conditions for the flow in the CND, as well as with improved radiation physics, such as implementing the self-shielding of gas and the local photoionization, which could both suppress cooling, and thus fragmentation.

Fiacconi *et al.* (2017a) made a first step in addressing the problem of the realism of the initial conditions by studying the nature and geometry of the gas flow in the nuclear region of a $z \sim 5-6$ massive galaxy in the state-of-the-art cosmological hydrodynamical simulation dubbed "PONOS". They found that the mass of the region within a few 100 pc from the center was varying in the range $10^7 - 10^8 M_\odot$, relatively comparable with the mass assumed in the CND models of Lupi *et al.* (2016). However, gas inflow rates were observed to be highly variable and anisotropic down to the sub-kpc scale of the nucleus (Figure 9). This is a generic expectation given the highly nonlinear and chaotic gas accretion pattern from the cosmic web to the disks, and is confirmed also in the simulations of more massive $z \sim 7-10$ protogalaxies in the MassiveBlackHR run (Capelo *et al.*, in prep.). This, perhaps combined with the resolution limit set by the gravitational softening of 40 pc, did not produce a long-lived CND-like structure as that assumed in Lupi *et al.* (2016). Another potential difference was that the effect of stellar and SN feedback was found to be much stronger than in the CND-scale simulations of Lupi *et al.* (2016), suppressing fragmentation and leading to a turbulent flow from tens of pc to kpc scales. The highly anisotropic and turbulent nature of the flow down to the nuclear region is qualitatively consistent with the fact that a coherent structure such as CND is not assembled (Figure 9). While the long term evolution of the angular momentum in the turbulent nuclear region was not studied, dissipation through shocks between colliding flows of turbulent eddies is expected to be capable of extracting angular momentum and sustain accretion to smaller (unresolved) scales. The mean inflow rates on scales of a hundred pc are indeed still high enough to be super-critical for a BH seed sitting in the center. The proposed chaotic accretion scenario (King and Pringle, 2006) might apply in this case.

5. Summary

We have reviewed the status of knowledge relevant to Super-Eddington accretion of BH seeds at high redshift. The latest developments of small scale simulations with detailed physics modeling accretion disks and envelopes in super-critical regimes were presented and discussed in the context of massive BH seeds. We then covered gas inflow rates in protogalaxies at $z > 15$ as well as in more massive galaxies in high sigma peaks at $z \sim 10$ and below. This is relevant to determine how efficiently the accretion disk or envelope is fed with fresh fuel from the surrounding ISM, and in particular if the inflow rate is compatible with the accretion rates found in small scale simulations of super-critical phases. We then concluded by describing results of simulations that attempt to bridge the scales by treating super-critical accretion with a sub-grid model. Currently only the SLIM disk solution has been considered in such sub-grid models. The main findings that we discussed are:

- Super-Eddington accretion in disks and envelopes is feasible and, if it happens, it reaches high amplitudes, from ten times to a few thousand times above the Eddington limit.
- The radiative efficiency characterizing optically-thick, super-critical flows is low, but whether it is 10 or 1000 times lower than the standard radiative efficiency in the conventional thin "α" disk model is still unclear, and seems to depend on the method adopted by simulations to solve for radiation transport.
- The ratio between radiative and kinetic luminosity emitted in a Super-Eddington accretion phase is also uncertain. In some cases it makes up for most of the emitted energy, which has important implications on how to model feedback from the BH seed or MBH onto the surrounding interstellar medium.
- Pc-scale gas inflow rates in protogalaxies at $z \sim 15-20$ or the first massive galaxies at $z < 8-10$ are always above the Eddington limit for light BH seeds ($M_{BH} \sim 100 - 1000 M_\odot$), but can be such for a wide range of black hole masses, up to $10^8 M_\odot$, only if we consider galaxies in high sigma peaks with halo masses $10^{11} - 10^{12} M_\odot$. Only in this latter case does prolonged Super-Eddington accretion appear feasible. Furthermore, galaxies in low mass halos might not have a long-lasting nuclear gas reservoir due to the strong effect of stellar and SN feedback, which might stifle BH growth via Super-Eddington accretion.
- The results of the SLIM disk accretion model can be easily parametrized as a (very low) modified radiative efficiency, with a weak dependence on the black hole spin. With the latter recipe both semi-analytical models and local, non-cosmological simulations of circumnuclear disks (CNDs) have obtained encouraging results, showing that super-critical accretion can occur steadily as well as episodically depending on the environmental conditions but always leads to rapid growth of light BH seeds, matching naturally the constraints imposed by the rapid emergence of high-z QSOs. However it is still unclear if the dense CND assumed in such simulations occurs in a realistic cosmological setting at high z.
- Once the conditions for a super-critical accretion flow are in place the actual growth rate will depend on the gas supply rate to the

nuclear envelope/accretion disk as well as on the nature and effectiveness of radiative and kinetic feedback from the growing BH. Other effects such as merging with other black holes in vigorously star forming CNDs or inside dense nuclear star clusters, might be important in high-redshift environments.

Future developments will rely heavily on the ability to carry out multi-scale simulations bridging the accretion disk/envelope domain with the galactic domain at nuclear scales to the very minimum. In parallel, small scale accretion simulations have to investigate a much wider range of initial conditions, in terms of black hole masses as well as boundary inflow rates and patterns. They also need to quantify more explicitly the dependence of results, in particular radiative efficiency and mean accretion rate, on the adopted radiation transport model. Finally, combining relativistic simulations with the most sophisticated radiation transport methods, such as VET, is among the technical challenges ahead.

Chapter 12

Current observational status

Dominik R.G. Schleicher
Astronomy Department,
Universidad de Concepción,
Barrio Universitario,
Concepción, Chile,
dschleicher@astro-udec.cl

In this chapter, we review the current observational status of the first supermassive black holes. It is clear that such a review can hardly be complete, due to the wealth of surveys that has been pursued, including different wavelengths and different observational techniques. This chapter will focus on the main results that have been obtained, considering the detections of $z \gtrsim 6$ supermassive black holes in large surveys such as SDSS, CFHQS and Pan-STARRS. In addition, we will discuss upper limits and constraints on the population of the first black holes that can be derived from observational data, in particular in the X-ray regime, as these provide additional relevant information for the comparison with formation scenarios.

1. Introduction

Having discussed the formation of the first black holes in Chapters 5 and 7, their initial mass function (Chapter 9) as well as the subsequent growth (Chapters 10 and 11), it is now time to review the current observational status regarding the first supermassive black holes. The main focus here is on the black holes themselves, not on their host galaxies, which have been discussed in detail for instance by Gallerani *et al.* (2017).

We will start presenting the main surveys that have identified quasars at $z \gtrsim 6$ in Section 2. Upper limits and constraints from X-ray surveys and stacking techniques are presented in Section 3, along with the constraints derived from them on the population of the first black holes.

2. Surveys of $z \gtrsim 6$ supermassive black holes

The currently known quasars at $z \gtrsim 6$ correspond to the most direct information that is available on the high-redshift black hole population, providing constraints on formation and accretion models through the masses of the supermassive black holes as well as the redshift at which they exist. In the following, we will provide an overview of the main surveys that have contributed to their detection.

2.1. *The Sloan Digital Sky Survey*

The Sloan Digital Sky Survey (SDSS)[a] [York *et al.* (2000)] is one of the largest and longest astronomical surveys running so far, since the year 2000. In the initial configuration over the first five years (SDSS-I), it carried out deep multi-color imaging in the u-, g-, r-, i- and z-band over 8000 square degrees and measured spectra of more than 700,000 astronomical objects (see Fukugita *et al.* (1996) for the description of the photometric system). Within the second period from 2005-2008 (SDSS-II), it completed the original survey goals to imagine half of the northern sky and to map the 3-dimensional clustering of one million galaxies and 100,000 quasars. In 2008-2014 (SDSS-III), after a major upgrade of the spectrographs and with two new instruments to execute a set of four surveys, it mapped the clustering of galaxies and intergalactic gas in the distant Universe through the BOSS survey. The current generation of the SDSS (SDSS-IV, 2014-2020) includes many programs, most relevantly here the SPectroscopic IDentification of EROSITA Sources (SPIDERS) that will provide a unique census of supermassive black hole and large scale structure growth, targeting X-ray sources from ROSAT, XMM and eROSITA.

Overall and especially in the beginning, SDSS has played a central role in identifying quasars at $z \gtrsim 6$. Many of them were discovered from the ~ 8000 deg^2 imaging data, where luminous quasars ($z_{AB} \leq 20$) were selected as i-dropout objects through optical colors. Near-infrared (NIR) photometry as well as optical spectroscopy were subsequently used to distinguish them from late-type dwarfs (Fan *et al.*, 2001). A first $z = 5.80$ quasar, SDSSp J104433.04-012502.2, was discovered in 2000 by SDSS (Fan *et al.*, 2000), and many more were found in subsequent years. Three additional objects, SDSSp J083643.85+005453.3 ($z = 5.82$), J130608.26+035626.3

[a]Webpage SDSS: http://www.sdss.org/

($z = 5.99$), and J103027.10+052455.0 ($z = 6.28$) were discovered in 2001 (see also Fig. 1 Fan *et al.*, 2001). These quasars were subsequently used to put constraints onto the end of reionization, and particularly the existence of a complete Lyman α Gunn-Peterson (GP) trough in the spectrum of the $z = 6.28$ quasar allowed to place strong limits (Fan *et al.*, 2002).

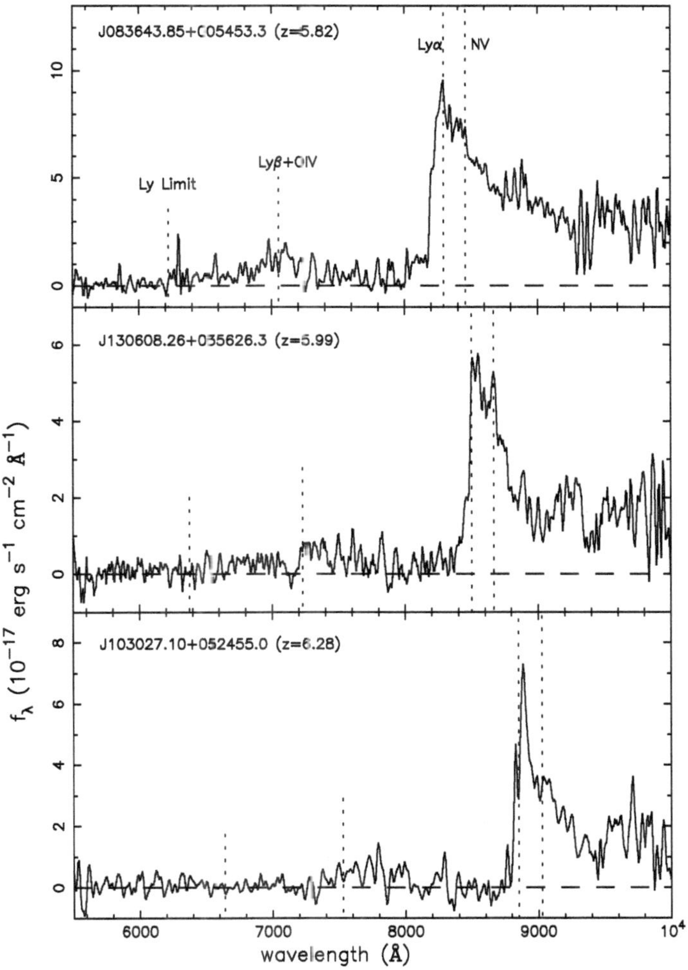

Fig. 1. The spectra of the three new $z > 5.8$ SDSS quasars discovered by Fan *et al.* (2001). Figure adopted from Fan *et al.* (2001), ©AAS. Reproduced with permission.

Three additional quasars, all at $z > 6$, have been discovered in 2003: J114816.64+525150.3 ($z = 6.43$), J104845.05+463718.3 ($z = 6.23$), and J163033.90+401209.6 ($z = 6.05$) (Fan *et al.*, 2003). These objects correspond to the bright-end slope of the quasar luminosity function, with $M_{1450} < -26.8$ (magnitude at 1450 Å). Their co-moving number density was thus estimated to be $(8 \pm 3) \times 10^{-10}$ Mpc^{-3}. The discovery of 5 additional objects at $z > 5.7$ has led to improved comoving number density estimates for objects with $M_{1450} < -26.7$ of $(6 \pm 2) \times 10^{-10}$ Mpc^{-3} at $z \sim 6$ (Fan *et al.*, 2004). 7 quasars were subsequently discovered at $z > 5.7$, with redshifts between 5.79 and 6.13. Two of them were at $z > 6$ (Fan *et al.*, 2006c). Based on such detections, a sample of 19 quasars at $z \sim 6$ could be established to probe reionization in 2006 (Fan *et al.*, 2006b).

With the discovery of 5 additional quasars at $z \sim 6$ selected from a 260 deg^2 field (one of them also independently discovered by the UKIRT Infrared Deep Sky Survey),[b] a complete flux-limited quasar sample at $z_{AB} < 21$ could be established, consisting of a total of 12 objects (Jiang *et al.*, 2008). This sample spans a redshift range of $5.85 \leq z \leq 6.12$ and a luminosity range of $-26.5 \leq M_{1450} \leq -25.4$. The co-moving number density per luminosity magnitude was estimated to be $(5.0 \pm 2.1) \times 10^{-9}$ Mpc^{-3} mag^{-1}. The sample allowed the first assessment of the bright-end quasar luminosity function at $z \sim 6$ as a power law, $\Phi(L_{1450}) \propto L_{1450}^{\beta}$, with $\beta = -3.1 \pm 0.4$. The subsequent detection of six additional quasars at $z \sim 6$ at at $21 < z_{AB} < 21.8$ essentially confirmed these conclusions (Jiang *et al.*, 2009).

The final sample based on SDSS discoveries alone consisted of 52 quasars with $5.7 \leq z \leq 6.4$, spanning a wide luminosity range of$-29.0 \leq M_{1450} \leq -24.5$ (Jiang *et al.*, 2016). The bright-end-slope is well constrained to be $\beta = -2.8 \pm 0.2$. Parametrizing the spatial density of luminous quasars as $\rho(M_{1450}) = \rho(z = 6) \times 10^{k(z-6)}$, they found that it drops rapidly from $z \sim 5$ to $z \sim 6$ with $k = -0.72 \pm 0.11$. Due to the small number density, it was shown that the observed population of quasars cannot provide enough photons to account for cosmic reionization, even though of course contributions from lower-luminosity quasars may be present. Overall, SDSS has opened up an important window to study the population of supermassive black holes at $z \sim 6$, including important objects such as the quasars at $z = 6.28$ and $z = 6.40$, which were included in many follow-up investigations. A particularly relevant recent discovery is the quasar SDSS J010013.02+280225.8

[b]Webpage UKIDSS: http://www.ukidss.org/

at $z = 6.30$ with an estimated black hole mass of 1.2×10^{10} $M_\odot$ (Wu *et al.*, 2015), and thus the most massive supermassive black holes at these redshifts so far.

2.2. *The Canada-France high-z quasar survey*

The Canada-France high-z quasar survey (CFHQS) makes use of optical imaging pursued in the Canada-France-Hawaii Telescope (CFHT) Legacy Survey.[c] It uses several different data sets for the search for high-redshift quasars: The majority of the sky area (~ 550 deg^2) is part of the RCS-2 survey (Yee *et al.*, 2007), and consists of MegaCam g$'$, r$'$, i$'$, z$'$ imaging data with exposure times in each filter of 240 s, 480 s, 500 s, 360 s, respectively. The CFHT Legacy Survey (CFHTLS) Very Wide covers several hundred square degrees, of which ~ 150 deg^2 are included in the CFHQS. The total exposure times for the Very Wide are comparable to RCS-2 with 540 s at i$'$ and 420 s at z$'$. The CFHTLS Wide is the intermediate depth per area component of the CFHTLS. It consists of 171 deg^2 at u$'$, g$'$, r$'$, i$'$, z$'$ with typical MegaCam integration times of 4300 s at i$'$ and 3600 s at z$'$. The CFHTLS Deep consists of four MegaCam pointings each of ~ 1 deg^2 at u$'$ g$'$ r$'$ i$'$ z$'$, with typical integration times of 250.000 s at i$'$ and 200.000 s at z$'$. Finally, the Subaru/XMM-Newton Deep Survey (SXDS) is a deep B, V, R, i$'$, z$'$ survey of ~ 1.2 deg^2 carried out at the Subaru 8.2m optical/infrared Telescope[d] (Furusawa *et al.*, 2008). SXDS employed typical integration times of 25.000 s at i' and 13.000 s at z' and is thus comparable to the CFHTLS Deep.

The first results of the survey yielded 24 quasar candidates at redshifts $5.7 < z < 6.4$ (Willott *et al.*, 2005), thereby yielding constraints and upper limits on the quasar luminosity function. In 2007, the survey discovered four quasars above redshift 6, including the highest-redshift quasar at that time, CFHQS J2329-0301 at $z = 6.43$ (Willott *et al.*, 2007). Six additional quasars above $z \geq 5.9$ were subsequently found, with luminosities $10 - 75$ times lower than the most luminous Sloan Digital Sky Survey quasars at this redshift (Willott *et al.*, 2009). The least luminous among them, CFHQS J0216-0455 at z $=$ 6.01 with an absolute magnitude $M_{1450} = -22.21$ is well below the expected break in the luminosity function where a transition from the steep power-law is expected to occur. Finally, nine additional $z \sim 6$ quasars were found in 2010, thus bringing the total number

[c]Webpage: http://www.cfht.hawaii.edu/Science/CFHTLS/
[d]Webpage Subaru: https://subarutelescope.org/

of CFHQS quasars to 19 (Willott *et al.*, 2010). Their binned luminosity function suggests a break at $M_{1450} \sim -25$, with a double power-law maximum likelihood fit to the data being consistent with the binned results. The CFHQS was thus particularly important to complement SDSS results through the detection of quasars at lower luminosity.

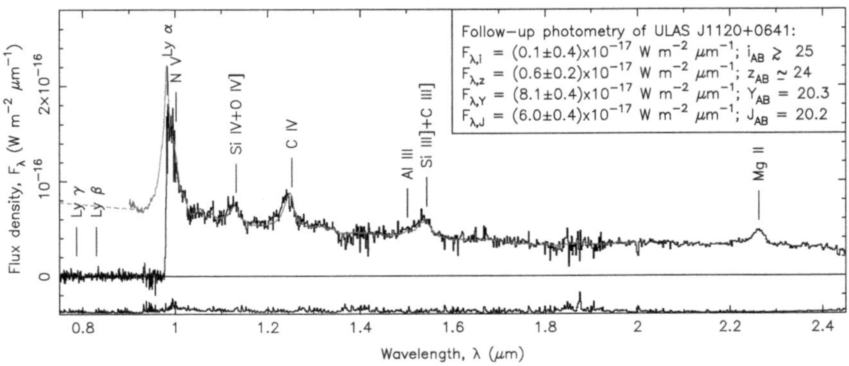

Fig. 2. Spectrum of the $z = 7.085$ quasar discovered by Mortlock *et al.* (2011). Figure from Daniel Mortlock, private communication.

2.3. *The UKIRT Infrared Deep Sky Survey*

While optical surveys were extremely important to find high-redshift quasars at ~ 6, the same surveys were insensitive to sources beyond $z = 6.5$, providing a strong limitation to find earlier objects. For the search of higher-redshift objects, near-infrared surveys are thus very important. The UKIRT Infrared Deep Sky Survey (UKIDSS)[e] was the first near-infrared survey with the capability to find quasars at $z > 6.5$. It is the successor to the 2MASS,[f] and began in 2005. Its goal is to survey 7500 deg^2 of the Northern sky, extending over both high and low Galactic latitudes. The depth is three magnitudes greater than 2MASS (K=18.3), thus forming the near-infrared counterpart to the Sloan Digital Sky Survey. The survey instrument is WFCAM on the UK Infrared Telescope (UKIRT)[g] in Hawaii. One of the scientific goals of the survey is finding the highest-redshift quasars at $z \sim 7$ (Lawrence *et al.*, 2007).

A first exploratory search for quasars at $z \sim 6 - 8$ has been performed in 2007 in the Early Data Release, initially yielding 34 candidates, which

[e] Webpage UKIDSS: http://www.ukidss.org/
[f] Webpage 2MASS: https://www.ipac.caltech.edu/2mass/
[g] Webpage UKRIT: http://www.ukirt.hawaii.edu/

were however shown to be brown dwarfs via spectroscopic follow-up (Glikman *et al.*, 2008). The first discovery of a luminous $z \sim 6$ quasar (ULAS J020332.38+001229.2) from near-infrared data was reported by Venemans *et al.* (2007), at a redshift of $z = 5.86$. Subsequently, the discovery of the $z = 6.13$ quasar ULAS J131911.29+095051.4 was announced by Mortlock *et al.* (2009). With ULAS J112001.48+064124.3, an even higher-redshift quasar, and for about 7 years the highest redshift quasar, was subsequently discovered at $z = 7.085$, i.e. about 770 million years after the Big Bang (Mortlock *et al.*, 2011). The quasar has a high luminosity of $6.3 \times 10^{13}\ L_\odot$, corresponding to a black hole mass of about $2 \times 10^9\ M_\odot$. The presence of such an object emphasizes the very limited time that is needed to grow such massive black holes, though requiring an average accretion rate of the order $2.6\ M_\odot\ yr^{-1}$. The spectra shown in Fig. 2 exhibit strong metal lines similar to the brightest quasars at lower redshift, and indicate that metal enrichment in its central region has been very efficient irrespective of the high redshift. Only recently, an even higher redshift supermassive black hole has been discovered, with a mass of $\sim 8 \times 10^8\ M_\odot$ at $z = 7.54$ (Bañados *et al.*, 2018b). The redshift of the source corresponds to a time of 700 million years after the Big Bang. This discovery was based on data from UKIDSS, combined with the AllWISE survey[h] (a successor of the WISE survey[i]), as well as DECam data (see Subsection 2.8 for more information on DECam).

2.4. *The Pan-STARRS distant $z > 5.6$ quasar survey*

The Panoramic Survey Telescope and Rapid Response System (Pan-STARRS)[j] is a system for wide-field astronomical imaging at the Institute for Astronomy of the University of Hawaii. Pan-STARRS1 (PS1) was the first part of Pan-STARRS to be completed. It used a 1.8 meter telescope along with a 1.4 Gigapixel camera to image the sky in five broadband filters (g, r, i, z, y), with 5σ limiting magnitudes of $(23.2, 23.0, 22.7, 22.1, 21.1)$. While PS1 is now completed, the focus has shifted towards PS2, where first light was achieved in 2013. The overall design foresees up to 4 possible telescopes.

The PS1 survey (Kaiser *et al.*, 2002, 2010) has imaged the whole sky above a declination of $-30°$ for about four years, which has led to a wealth

[h] Website AllWISE: http://wise2.ipac.caltech.edu/docs/release/allwise/
[i] Webpage WISE: https://www.nasa.gov/mission_pages/WISE/main/index.html
[j] Webpage Pan-STARRS: https://panstarrs.stsci.edu/

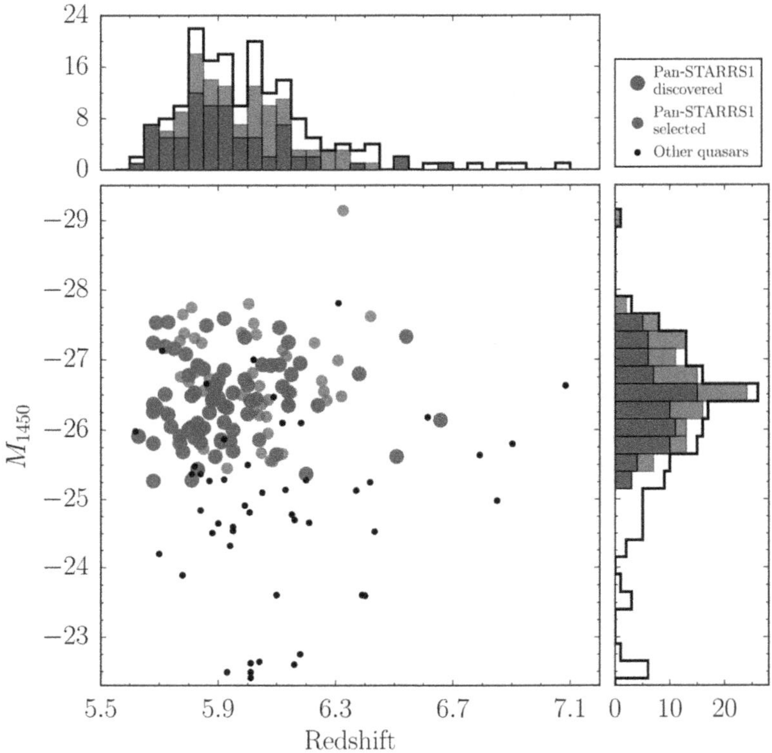

Fig. 3. Redshift and absolute UV magnitude (M_{1450}) distribution of all the known $z > 5.6$ quasars as of 2016 March (Bañados *et al.*, 2016). Figure adopted from Bañados *et al.* (2016), ©AAS. Reproduced with permission.

of data and a large amount of new discoveries. The first high-redshift quasar discovered through the survey was announced in 2012, with a redshift of $z = 5.73$ and an estimated black hole mass of 6.9×10^9 $M_\odot$ (Morganson *et al.*, 2012). Bañados *et al.* (2014) subsequently discovered 8 new quasars at $5.7 \leq z \leq 6.0$ with a range of luminosities, $-27.3 \leq M_{1450} \leq -25.4$, increasing the number of known quasars at $z > 5.7$ by 10% at that time. Venemans *et al.* (2015) subsequently announced 3 new quasars at $z = 6.50$, 6.52, and 6.66, thereby substantially adding to the number of $z > 6.5$ objects, including the brightest such object reported so far with $M_{1450} = -27.4$, the quasar PSO J036.5078 + 03.0498.

With improved selection criteria, Bañados *et al.* (2016) identified 63 additional quasars at $5.6 \leq z \leq 6.7$, leading to a total of 77 quasars that were identified through PS1. The total PS1 sample, which includes quasars previously discovered through other surveys, comprises a total of 124 quasars and spans a factor of $\sim$ 20 in quasar luminosity. As such, it has considerably extended the number of known quasars at such redshift. The redshift and UV magnitude distribution of the currently known sample is given in Fig. 3. Recently, Mazzucchelli *et al.* (2017) discovered 6 additional $z \gtrsim 6.5$ quasars via Pan-STARRS, and obtained a sample of 15 $z \gtrsim 6.5$ quasars to perform a homogeneous and comprehensive analysis. They derived typical black hole masses of $(0.3-5)\times 10^9$ $M_\odot$ and Eddington ratios of at least 0.39 or higher. In addition, the majority of $z \gtrsim 6.5$ quasars show large blueshifts of the broad C IV λ 1549 emission line compared to the systemic redshift of the quasars, with a median value $\sim 3\times$ higher than a quasar sample at $z \sim 1$.

2.5. *The Subaru High-z Exploration of Low-Luminosity Quasars survey*

The Subaru High-z Exploration of Low-Luminosity Quasars (SHELLQs) survey is the first 1000 deg^2 class survey for high-z quasars with an 8 m class telescope like Subaru.[k] The survey exploits multiband photometry data produced by the Subaru Hyper Suprime-Cam (HSC) (Miyazaki *et al.*, 2012), as part of the Subaru Strategic Program (SPP) survey. The HSC-SPP survey is a large collaborative project including researchers from Japan, Taiwan and Princeton. It started in early 2014, and will include 300 observing nights, lasting until 2019. It uses the HSC, a wide-field camera equipped with 116 2K $\times$ 4K Hamamatsu fully depleted CCDs, of which 104 are used to obtain science data. Within the HSC-SPP, the Wide layer aims to observe 1400 deg^2 mostly along the equator through five broad-band filters, aiming to reach a 5σ limiting magnitude of $g = 26.5$, $r = 26.1$, $i = 25.9$, $z = 25.1$, $y = 24.4$. mag within 2.″0 apertures. The Deep and Ultra-Deep layers observed 27 and 3.5 deg^2, using 5 broad-band and 4 narrow-band filters, aiming to reach a 5σ limiting depth of $r = 27.1$ mag (Deep) or 27.7 mag (Ultra-Deep).

SHELLQs exploits the HSC data to search for low-luminosity quasars at high redshift. The filter set is sensitive to quasars with redshifts up to $z \sim 7.4$, i.e. beyond the current quasar redshift record, even though the detection capability sharply drops at $z > 7$ as the Gunn-Peterson effect, so that the survey is limited to very luminous objects at those redshifts. The

[k]Webpage Subaru: https://subarutelescope.org/

survey discovered 15 quasars and bright galaxies at redshifts $5.7 < z < 6.9$, of which 6 were determined to be likely quasars (Matsuoka *et al.*, 2016). Subsequently the survey presented the spectroscopic identification of 32 new quasars and luminous galaxies at $5.7 < z < 6.8$, of which 24 were identified as quasars. As the survey already includes quasars with $M_{1450} \sim -22$ mag, extending the quasar luminosity function to this magnitude is now likely within reach.

2.6. *The VST ATLAS survey*

The VST ATLAS survey[l] (Shanks *et al.*, 2015) is targeting 5000 deg^2 of the Southern sky (Chehade *et al.*, 2018). It is pursued through the Very Large Telescope (VLT),[m] which consists of four Unit Telescopes with main mirrors of 8.2 m diameter and four movable auxiliary telescopes. The survey is conducted in the U, V, R, I and Z filters at depths comparable to SDSS. However, it is deeper in the z band, with a mean 5σ limiting magnitude of 20.89, and has considerably better seeing, in the range $0.''8 - 1.''0$ for all five bands.

While the main goal of the survey is to measure baryonic acoustic oscillations, it contributes to the detection of bright $z > 6$ quasars. In particular, Carnall *et al.* (2015) reported the discovery of two $z > 6$ quasars (selected as i-band dropouts). The first of the quasars in their sample has a redshift $z = 6.31 \pm 0.03$ with $M_{1450} = -27.8 \pm 0.2$, thus making it the joint second most luminous quasar known at $z > 6$. The second quasar corresponds to a redshift of $z = 6.02 \pm 0.03$ with magnitude $M_{1450} = -27.0 \pm 0.1$. The detection shows the potential of the survey to still discover new very bright quasars at high redshift.

2.7. *The VISTA surveys*

The VISTA surveys[n] are conducted via the Visible and Infrared Survey Telescope for Astronomy (VISTA).[o] VISTA is a 4-m class wide field survey telescope at Paranal in the southern hemisphere, equipped with a near infrared camera (1.65 degree diameter field of view) containing 67 million pixels, available broad band filters at Z,Y,J,H,Ks and a narrow band filter

[l]Webpage VST surveys: https://www.eso.org/public/teles-instr/paranal-observatory/surveytelescopes/vst/surveys/
[m]Webpage VLT: http://www.eso.org/public/teles-instr/paranal-observatory/vlt/
[n]Webpage VISTA surveys: https://www.eso.org/public/teles-instr/paranal-observatory/surveytelescopes/vista/surveys/
[o]Webpage VISTA: http://www.vista.ac.uk/

at 1.18 μm. For the search for high-redshift quasars, the relevant VISTA surveys are UltraVISTA and VIKING. UltraVISTA is the deepest and narrowest VISTA survey, imaging one patch of sky over and over again to unprecedented depths. The VISTA Kilo-Degree Infrared Galaxy Survey (VIKING) is imaging 1500 deg^2.

VIKINGS contributed in particular to extend the number of $z > 6.4$ quasars, reporting detections at $z = 6.60$, 6.75, and 6.89, with black hole masses of $1 - 2 \times 10^9$ $M_\odot$ (Venemans *et al.*, 2013). Within the redshift range $6.44 < z < 7.44$, they established a lower limit on the black hole mass density of $\rho(M_{BH} > 10^9\ M_\odot) > 1.1 \times 10^{-9}$ Mpc^{-3}. Jointly with the Kilo-Degree Survey (KiDS),[p] which is pursued by the VLT in the Southern sky, four additional quasars were subsequently discovered at $5.8 < z < 6.0$, with magnitudes $-26.6 < M_{1450} < -24.4$ (Venemans *et al.*, 2015), thus corresponding to relatively faint objects, and it may be possible to find 30 objects of similar luminosity during the further course of the survey. Discoveries based on joint data from VISTA and the Dark Energy Survey will be described further below.

2.8. *The Dark Energy Survey*

The Dark Energy Survey (DES)[q] consists of a 5000 deg^2 area of the Southern sky (roughly 1/8 of the total sky), which will be observed over 525 nights using the new Dark Energy Camera (DECam) mounted on the Blanco 4-meter telescope[r] at the Cerro Tololo Inter-American Observatory in the Chilean Andes. While the main science goal is the detection of thousands of supernovae to probe the history of cosmic expansion and thereby the nature of dark energy, the data of the survey have also contributed to the discovery of additional quasars at high redshift.

With DES J0454-4448, the first luminous quasar discovered through the Dark Energy Survey was announced in 2015 (Reed *et al.*, 2015), with a redshift of $z = 6.09 \pm 0.02$ and $M_{1450} = -26.5$, thus corresponding to a rather bright object. Overall, the survey is expected to discover $50 - 100$ new quasars with $z > 6$ as well as a few with $z > 7$. Indeed, the detection of 8 new quasars, based both on data from DES, VISTA and the Wide-field Infrared Survey Explorer (WISE),[s] was reported in 2017 by Reed *et al.* (2017). This includes the $z = 6.5$ quasar VDES J0224-4711, the second

[p]Webpage KiDS survey: http://kids.strw.leidenuniv.nl/
[q]Webpage DES: https://www.darkenergysurvey.org/
[r]Webpage Blanco telescope: http://www.ctio.noao.edu/noao/node/9
[s]Webpage WISE: https://www.nasa.gov/mission_pages/WISE/main/index.html

most luminous quasar known with $z \geq 6.5$ ($M_{1450} \gtrsim -27$). The overall redshifts range between 6.0 and 6.5, and the magnitude extends to $M_{1450} \lesssim -25$.

3. Upper limits and constraints from X-ray surveys and observations

The role and importance of observations in the X-ray regime can hardly be overemphasized to follow the evolution of the supermassive black hole population over cosmic history (Treister *et al.*, 2010), as well as the identification of heavily obscured objects that would not be visible at other wavelengths (Treister *et al.*, 2009). Surveys pursued with the Chandra satellite,[t] XMM-Newton[u] and NuSTAR[v] have thus greatly advanced our knowledge on the high-redshift black hole population and demographics, with newly discovered quasars up to redshifts of $z \sim 5$ [Ueda *et al.* (2014)]. An excellent review on these discoveries has been provided by Brandt and Alexander (2015), to which we refer here for a description of the demographics until $z \sim 5$. We also emphasize here that the ATHENA X-ray Observatory[w] has the potential to overcome this boundary, thus potentially providing detections of the first quasars at very high redshift in the future (Nandra *et al.*, 2013). In addition, X-ray observations can be useful to increase available information on already known quasars at $z \gtrsim 6$, as demonstrated by Bañados *et al.* (2018a). In the following, we will discuss how existing X-ray data are already providing strong relevant constraints about the first supermassive black holes.

3.1. *Constraints from the unresolved X-ray background*

The cosmic X-ray background is a radiation background resulting from many individual sources, many of which were initially unresolved, while substantial progress from X-ray surveys has allowed to resolve a large fraction of that background. As there is no anticipated cosmological or primordial origin, it is expected that the remaining unresolved background results from low-luminosity sources at low to moderate redshifts, or potentially from higher-luminosity objects at very high redshift, including the population of the very first quasars. With realistic assumptions on the X-ray

[t] Webpage Chandra: http://chandra.harvard.edu/
[u] Webpage XMM-Newton: https://www.cosmos.esa.int/web/xmm-newton
[v] Webpage NuSTAR: https://www.nustar.caltech.edu/
[w] Webpage ATHENA: http://www.the-athena-x-ray-observatory.eu/

spectra of the first quasars, the unresolved X-ray background thus translates into a constraint for the first supermassive black holes (Dijkstra *et al.*, 2004; Salvaterra *et al.*, 2005).

To provide constraints on the first quasars, the most relevant component of the unresolved X-ray background is the soft X-ray band at energies of 0.5 – 2 keV, which corresponds to hard X-rays at redshifts $z \geq 6$. These are well within the regime of energies where the intergalactic medium is optically thin. Moretti *et al.* (2003) have studied the X-ray background in that energy range, finding an integrated energy flux of $\int f_E dE \sim 7.53 \pm 0.35 \times 10^{-12}$ erg cm^{-2} s^{-1} deg^{-2} at an energy of 1 keV, where they combined data from 6 surveys performed by 3 satellites, ROSAT,[x] Chandra, and XMM-Newton. They further included deep pencil beam surveys together with wide field shallow surveys to determine the flux resulting from resolved sources, with individual sources having fluxes of $2.44 \times 10^{-17} - 1.00 \times 10^{-11}$ erg cm^{-2} s^{-1}. Overall, their analysis could show that 90^{+6}_{-7}% of the soft X-ray background consists of discrete sources.

Point-like sources are however not the only contribution to the X-ray background. Extended emission can originate for instance in clusters and groups of galaxies throughout the Universe, and the expected resulting background was calculated by Wu and Xue (2001) using the observed X-ray luminosity function, finding an expected contribution of 1.18×10^{-12} erg cm^{-2} s^{-1}, i.e. about 16% of the total soft X-ray background. As argued by Dijkstra *et al.* (2004), the contribution coming from groups is rather uncertain, so a more conservative estimate may correspond to about 60% of that value, the contribution coming from clusters of galaxies, showing that the current data are not fully consistent. A recent analysis based on the Chandra COSMOS Legacy Survey (Cappelluti *et al.*, 2017) provides a 1 keV normalization of the unresolved X-ray background of ~ 1.37 keV cm^{-2} s^{-1} sr^{-1} keV^{-1}, stating that unresolved sources contribute 8 – 9% of the soft background. Even from the unresolved fraction, one would expect that only a small part is really due to the first black holes, thus potentially resulting in a strong upper limit.

In their analysis, Dijkstra *et al.* (2004) determined that, depending on how to sum up the individual contributions and regarding different scenarios concerning the treatment of the error bars, the unresolved flux is between $0.35 - 1.23 \times 10^{-12}$ erg cm^{-2} s^{-1} deg^{-2}, which we will adopt here for definiteness and for comparison with their work. It is however clear

[x]Webpage ROSAT: https://heasarc.gsfc.nasa.gov/docs/rosat/rosat3.html

that the resulting constraints can be rescaled for different values of the unresolved X-ray background. An additional important ingredient to derive constraints on the first black holes is a quasar spectrum. For this purpose, they adopted the characteristic spectrum derived by Sazonov *et al.* (2004), who computed the characteristic angular-integrated, broad-band spectral energy distribution for average quasars. This spectrum scales approximately as $f_E \propto E^{-1.7}$ for $E > 13.6$ eV, but becomes much shallower beyond 2 keV, where it scales as $f_E \propto E^{-0.25}$. Through their analysis, Dijkstra *et al.* (2004) have shown that the contribution of quasars to the epoch of reionization is strongly constrained via the unresolved X-ray background, which limits the production rate of ionizing photons. As a result, it became clear that quasars are not the main sources driving reionization, unless their spectra were considerably different at such early times. Partial contributions however still seemed possible, with fractions up to 50% in more optimistic scenarios. Using a similar approach, Salvaterra *et al.* (2005) derived constraints on the mass density of intermediate mass black holes at $z \geq 6$, finding an upper limit of $\rho_{BH} < 3.8 \times 10^4$ $M_\odot$ Mpc^{-3}, corresponding to about one intermediate mass black hole with 1 Mpc^{-3}.

3.2. *Constraints via stacking techniques*

While X-ray observatories have not yet provided any new discoveries of supermassive black holes at $z > 6$ (though already known quasars at redshifts $z \gtrsim 6$ have been detected with Chandra or XMM, see Bañados *et al.* (2018a)), the results of their surveys can nevertheless be employed to place upper limits on supermassive black holes in typical galaxies. In particular, we recall that the detections in optical surveys are mostly in the higher-luminosity tail of the quasar luminosity function at $z \sim 6$. As a result, we obtain statistical information on the properties of the brightest and most massive objects. While these may be interesting as extreme cases to probe potential formation scenarios, it is also important to understand the more typical outcome of black hole and galaxy formation, and to derive constraints on the black hole masses in typical galaxies.

To accomplish this, stacking techniques have been developed by Treister *et al.* (2011) to combine X-ray data at positions where star-forming galaxies have been found via optical surveys, thus efficiently increasing the signal-to-noise ratio by adding up the data at the different positions. While this of course will not provide information on individual galaxies, it can provide information on the average properties of such star-forming galaxies, either in the form of the detection of a mean signal, or through an upper limit that constrains their X-ray activity. The technique is however technically complex, and was thus subsequently improved by Willott *et al.* (2011) and Cowie *et al.* (2012).

Using this technique, Treister *et al.* (2013) employed the 4 Ms Chandra observations of the Chandra Deep Field-South (CDF-S) (Xue *et al.*, 2011), the deepest X-ray observations taken so far, and combined them with additional survey data with galaxy detections at other wavelengths. Within the CDF-S, Bouwens *et al.* (2011) detected 66 $z \sim 7$ galaxies and 47 at $z \sim 8$ through the Hubble Space Telescope (HST)[y] using the WFC3 camera. In addition, HST/WFC3 observations by Finkelstein *et al.* (2012) in the CANDELS fields[z] yielded a sample of 223 galaxies at $z \sim 6$, 80 at $z \sim 7$ and 33 at $z \sim 8$. While none of these galaxies are detected individually in X-rays, count rates for their positions are nevertheless available from the 4 Ms Chandra observations by Xue *et al.* (2011), which can be stacked at the respective positions to increase the signal-to-noise ratio. Using this procedure, Treister *et al.* (2013) obtained an upper limit of the X-ray luminosity of 2.6×10^{41} erg s^{-1} for the $z \sim 6$ sources in the soft band, and 1.6×10^{42} erg s^{-1} in the hard band. For $z \sim 7$ galaxies, these limits correspond to 6.8×10^{41} erg s^{-1} and 5.3×10^{42} erg s^{-1}, as well as 1.5×10^{42} erg s^{-1} and 9.8×10^{42} erg s^{-1} for $z \sim 8$ galaxies. All of these are below the standard threshold for active galaxies of $\sim 10^{42}$ erg s^{-1} (Szokoly *et al.*, 2004), implying relatively low activity on average.

Upper limits on the X-ray luminosity function can be translated into upper limits on the accreted black hole mass density using Soltan's argument (Sołtan, 1982), considering the link between mass accretion and radiation energy production in the accretion process. Applying this analysis to the results from the stacking technique implies upper limits of 990, 1142 and 1263 M$_\odot$ Mpc^{-3} at $z \sim 6$, 7 and 8, respectively. These limits assume a radiative efficiency of $\epsilon = 10\%$, and are even tighter than the ones derived via the unresolved soft X-ray background in Section 3.1, due to the combination of information from a large number of sources.

Translating this into properties of individual galaxies requires further assumptions on the bolometric correction, the Eddington ratio as well as the fraction of active black holes. Assuming 100% activity, a canonical bolometric correction of 10% for hard X-rays and a 10% Eddington ratio, Treister *et al.* (2013) derived an upper limit of the black hole mass of

[y] Webpage HST: https://www.nasa.gov/mission_pages/hubble/main/index.html
[z] Webpage CANDELS: http://candels.ucolick.org/

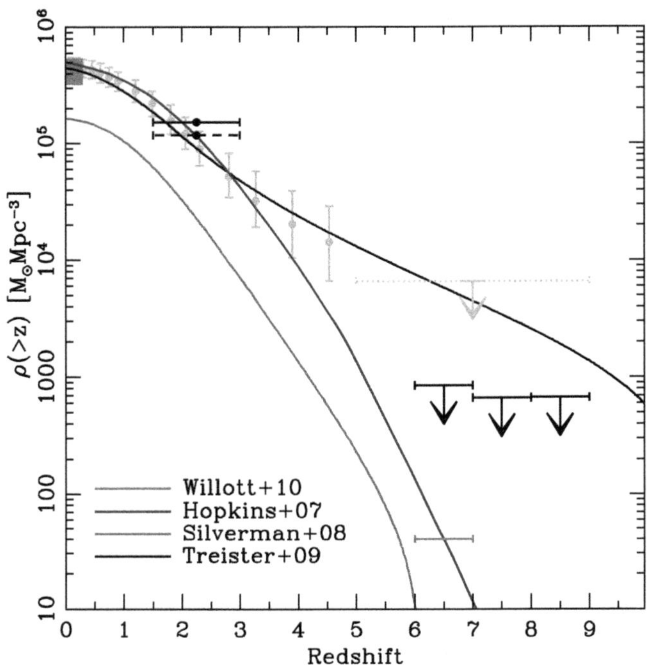

Fig. 4. Comparison between the observed accreted mass density in supermassive black holes and expectations from the observed luminosity functions, combined with the upper limits from stacking (black solid) and the unresolved soft X-ray background (grey dashed). Adopted from Treister *et al.* (2013), ©AAS. Reproduced with permission.

about 3×10^6 $M_\odot$ for a typical $z \sim 6$ galaxy. This can be higher if not all of them are active, or if the Eddington ratio is further reduced. While the constraints are tighter than originally anticipated, recent research suggests that the limits may still be consistent with total stellar mass - black hole mass relations obtained through calibrations within the local Universe (Volonteri and Reines, 2016). It is nevertheless important to emphasize that black hole formation models need not only reproduce the most luminous quasars discovered via optical surveys, but also have to comply with the constraints on the average population that is available from X-ray data.

While the current data do not yet strongly constrain models, the upper limits are important to prevent an overproduction of black holes in certain scenarios. In addition, the main challenge may consist in explaining the most massive quasars that have been observed, through massive seeds, strong subsequent accretion or both. In the two remaining chapters, we will now address the future prospects, both through the prospects of gravitational wave observatories (Chapter 13) as well as future observational prospects (Chapter 14).

Chapter 13

Probing the formation of the seeds of supermassive black holes with gravitational waves

Monica Colpi[1,2]

[1]*Department of Physics G. Occhialini, University of Milano Bicocca*
[2]*National Institute of Nuclear Physics, Milano Bicocca*
Piazza della Scienza 3, I 20123 Milano, Italy
monica.colpi@unimib.it

The existence of black holes in the intermediate-mass interval between $10^2\,M_\odot$ and $10^5\,M_\odot$, filling the *gap* between the stellar and the supermassive black holes is a key prediction to explain the origin of luminous QSOs at redshifts z as large as 7. There is a sheer difficulty in forming a $10^9\,M_\odot$ black hole in less than one billion years. This has led to the concept of *seed black holes*. They are high-z intermediate-mass black holes that formed during the dark ages of the universe. Seeds are a transient population, which later grew massive through episodes of accretion and mergers. In this chapter we explore the possibility of discovering seed black holes and track their growth across all cosmic epochs, by detecting the gravitational wave signal they emit at the time of their coalescence, if/when they pair to form close binaries. We show that the ESA's LISA mission for the detection of low-frequency gravitational waves will be paramount in granting this insight. Gravitational waves travel unimpeded through the cosmos and carry exquisite information on the masses and spins of the merging black holes. To this purpose we introduce key concepts on the gravitational wave emission from binaries, describing briefly their formation pathway during halo-halo mergers and galaxy collisions.

1. Paving the way to unveil black holes in the high redshift universe

The simplest object in Nature is a *black hole* (BH)* exact solution of Einstein's field equations representing a spinning, point mass M, in otherwise empty space. BHs are believed to power many astrophysical sources, from the luminous quasi-stellar objects (QSOs) to the galactic X-ray sources, through the conversion of gravitational energy into electromagnetic (EM) radiation. In this chapter we show that BHs in close binaries are among the most powerful sources of gravitational waves (GWs) in the universe.

EM observations reveal the occurrence of BHs in two different flavors: the *stellar-origin black holes* (sBHs) and the *supermassive black holes* (SMBHs). sBHs are widespread in all the galaxies of the universe, as they are the relic of massive stars. They started to form in pristine, dark matter halos as early as redshift $z \sim 20$ from the collapse of massive population III stars (Madau and Rees, 2001), and continue to form in the galaxies as stars do (see also chapter 4). When a sBH is bound to a star in a close binary, it lights up as an X-ray source, but when bound to a sBH, the system emits GWs. The GW signal is initially weak, but it grows loud when the two sBHs are in the process of merging to form a new sBH. GW150914 (Abbott *et al.*, 2016b), the first cosmic source of GWs hosting two coalescing sBHs, is a chief example.

Prior to the discovery of GW150914, the measurements of sBH masses in the Milky Way Galaxy were indicating a narrow distribution around $7.8 \pm 1.2\,\mathrm{M}_\odot$ (Özel *et al.*, 2010). The discovery of "heavy" sBHs, with mass up to $\sim 36^{+5}_{-4}\,\mathrm{M}_\odot$, in GW150914, has shown that the sBH mass spectrum is wider than EM observations revealed so far. With the recent detection of GW170608 which hosts a sBH of $7^{+2}_{-2}\,\mathrm{M}_\odot$, the mass function of coalescing sBHs extends, bracketing uncertainties, between about $5\,\mathrm{M}_\odot$ and about $40\,\mathrm{M}_\odot$[Abbott *et al.* (2017b,d,c)].

Theoretical studies (Heger and Woosley, 2002) show that the mass of a sBH, $M_\bullet^{\mathrm{sBH}}$, depends not only on the mass of the progenitor star on the zero-age main sequence m_{zams}, but also on the metallicity Z, and degree of (differential) rotation. The metallicity controls the extent of mass loss by

*Through this chapter, BHs are described by the Kerr family of solutions and carry *no hair*. The modulus of the spin vector $\mathbf{S}$ is expressed in terms of the dimensionless spin parameter $\chi_{\mathrm{spin}} \equiv |\mathbf{S}|c/(GM^2)$, where M is the BH mass. According to the cosmic censorship conjecture $|\chi_{\mathrm{spin}}| \leq 1$. If this condition is violated, a naked singularity appears and General Relativity would not be able to describe the exterior solution nor the dynamics of BHs in Nature.

Abbreviation	Full name
BH	black hole
QSO	quasi-stellar object
EM	electromagnetic
GW	gravitational waves
sBH	stellar-origin black hole
SMBH	supermassive black hole
IMBH	intermediate-mass black hole
MBH	massive black hole
PPI	pulsational pair-instability
PISN	pair instability supernova
AGN	active galactic nuclei
MDO	massive dark object
GR	General Relativity
ISCO	innermost stable orbit
BBH	binary black hole
PN	post-Newtonian
NR	Numerical Relativity
SNR	signal-to-noise ratio
IMRI	intermediate mass ratio inspirals
EMRI	extreme mass ratio inspiral
IMF	initial mass function

winds during stellar evolution (Vink, 2008). The sBH may form through the fall back of bound material onto a hot-neutron star driven unstable above its maximum mass during the supernova explosion, or through direct collapse. Consensus is rising that heavy sBHs form in sub-solar metallicity environments, and that in general heavier, lower metallicity stars release heavier sBHs (Belczynski *et al.*, 2010; Spera *et al.*, 2015; Belczynski *et al.*, 2016b). However, at increasingly larger stellar masses, pair instabilities and pair instability supernovae create an *edge* in the sBH mass distribution around $40-60\,\mathrm{M}_\odot$ (Giacobbo *et al.*, 2018), and possibly a *gap* between about $40-60\,\mathrm{M}_\odot$ and $120\,\mathrm{M}_\odot$.[†] For stars with $m_{\rm zams} > 220\,\mathrm{M}_\odot$ and

[†]In metal-poor stars with $Z < 0.7\,\mathrm{Z}_\odot$ (with $\mathrm{Z}_\odot = 0.01524$) and mass $70 < m_{\rm zams}/\,\mathrm{M}_\odot < 150$, pulsational pair-instabilities (PPIs) cause substantial mass loss during explosive oxygen/silicon burning, limiting the maximum mass of the sBH to about $M_\bullet^{\rm sBH} \simeq 40$-$60\,\mathrm{M}_\odot$(Woosley, 2017; Belczynski *et al.*, 2016a; Spera and Mapelli, 2017). Pair instability supernovae (PISNae) appear at lower metallicities, $Z \leq 8\times10^{-3} \simeq 0.5\,\mathrm{Z}_\odot$. In the mass

metallicities below $0.07\,Z_\odot$, the sBH mass function resumes and sBHs with masses in excess of about $120\,\mathrm{M}_\odot$ can form, if very massive stars form in Nature. The mass spectrum of currently observed sBHs is sketched in Fig. 1.

Supermassive black holes (SMBHs) are observed as luminous QSOs and active galactic nuclei (AGN), fed by accretion of gas (Merloni, 2016), or as massive dark objects (MDOs) at the center of quiescent galaxies by tracing the stellar and/or gas dynamics in the nuclear regions (Kormendy and Ho, 2013). The SMBH mass spectrum currently observed extend from about $5\times 10^4\,\mathrm{M}_\odot$ (the SMBH in the galaxy RGG118 (Baldassare *et al.*, 2015)) up to about $1.2\times 10^{10}\,\mathrm{M}_\odot$ (SDSS J0100+2802 (Wu *et al.*, 2015)), as illustrated in Fig. 1.

Observations suggest that SMBHs have grown in mass through repeated episodes of gas accretion and (to a minor extent) through mergers with other BHs. This complex process initiates with the formation of a *seed* BH of yet unknown origin. The concept of seed has emerged to explain the appearance of a large number of SMBHs of billion suns at $z \sim 6$, shining when the universe was only 1 Gyr old (Jiang *et al.*, 2016). In addition, the discovery of the correlation (Kormendy and Ho, 2013) between the SMBH mass $M_\bullet$ and the stellar velocity dispersion σ in nearby spheroids, extended to low values of the velocity dispersion, predicts BH masses of $M_\bullet \sim 10^{4.6}\,\mathrm{M}_\odot$, at σ as low as $40\,\mathrm{km\,s}^{-1}$ (typical of nuclear star clusters) (van den Bosch, 2016). Furthermore, the comparison between the local SMBHs mass density, as inferred from the $M_\bullet - \sigma$ relation, with limits imposed by the cosmic X-ray background light (see chapter 12), resulting from unresolved AGN powered by SMBHs in the mass interval between $10^{8-9}\,\mathrm{M}_\odot$, indicates that radiative efficient accretion played a large part in the building of SMBHs below $z \sim 3$, and that information is lost upon their initial mass spectrum (Marconi *et al.*, 2004). Thus, SMBHs are believed to emerge from a population of seeds of yet unconstrained initial mass, in the range from about $100\,\mathrm{M}_\odot$ to about $10^{4-5}\,\mathrm{M}_\odot$. BHs in this mass interval are customarily referred to as *intermediate mass black holes* (IMBHs).[‡]

The word *seed* embodies two concepts. Seeds are IMBHs that form "early" in cosmic history (at redshift $z \sim 20$, when the Universe is only 180

range about $150 < m_{\mathrm{zams}}/\mathrm{M}_\odot < 220\,\mathrm{M}_\odot$, PISNae lead to the complete disruption of the star, leaving no remnant.

[‡]Here on, we refer to as *seed BHs* those IMBHs that grow to give origin to the population of SMBHs. With *BH seeds* (or better BH progenitors) we refer instead to those stellar configurations, which through core collapse, evolve into IMBHs (or seed BHs) which terminate their evolution as SMBH.

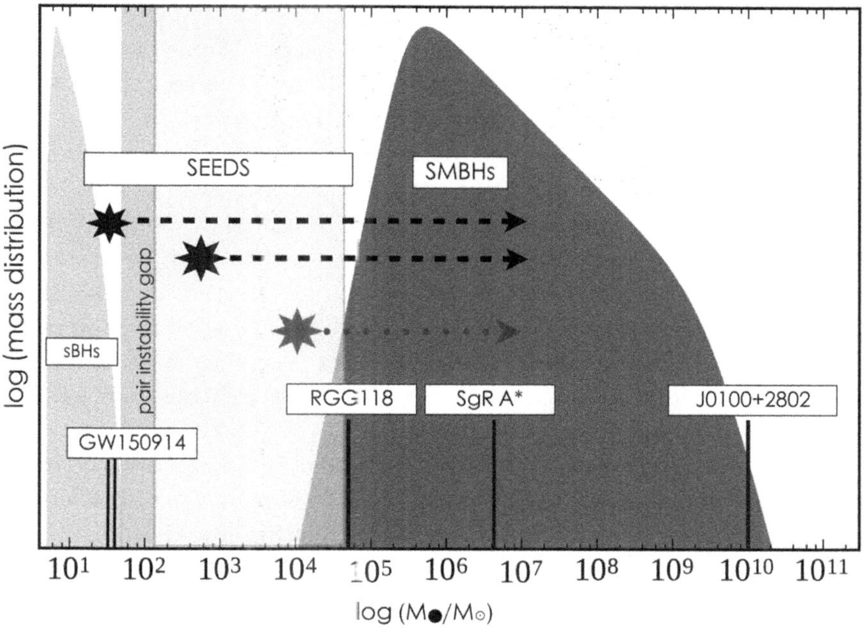

Fig. 1. (Color online) Schematic illustrating the BH mass spectrum encompassing the whole astrophysical relevant range, from sBHs to SMBHs, through the unexplored (light-green) zone where seed BHs are expected to form and grow. Vertical black-lines denote the two sBH masses in GW150914 (Abbott *et al.*, 2016b), the mass $M_\bullet$ of RGG118 (the lightest SMBH known as of today in the dwarf galaxy RG118), of SgrA* in the Milky Way (Genzel *et al.*, 2010), and of J0100+2802 (the heaviest SMBH ever recorded). The mass distribution of sBHs, drawn from the observations of the Galactic sBH candidates, has been extended to account for the high-mass tail following the discovery of GW150914, which host sBHs of $\sim 30\,M_\odot$, in the binary prior to merging. The minimum (maximum) sBHs is arbitrarily set equal to $3\,M_\odot$ ($60\,M_\odot$), and the theoretically predicted pair-instability gap is depicted as narrow darker-grey strip. The SMBH distribution has been drawn scaling their mass according to the local galaxy mass function and $M_\bullet$-σ correlation. The decline below $\sim 10^5\,M_\odot$ is arbitrarily set: BH of $\sim 10^{4-5}\,M_\odot$ may not be ubiquitous in low-mass galaxies as often a nuclear star cluster is in place in these galaxies, which may or may not host a central IMBH (Graham and Spitler, 2009). The black stars and dashed tracks illustrate the possibility that a SMBH forms as sBH-only (born on the left side of the sBH gap) or as light seed (on the right of the gap) which then grows through phases of super-Eddington accretion (Lupi *et al.*, 2016). The red star and dotted track illustrates the possibility of a *genetic* divide between sBHs and SMBHs, and that a heavy seed forms through the direct collapse of a supermassive isolated cloud/protostar in a metal free, atomic-hydrogen cooling, dark-matter halo (Latif *et al.*, 2013c; Schleicher *et al.*, 2013). The seed later grows via gas accretion and mergers with other halos. LIGO-Virgo, the third generation of Ground Based Interferometers, and the LISA space mission will shed light into the physical mechanisms leading to the formation of SMBHs in the unexplored range between sBHs and SMBHs.

Myr old). They form in metal-poor environments under rare conditions, and grow over cosmic time by accretion and mergers. Some of these seeds are able to grow to giant sizes, of up to $10^8 - 10^{10}\,M_\odot$, enabling the presence of luminous QSOs during the epoch of cosmic reionization.

Different formation channels have been proposed for the seeds (Volonteri, 2010; Schleicher *et al.*, 2013; Latif and Ferrara, 2016). *Light seeds* refer to IMBHs of about $100\,M_\odot$ and up to $\sim 10^3\,M_\odot$. Seeds of about $100\,M_\odot$ (above the instability gap) are the relics of the first generation of massive stars of zero or very low metallicity (see Chapter 4), while seeds of $\sim 10^3\,M_\odot$ likely arise from runaway collisions of massive stars in dense star clusters of low metallicity (Mapelli, 2016; Devecchi *et al.*, 2012, see Chapter 7), or from repeated mergers of sBHs in star clusters subjected to gas-driven evolution (Lupi *et al.*, 2014).

Heavy seeds refer instead to IMBHs of about $10^{4-5}\,M_\odot$ resulting from the monolithic collapse of massive gas clouds, forming in $T_{\rm vir} \gtrsim 10^4$ K metal-free halos exposed to an intense H_2 photo-dissociating ultraviolet flux (Latif *et al.*, 2013c; Dijkstra *et al.*, 2014; Habouzit *et al.*, 2016a; Regan *et al.*, 2017, see Chapter 5).[§] The seeds of the first SMBHs are still elusive to most instruments that exist today, preventing us from setting constraints on their nature.

In the current cosmological framework, and according to recent studies (Valiante *et al.*, 2018) that we take here to set a logical flow, light and heavy seeds form around $15 < z < 18$, inside halos of comparable masses of about $10^8\,M_\odot$, and evolve in isolation for their first 150-200 Myrs. Over this lapse time, the halos are enriched with baryons through gas inflows from the external medium and seeds grow via Bondi accretion (with accretion rates $\dot{M} \propto M_\bullet^2$) or at (super-) Eddington rates ($\dot{M} \propto M_\bullet$). Then halo-halo multiple mergers begin at $z \sim 10$. Seed BHs are necessarily a *transient* population of objects and inferring their initial mass function from observations is possible only as long as they preserve some features acquired at birth. As an illustration of this concept, Figure 2 shows the mass distribution of both light and heavy seeds before halos experience multiple mergers (Valiante *et al.*, 2018). Isolated heavy seeds are accreting gas more efficiently than their lighter counterparts. Both populations grow, but as heavy seeds grow faster, the two populations remain distinct over time, and retain information of their origin. As a result of efficient growth and radiative feedback, the host galaxies of heavy seeds experience a less

[§]Heavy seeds might also form in major gas-rich galaxy mergers over a wider range of redshifts, as mergers trigger massive nuclear inflows (Mayer *et al.*, 2015).

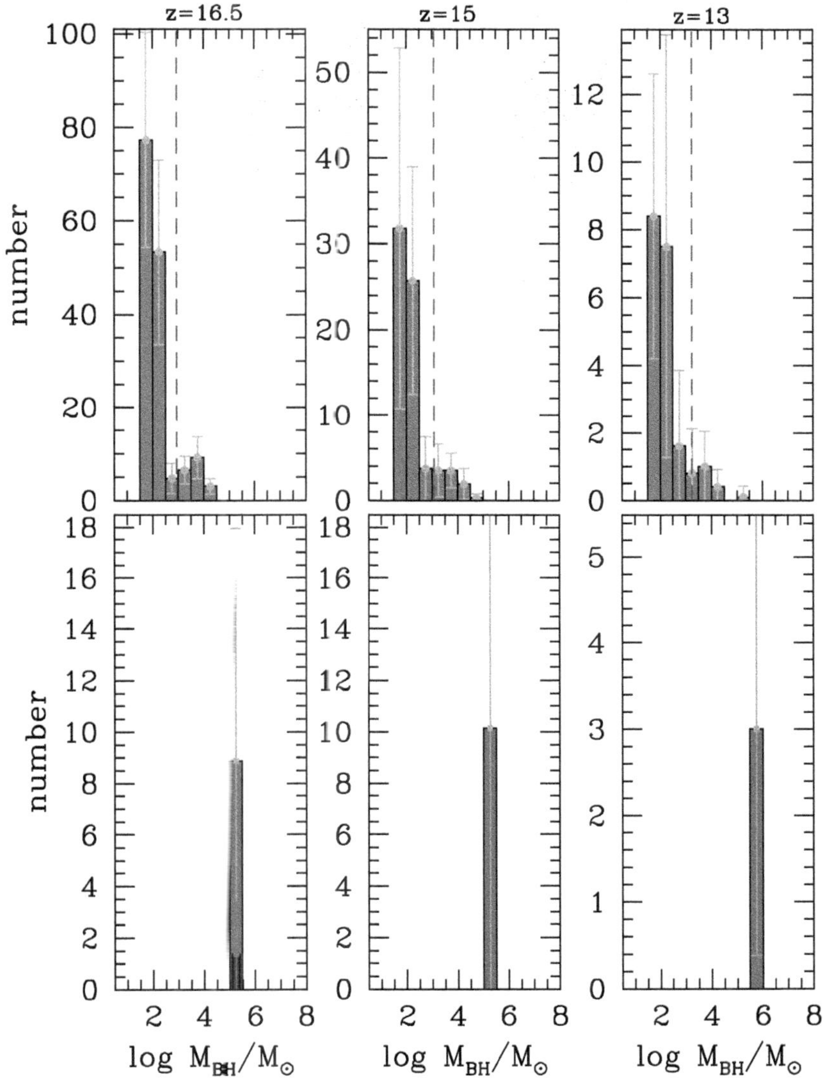

Fig. 2. Mass distribution of *light seeds* (top panel) and *heavy seeds* (bottom panel) born in dark matter halos of $\sim 10^8\,M_{\odot}$ at $z > 16.5$, at three different redshifts $z = 16.5, 15, 13$, in a biased region of the universe selected to explain the existence of SMBHs at $z \sim 7$. The seeds live in isolated halos for about 200 Myrs, before halo-halo mergers occur. During this lapse time, heavy seeds grow faster than light seeds that do not modify their mass spectrum significantly. The dashed lines in the upper panels indicate the average mass of isolated light seed at the corresponding z. Courtesy of R. Valiante and R. Schneider (Valiante *et al.*, 2013).

intense star formation activity, and have lower stellar content and lower metallicities. Below $z < 10$, the fraction of isolated systems dramatically decreases to less than 2% (20%), for heavy (light) seed hosts, as both major and minor mergers occur among halos. Thus, below this redshift *binary seed BHs* might form through halo-halo mergers. As seeds grow via mergers and accretion, it is extremely difficult to infer their initial mass spectrum, as they evolve into massive BHs (MBHs) of $10^6 - 10^7\,\mathrm{M}_\odot$, and some to SMBHs ($10^8 - 10^{10}\,\mathrm{M}_\odot$).

To unveil seeds and track their early growth we need to have access to an overwhelmingly wide cosmological volume and have the capability of measuring at the earliest possible redshifts, their masses and spins. How can this be possible?

According to General Relativity (GR), *BHs of any flavor captured in binaries are loud sources of GWs at the time of their merging. Their GW signal contains direct information on the individual masses and spins of the BHs, and on the source luminosity distance.* Thus, detecting binary BHs through their GW signal over a wide range of masses offers a unique way to learn about how and when seeds form and pair in binaries.

The *Gravitational Wave Universe* is the universe we can sense using GWs as messengers (Colpi and Sesana, 2017; Amaro-Seoane *et al.*, 2017). By observing this universe over a wide interval of GW frequencies, from mHz to kHz, will enable us answering a number of key questions:

- Are sBHs the only elementary building blocks of SMBHs, or is there a "genetic" divide between sBHs and SMBHs, with heavy seeds being the only path to SMBH formation?
- Are there yet undiscovered channels of SMBH formation, calling for the existence of unforeseen, unstable MDOs?
- Can we predict from cosmological models and numerical simulations the mass, spin and mass ratio distributions of seed BHs?
- How fast do seeds grow hand-in-hand with the growth of cosmic structures?
- Can we distinguish at statistical level from the GW data the nature of the seeds, whether light or heavy, and infer if one population dominates over the other?
- How fast has information on the two seed populations been corrupted by the cosmic growth of structures?
- Do seed BHs merge when halos merge?
- How do seed BHs pair in pristine dark matter halos? Is their pairing driven *in-situ* from local dynamics, or induced by halo-halo mergers only?

• What do low redshift MBHs at the centre of dwarf galaxies or IMBHs in dense star clusters tell us about seeds?

This chapter is organized as follows: In § 2 we briefly describe how gas accretion shapes the mass and spin of BHs. § 3 describes shortly the nature of GWs, while § 4 gives an overview of the dynamics of BH binaries and the GW signal they emit. § 5 describes how seed BHs can be detected as GW sources by LISA, and IMBHs by the LIGO-Virgo Observatories in their advanced configurations. In § 6 we show LISA's unique capabilities of detecting BHs in the universe over a yet-uncharted interval of masses and redshifts, while in § 7 we describe shortly the formation channels of binary BHs. Coalescence rates for heavy seeds are given in § 8. Before concluding, we explore briefly in § 9 the possibility of discovering IMBHs of $10^2 - 10^3\,\mathrm{M}_\odot$ retained in globular clusters or/and in nuclear star clusters, and MBHs between $10^4 - 10^6\,\mathrm{M}_\odot$ in dwarf galaxies by detecting their multiband GW signal.

2. Accreting black holes

Seed BHs are *transitional* objects as they are expected to undergo episodes of accretion and mergers that change their mass and spin. Accretion may occur via an optically thick, geometrically thin disc which carries mass and angular momentum, transferred to the BH at the innermost stable circular orbit (ISCO) close to the BH event horizon. The BH spin thus changes in norm and direction (Bardeen and Petterson, 1975). Phases of super-Eddington accretion from a thick disc are possible, which bring to fast mass growth of the seeds in gas-rich gravitational potential wells (Lupi *et al.*, 2016). Accretion at a luminosity L equal to a fraction f_{E} of the Eddington luminosity L_{E} enhances the BH mass on an e–folding time $\tau_{\mathrm{BH,acc}} = 4.7\times 10^8\eta/[f_{\mathrm{E}}(1-\eta)]$ yr, where η is the accretion efficiency, which varies from 0.057 (for $\chi_{\mathrm{spin}} = 0$) to 0.151 (for $\chi_{\mathrm{spin}} = 0.9$) and 0.43 ($\chi_{\mathrm{spin}} = 1$, for a maximally rotating BH). Mass accretion and spin are tightly coupled. A non-rotating (maximally-rotating) BH is spun-up (spun-down) to $\chi_{\mathrm{spin}} = 1$ (to spin zero) after increasing its mass by only a factor $\sqrt{6}$ ($\sqrt{3/2}$) for prograde (retrograde) accretion. If accretion comes in clumps of matter randomly oriented, after a sufficiently long time the BH spin is expected to be low (as retrograde accretion carries a higher angular momentum then prograde accretion). Instead, episodes of coherent accretion in a stable direction leave behind a highly spinning BH (King *et al.*, 2005; Perego *et al.*, 2009). Measuring the spin of coalescing seeds,

IMBHs and MBHs in close binaries bring precious information on the way BHs have accreted gas in the host galaxy (Sesana *et al.*, 2014).

3. Gravitational waves in a nut-shell

According to GR, a GW is a perturbation of the metric tensor $h_{\mu\nu}$ in the 4-dimensional spacetime, which propagates at the speed of light over a stationary (flat) background. In the traceless and transversal gauge, the perturbation $h_{\mu\nu}$, orthogonal to the propagation direction, has vanishing space-time and time-time components so that the GW is characterized by only two linearly independent propagating degrees of freedom, called the cross and plus polarizations, and denoted as $h_\times$ and h_+ respectively. A single detector can measure a linear combination, called *strain amplitude*, $h(t) = F_+h_+(t) + F_\times h_\times(t)$, where F_+ and $F_\times$ are the antenna patterns of the detector, which depend on the source sky position (Sathyaprakash and Schutz, 2009).

Gravitational radiation is sourced by non-symmetrically accelerated masses (the "charges" of gravity), and tracks the large scale coherent motions of cosmic bodies. Thus, there is a similarity with the process of emission of EM radiation from accelerated elementary charged particles. But, no gravitational dipole radiation is emitted, as the time derivative of the mass dipole moment $\mathbf{D} = \sum m_i\mathbf{r}_i$ of a closed system is the total linear momentum which is conserved, i.e. $\ddot{\mathbf{D}} = 0$. Hence, at the leading order in the multipole expansion, gravitational radiation is *quadrupolar*, where the quadrupole moment is of order $Q \sim \varepsilon MR^2$ for a source of mass M, size R, with ε describing the extent of the non-symmetric deformation. Einstein's linearized field equations establish a relation between Q and the strain amplitude $h \propto \ddot{Q}$: a back-of-the-envelope calculation gives $h \sim (R/d)(R_{\rm G}/R)(v/c)^2$, where $R_{\rm G} = 2GM/c^2$ is the Schwarzschild radius associated with the mass M, d the source distance, and v the velocity of the asymmetric mass motion (G and c denote the gravitational constant and the speed of light). A compact self-gravitating source ($R \gtrsim R_{\rm G}$) moving at relativistic speed ($v \lesssim c$) has $h \sim GM/(c^2d)$. Binary black holes (BBHs) at the time of their coalescence have separations R comparable to about $2R_{\rm G}$ (with $M_{\rm B}$ the total mass of the binary) and thus $h \sim 2GM_{\rm B}/(c^2d)$. In typical astrophysical conditions, $h \sim 10^{-22} - 10^{-19}$. The frequency of the GW is related to the inverse of the dynamical time associated to the non-symmetric motion, $f_{\rm gw} \sim \sqrt{R_{\rm G}/R}\,(c/R)$. In the case of a BBH coalescence, $f_{\rm gw}$ is close to $c^3/(GM_{\rm B})$, as described in Section 4.

4. Linking the black hole binary dynamics to the GW signal

In GR, a binary system comprising two masses emits GWs and as GWs carry away orbital energy and angular momentum, the binary is fated to coalesce. Calculating the dynamics and radiative properties of two coalescing BHs requires solving Einstein's non-linear field equations in the framework of the so called "geometro-dynamics" where space-time is foliated into a series of surfaces that allow the notion of time (Lehner and Pretorius, 2014).

The coalescence of a BH binary in vacuum has the beauty of being one of the elementary processes in Nature. It is a well posed problem characterized by a relatively small set of intrinsic parameters: the mass ratio $q \equiv m_2/m_1$, where m_1 and m_2 are the masses of the two BHs in the binary, and the BH individual spins are $\mathbf{S}_1$ (equal to $\chi^1_{\rm spin} G m_1^2/c$) and $\mathbf{S}_2$ defined at large (initial) separations. The initial semi-major axis of the relative orbit (or total energy $E_{\rm B}$), and the modulus of the orbital angular momentum $\mathbf{L}$ (whose direction can be taken as reference axis due to the isotropy of space) define the initial condition and the nature of the orbit (whether it is a head-on collision or a gentle inspiral from an initially bound orbit). The merger product is a new Kerr black hole of mass $M_{\rm f}$ and spin $\mathbf{S}_{\rm f}$. There is no intrinsic scale in vacuum Einstein's gravity. Thus any solution can be re-scaled knowing the binary mass $M_{\rm B} = m_1 + m_2$ under study.

Coalescing BHs produce a universal GW signal which contains unique information on the BH individual masses and spins that source the gravitational field and its vibrations. The GW signal is as a progression of 3+1 phases:

• The *late-inspiral* refers to the phase in which the two BHs can still be considered as structureless and their dynamics and emission can be described by Post Newtonian (PN) theory (Blanchet, 2014). The two BHs in this phase have small orbital velocities compared to the speed of light c. This phase, which is the longer lasting, is crucial in obtaining first estimates of the binary system's parameters. If the BHs carry high spins, spin-orbit and spin-spin couplings lead to in-plane and off-plane spin precessions around the total angular momentum (Schnittman, 2004), and the GW encodes these changes (Santamaría *et al.*, 2010).

• The *merger* refers to the non-linear plunge-in phase, coalescence and early post-merger. Moving at around one third of the speed of light, the two highly deformed BH space-times are described in the realm of Numerical Relativity (NR) as the faithfulness of the PN expansion decreases. The plunge refers to the rapid increase in the magnitude of the inward radial velocity leading to merger, and initiates when the two BHs attain a distance comparable to ISCO, defined here as $R_{\rm ISCO} = 6GM_{\rm B}/c^2$. NR simulations are successful in tracing the non-linear dynamics of the highly distorted vacuum space-time (Lehner and Pretorius, 2014).

• The *ring-down* is the phase describing the rapid settling of the new BH into a quiescent, stationary state described by the Kerr family which is attained after the GWs carry away any residual space-time deformation. The radiation consists of a superposition of quasi-normal modes whose frequencies and decaying timescales depend only on the mass and spin of the BH, according to the "no-hair" theorem (Berti *et al.*, 2016). This phase is tracked using NR simulations and/or perturbation theory.

• A fourth phase of *recoil* exists, as the new BH that forms experiences a natal kick (Zlochower *et al.*, 2011). The gravitational recoil emerges when the two BHs are not symmetric, as in this case GWs carry away linear moment. The asymmetry can be due to unequal masses, unequal spins, or a combination of the two. For non-spinning BHs, the maximum recoil is of $\sim 175\,{\rm km\,s^{-1}}$ when the mass ratio is $q \sim 0.195$. But generic binaries with in-plane spin components may lead to much higher recoil velocities, up to $\sim 4000\,{\rm km\,s^{-1}}$ which arise when equal-mass, maximally spinning BHs merge with spins in the orbital plane equal in magnitude and opposite in direction.

In the next section we focus on key properties of the GW signal in the inspiral phase, particularly relevant when searching for high redshift seed BHs.

4.1. *The inspiral and chirp of black holes in binaries*

The GW signal during the inspiral phase can be represented as a sequence of quasi-closed orbits where the semi-major axis a and eccentricity e both decrease with time, as in GR, orbital energy is carried away by GW radiation faster than orbital angular momentum. Little eccentricity is indeed left at the time of coalescence, unless binaries form in triple interaction (Bonetti *et al.*, 2017). For this reason and for quest of simplicity, we focus only on *circular* binaries. Under this assumption, the GW signal is, to leading order, a monochromatic wave of frequency $f_{\rm gw} = 2f_{\rm K} = \pi^{-1}(GM_{\rm B}/a^3)^{1/2}$ in the source rest frame. $f_{\rm gw}$ depends on the square root of total mass of the binary $M_{\rm B}$ and on $a^{-3/2}$. Thus, depending on the separation a,

BBHs can emit at arbitrarily low frequencies, and only when approaching coalescence, i.e. when a approaches $R_{\rm ISCO}$, the frequency of the GW attains its maximum.

The orbital energy of a binary is a function of a, and can be expressed in terms of $f_{\rm gw}$: $E_{\rm B} = -\nu G M_{\rm B}^2/2a = -[\pi^{2/3}/2]\nu M_{\rm B}(GM_{\rm B})^{2/3}(f_{\rm gw}^{2/3})$, where ν is the *symmetric mass ratio*, $\nu = m_1 m_2/M_{\rm B}^2 = q/(1+q)^2$ (equal to 1/4 for an equal mass binary and invariant under the change of the mass ratio $q = m_2/m_1$ in q^{-1}). As GWs carry away energy, $\dot{E}_{\rm B} < 0$, the binary contracts. During inspiral, the semi-major axis decreases, the Keplerian frequency increases and thus $f_{\rm gw}$. To leading order, the key (orbit-averaged) equation that rules the binary contraction during the quasi-adiabatic, long-lived inspiral, is

$$\dot{E}_{\rm B} = -P_{\rm GW} \equiv -\frac{32}{5}\frac{c^5}{G}\left(\frac{GM_{\rm c}}{c^3}\pi f_{\rm gw}\right)^{10/3} \tag{1}$$

where the right-hand side gives the radiated power in GWs, $P_{\rm GW}$, for a circular binary.[¶] Equation (1) contains the so called binary *chirp* mass

$$M_{\rm c} \equiv \frac{(m_1 m_2)^{3/5}}{(m_1+m_2)^{1/5}} = \nu^{3/5} M_{\rm B} = \frac{q^{3/5}}{(1+q)^{6/5}} M_{\rm B}^{2/5}, \tag{2}$$

i.e. a combination of the individual BH masses, expressed in terms of the symmetric mass ratio ν.

As $E_{\rm B}$ and $f_{\rm gw}$ can be expressed in terms of the semi-major axis a, equation (1) can be recast to give the rate of change of $\dot{a}$. It is then a simple exercise to determine the *coalescence time*, defined as the time it takes a BH binary, driven by GW losses, to reach null separation (in the point mass approximation), knowing that the binary formed with an initial semi-major axis $a_\circ$, set by "astrophysical" conditions. For a circular binary

$$t_{\rm coal} = \frac{5}{256}\frac{c^5}{G^3}\frac{a_\circ{}^4}{\nu M_{\rm B}^3} = \frac{5\cdot 2^4}{256}\frac{1}{\nu}\frac{GM_{\rm B}}{c^3}\tilde{a}_\circ^4, \tag{3}$$

where $\tilde{a}_\circ$ is the binary semi-major axis in units of $R_{\rm G} = 2GM_{\rm B}/c^2$. [If the binary is eccentric with initial eccentricity $e_\circ$, the coalescence time is shorter by a factor $G(e_\circ)(1-e_\circ^2)^{7/2}$, with $G(e_\circ)$ varying from 1 (for $e_\circ = 0$) to 1.80 (for $e_\circ = 1$).]

Due to the weakness of the gravitational coupling constant and the steep dependence of $\tau_{\rm coal}$ on $a_\circ$, the coalescence time can be very long,

[¶]We defer the reader to the book by M. Maggiore, *Gravitational Waves* (Oxford University Press), for a detailed derivation of all the equations presented in this Chapter.

unless the binary is initially extremely eccentric. A (circular) binary of $M_{\rm B} = 10^5\,{\rm M}_\odot$ ($10^3\,{\rm M}_\odot$) reaches coalescence in 0.27 Gyrs, corresponding to the cosmic time at redshift $z \sim 15$, if the two BHs have a separation of $a_\circ \sim \nu^{1/4}\,1.5 \times 10^4 R_{\rm G}$ ($\nu^{1/4} 4.84 \times 10^4 R_{\rm G}$), corresponding to $\nu^{1/4}$ 30 AU ($\nu^{1/4}$ 0.1 AU). Reaching these separations requires the presence of non-GW dissipative processes, active when the binary interacts with the environment or when the progenitor stars undergo a fine-tuned evolution, as discussed in Section 7.

Coalescing seed BHs are at cosmological distances. Thus the GW signal is affected by the expansion of the universe. If z is the cosmological redshift of the source of GWs, it is possible to show that (i) the observed frequency, denoted for simplicity as f here-on, is redshifted with respect to the frequency $f_{\rm gw}$ as measured in the source frame, $f = f_{\rm gw}/(1+z)$; (ii) the chirp mass in the source frame $M_{\rm c}$ is replaced by the redshifted chirp mass $\mathcal{M}_{\rm c} = (1+z)M_{\rm c}$; (iii) the source distance d is identified with the luminosity distance d_L; and (iv) the time to coalescence as measured in the observer frame is $\tau_{\rm coal} = (1+z)t_{\rm coal}$. Note that the combination $\mathcal{M}_{\rm c} f = M_{\rm c} f_{\rm gw}$ is redshift independent.

Equation (1) can be recast to determine the rate of change of the frequency of the GW, $f_{\rm gw}$, during the inspiral phase, in the source frame. If one includes PN terms and refers to the rate of change of f in the observer frame, the frequency is chirping at a rate

$$\dot{f} = \frac{96}{5}\pi^{3/8}\left(\frac{G\mathcal{M}_{\rm c}}{c^3}\right)^{5/3} f^{11/3}\left[1+\mathcal{D}^{\rm PN}\right], \tag{4}$$

where $\mathcal{D}^{\rm PN}$ is the PN correction, up to the desired order higher than leading (Blanchet, 2014). $\mathcal{D}^{\rm PN}$ depends on the initial phase and coalescence time. It can be expanded analytically in powers of ν and of $\mathcal{M}_{\rm c} f$. Spin effects can also be inserted in $\mathcal{D}^{\rm PN}$ [Blanchet (2014)].

The frequency evolution of the GW is determined by the chirp mass $\mathcal{M}_{\rm c}$, according to equation (4). Thus, measuring $\dot{f}$ provides directly the measure of the redshifted chirp mass associated to the binary. The PN corrections present in $\mathcal{D}^{\rm PN}$ help lifting the degeneracy in the intrinsic parameters of the source present in the signal, and in particular the individual masses in the observer frame.

If we neglect for simplicity the PN term in equation (4), we can estimate the residence time of an inspiraling BBH measured by the observer's clock, when it enters the sensitivity window at a frequency $f_{\rm in}$ and exits at $f_{\rm out}$:

$$\tau_{\rm det} = \frac{5}{256\,\pi^{8/3}}\frac{1}{\nu}\left(\frac{c^3}{G\mathcal{M}_{\rm c}}\right)^{5/3}\left(\frac{1}{f_{\rm in}^{8/3}} - \frac{1}{f_{\rm out}^{8/3}}\right). \tag{5}$$

Equation (5) shows that an unequal mass binary with $\nu \ll 1$, and/or a binary with a smaller chirp mass stay in band much longer covering many cycles. A back-of-the-envelope calculation shows that the number of cycles that a BBH covers before merging scales as $\mathcal{N}_{\rm cycles} \propto \nu^{-1} M_{\rm B}^{-5/3} f_{\rm in}^{-5/3}$.

The chirp signal contains further information on the binary, as it is accompanied by an increase of the amplitude h of the GW. According to GR, the GW signal from an inspiraling binary can be decomposed in its plus and cross polarization states, which take the following form in the Fourier domain:

$$\tilde{h}_+(f) = \left(\frac{5}{12}\right)^{1/2}\frac{1}{\pi^{2/3}}\frac{c}{d_L}\left(\frac{G\mathcal{M}_{\rm c}}{c^3}\right)^{5/6} e^{i\Phi_+^{\rm PN}(f)}\frac{1}{f^{7/6}}\left(\frac{1+\cos^2\iota}{2}\right) \tag{6}$$

$$\tilde{h}_\times(f) = \left(\frac{5}{12}\right)^{1/2}\frac{1}{\pi^{2/3}}\frac{c}{d_L}\left(\frac{G\mathcal{M}_{\rm c}}{c^3}\right)^{5/6} e^{i\Phi_\times^{\rm PN}(f)}\frac{1}{f^{7/6}}\cos\iota \tag{7}$$

where $\Phi_+^{\rm PN}(f) = \Phi_\times^{\rm PN}(f) - \pi/2$ are the phases, computed at the desired PN order, and ι is the binary inclination angle, i.e. the angle between the orbital angular momentum $\mathbf{L}$ and the line of sight. The polarization of the GW has a direct relationship to the motions of the BHs projected on the observer's sky plane, and the $h_+/h_\times$ gives directly the inclination angle ι. When the binary is seen edge-on ($\iota = \pi/2$), $h_\times = 0$, radiation has $+$ polarization only, since from the observer's view the motion of the two BHs projected on the sky is linear. When $\iota = 0$, the binary is seen face-on and the BHs execute a circular motion in the sky. In this case both polarization components have equal amplitude and are out of phase by $\pi/2$.

If the GW detector (or a network of detectors) enables the measure of the two independent polarization states of a GW, from the chirp signal (equations 4, 6, and 7) one can infer the luminosity distance d_L of the source. From the ratio $h_+/h_\times$ we obtain the inclination angle $\cos\iota$, from the frequency chirp the redshifted chirp mass, so that from the measured value of h_+ (or $h_\times$) we can read off d_L. Coalescing BBHs are gravitational standard candles, or as often said, *standard sirens*. The application of the luminosity distance-redshift relation (d_L versus z) from the current cosmological model provides the redshift of the source $z_{\rm source}$ and the inference of the source-frame physical parameters. But, if an EM counterpart is associated to and identified with the GW source, it is then possible to have a direct measure of $z_{\rm source}$ and this allows for an independent measure of the Hubble expansion parameter to a certain degree of accuracy (Tamanini *et al.*, 2016).

4.2. *The merger and the newly formed black hole*

The *merger* occurs in the strong field regime when the space-time of the two plunging BHs is violently changing, and NR simulations are the only tool to track the dynamics and extract the GW signal in the far field. The observation of the first coalescing BBHs (Abbott *et al.*, 2016b) has shown that the GW signal is extremely simple at merger: after the chirp, the amplitude of the GW is observed to decay rapidly, as predicted by NR simulations and perturbation theory. The frequency derivative is no longer described by equation (4), as it coasts to a finite value, when the new BH forms. This maximum frequency, derived from NR simulations, is close to twice the Keplerian frequency at the ISCO for a binary of total mass $M_{\rm B}$,

$$f_{\rm gw,max} \sim \frac{c^3}{\pi 6^{3/2} G M_{\rm B}} = 4.4 \times 10^3 \left(\frac{{\rm M}_\odot}{M_{\rm B}}\right) \,{\rm Hz}. \qquad (8)$$

The simple scaling of $f_{\rm gw,max}$ with $M_{\rm B}$ shows that detecting coalescing sBHs of $10\,{\rm M}_\odot$ requires interferometers operating at frequencies 10^4 times higher than those requested to detect IMBHs in a binary with $M_{\rm B} = 10^5\,{\rm M}_\odot$.

The merger of BBHs carries tremendous luminosity on the order of $L_{\rm GW} \approx \nu\,(c^5/G) \sim 3.6 \times 10^{59}\,\nu$ erg s^{-1} which is independent on $M_{\rm B}$, as in GR mass $M_{\rm B}$ is a length, and time has the same dimension. This huge luminosity is emitted for a short time lapse, in the immediate vicinity of the merger proper. What depends on the mass scale is the total energy radiated away in GWs. It arises from two sources: the gravitational binding energy liberated during the inspiral and plunge, and the energy present in the geometry of the deformed, rotating relic before it settles into a new Kerr BH. As a rule of thumb, the radiated energy is approximately equal to the binding energy at ISCO of a particle with reduced mass μ (where $\mu = \nu M_{\rm B}$), and can account for $3.8\% - 42\%$ of μc^2 depending on the spin of the primary (i.e. heavier) BH and whether the orbit is prograde or retrograde, relative to the spin direction.

If the inspiral, merger and ringdown of a coalescing binary is fully in band and the source is detected in both polarization states with a high signal-to-noise-ratio, then it is possible to carry on highly accurate measurements of the intrinsic and extrinsic parameters of the source, as the luminosity distance, the individual masses and spins of the BHs, and the final mass and spin of the new BH, in the source frame.‖ Fitting formulae instructed by NR simulations exist (Barausse *et al.*, 2012) that help inferring the final mass of the new BH, given approximately by the difference between the initial mass $M_{\rm B}$ and the energy carried away by the GWs. Likewise, the final spin $\mathbf{S}_{\rm f}$ can be viewed to lowest order as the sum of the orbital angular momentum $\mathbf{L}$ at ISCO plus the intrinsic angular momenta of the two BHs. For equal-mass non spinning BBHs, the final spin is $\chi_{\rm spin,f} = 0.6865$.

4.3. *Extracting the signal from the noise*

Match filtering is the technique used to extract the source signal from the noise. The signal-to-noise ratio (SNR) (or ρ) of a source is computed as

$$({\rm SNR})^2 = \int_0^\infty d\ln f \, \frac{|2\tilde{h}(f)\sqrt{f}|^2}{S_{\rm noise}} \tag{9}$$

where $S^{1/2}_{\rm noise}$ is the *spectral strain sensitivity* or *spectral amplitude* (with dimension $\mathrm{Hz}^{-1/2}$) describing the noise in the detector, and $\tilde{h}(f) = F_+\tilde{h}_+(f) + F_\times\tilde{h}_\times(f)$, (in units of Hz^{-1}). The SNR accumulates over the observing time, until the source rapidly fades away, as in the case of a coalescing BH binary, or exits from the observing bandwidth. Customarily, $(fS_{\rm noise})^{1/2}$ and $f\tilde{h}(f)$ are plotted in the same diagram as function of frequency f, to describe the sensitivity window of an interferometric experiment and the strength of the source signal, respectively. During BBH inspiral $f\tilde{h}(f) \propto f^{-1/6}$ decreases with increasing f, as the source completes fewer cycles in a given frequency interval. But, the decrease in $f\tilde{h}$ is compensated by the increase in the SNR, which accumulates over time up to $f_{\rm max}$ or $f_{\rm out}$. Typically, merging BBHs cover from a few and up to many cycles $\mathcal{N}_{\rm cycles} \sim O(10) - O(1000)$ depending on their mass and on the value of the frequency $f_{\rm in}$ at the time of entrance in the interferometer.

5. High redshift binary seed black holes and low redshift intermediate mass black holes

The *Gravitational Wave Universe* is the universe we can explore using GWs as new messengers over a wide range of frequencies. Coalescing *heavy* seeds

‖A current and powerful tool used to carry on parameter estimation is the Effective One Body (EOB) theory, where re-summed PN inspiral waveforms are smoothly attached to quasi-normal ringdown modes via a transition function calibrated by NR (Buonanno and Damour, 1999; Buonanno and Sathyaprakash, 2014).

and MBHs of up to a few $10^7\,M_\odot$ are the targets of the Laser Interferometer Space Antenna (LISA) operating in the 0.1 mHz-0.1 Hz interval [Amaro-Seoane *et al.* (2013, 2017)]. *Light* seeds are the targets of the third generation of ground-based detectors (III GBI) as the Einstein Telescope [Sathyaprakash *et al.* (2012); Abbott *et al.* (2017a)], sensitive between about 1 Hz (or a few Hz) and several 10^3 Hz, that will let us detect the signal from sBHs and IMBHs in binaries out to high $z \sim 5-6$. sBH binaries with total masses between $60\,M_\odot$ and $300\,M_\odot$ can sweep through the LISA frequency domain years to months before coalescing in the kHz bandwidth [Sesana (2016)]. Advanced LIGO and Virgo interferometers operating in the 10–10^3 Hz interval [Abbott *et al.* (2016c)] will have the capability of detecting binary IMBHs of $300\,M_\odot$ out to $z \sim 1$, setting the conditions for the early search of IMBHs in the universe.

Figure 3 shows the GW signal $f\tilde{h}$ from seed BHs in circular, equal-mass binaries with (source-frame) total mass $M_B = 10^5\,M_\odot$ (solid lines) and $10^4\,M_\odot$ (dashed lines) in blue and light blue colors as they sweep across the LISA band. The signal is shown varying the redshift of the source from $z \sim 13$ down to $z \sim 3$. For BH binaries with $M_B = 10^5\,M_\odot$, the coalescence signal is in band enabling accurate source parameter estimations. For BH binaries of $10^4\,M_\odot$, LISA can only detect the inspiral phase and measure the chirp mass and precession effects if spin are high and misaligned. These seeds are sufficiently light to attain their maximum frequency out of the LISA sensitivity band. Their signal stands above the bucket of the sensitivity curve with accumulated SNR of $\lesssim$50, typically. Figure 3 shows also the signal (solid lines in yellow and dark green colors) from circular, equal-mass binaries with source-frame total mass $M_B = 300\,M_\odot$ as they transit across the LISA sensitivity band during the inspiral phase and enter the LIGO-Virgo band where they merge. Only relatively close-by ($z < 1$) IMBHs in equal mass binaries with individual mass of $\sim 150\,M_\odot$ (just above the pair-instability gap) can be detected both by LISA and LIGO-Virgo. Figure 4 shows the range where both space and Earth-based interferometers can operate jointly in the search of heavy sBHs and IMBHs, in the mass interval between $10\,M_\odot$ up to several $10^3\,M_\odot$. At the redshifts $z \lesssim 1$, IMBHs may not represent unevolved light seeds, and may describe a population of IMBHs that form in dense star clusters that build up their present mass, during the cosmic evolution of their host galaxies.

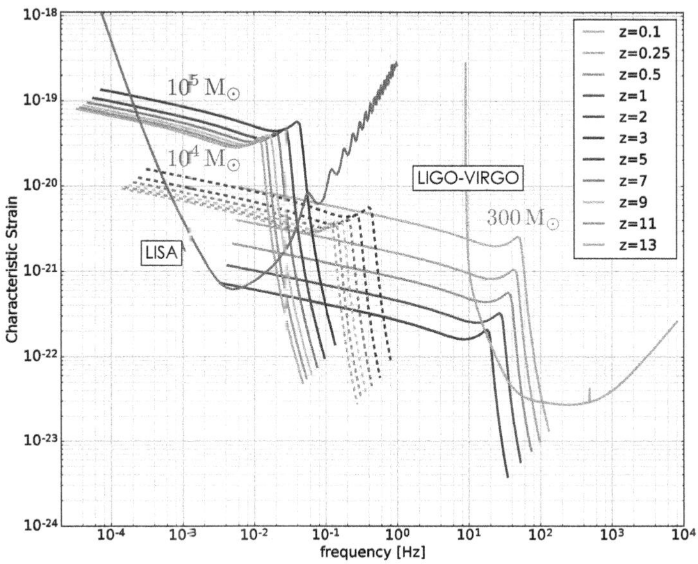

Fig. 3. (Color online) The curve labeled LISA depicts the sensitivity curve of the LISA observatory in its current design (Amaro-Seoane *et al.*, 2017), expressed in terms of the strain noise $(f S_{\rm noise})^{1/2}$ as a function of the observer-frame frequency f. The GW signal from coalescing seed BHs in circular binaries at selected redshifts is described in terms of the sky and inclination averaged characteristic strain $f\tilde{h}(f)$ (in dimensionless units) inferred using non-precessing waveform models (Santamaría *et al.*, 2010): upper solid and dashed lines in blue and light blue colors refer to signals from BBHs with mass ratio $q = 1$ and total rest-frame mass $M_{\rm B} = 10^5\,{\rm M}_\odot$ and $10^4\,{\rm M}_\odot$, respectively. The two family of lines refer to BH binaries detected at different redshift, as indicated in the inset, from $z = 3$ up to $z = 13$. Lower yellow and dark-green solid curves refer to IMBH binaries with total mass $M_{\rm B} = 300\,{\rm M}_\odot$ and $q = 1$; colors are associated to different redshifts varying between 0.1 and 2, as indicated in the inset. Notice that for each BBH of given mass $M_{\rm B}$, the frequency at which $f\tilde{h}(f)$ is maximum (i.e. near coalescence) shifts to lower values with increasing redshift due to cosmic expansion. Seeds of $10^4\,{\rm M}_\odot$ do not coalesce in the LISA band, and can be detected during their inspiral phase only, out to very large redshifts. IMBH binaries with mass of $300\,{\rm M}_\odot$ are detectable first in the LISA band, during their slow adiabatic contraction, and later at the time of coalescence in the LIGO-Virgo band, only if they are at low redshift $z \lesssim 1$ (as also described in Figure 4).

6. The black holes of the gravitational wave universe

Figure 5 illustrates the vastness of the cosmic horizon probed by the LISA observatory, in the hunt of heavy seed BHs and of MBHs. Contour lines of constant (sky averaged) SNR are shown in the $(M_{\rm B} - z)$ plane, for spinning

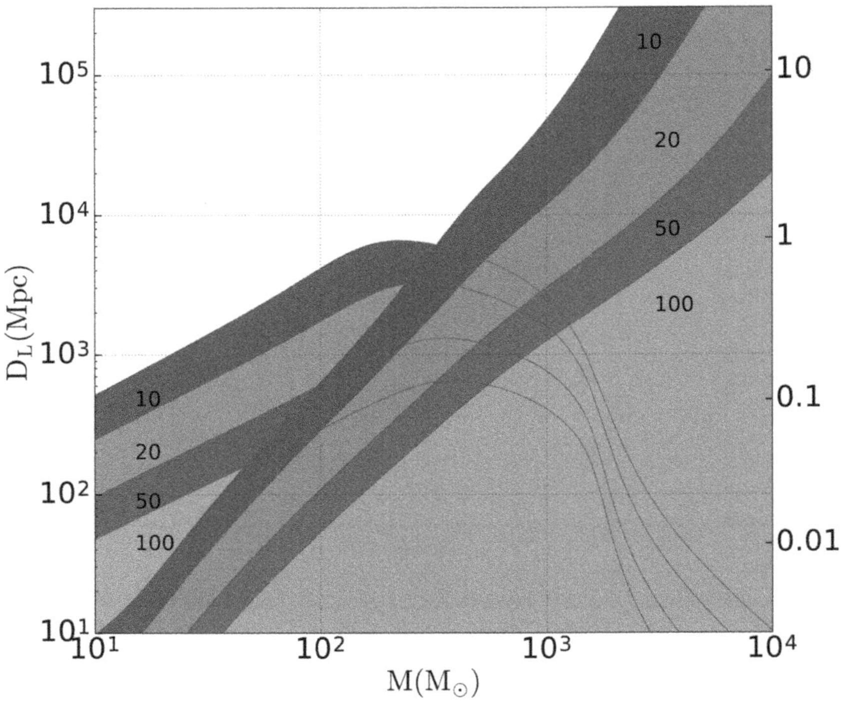

Fig. 4. $D_L - M$ plane, where on the left y-axis D_L is the luminosity distance (in units of Mpc), on the right $y-$axis, the redshift z, and on the $x-$axis, the total mass of the binary $M_{\rm B}$ in the the source-frame. Lines of constant sky-and-inclination averaged SNR are drawn for non-spinning, equal mass binaries hosting sBHs and IMBHs. The SNR is computed using the waveform models of Santamaría *et al.* (2010). The figure illustrates the synergy between LISA and advanced LIGO in the common effort to search for low redshift sBH and IMBH binaries. LISA detects the sources in the inspiral phase, whereas LIGO in the merger proper.

(non-precessing) BHs in binaries with mass ratio $q = m_2/m_1 = 0.2$. We recall that the SNR scales linearly with $\nu = q/(1+q)^2$, thus binaries of equal mass carry the highest SNR. Astrophysical considerations on BH sinking in galactic mergers suggest that the likely interval for q is between 0.1 and 1 (Colpi, 2014).

As shown in Figure 5, LISA is dominated by high SNR sources, and can reveal BBHs up to $z \sim 20$ in a large mass interval between a few $10^3\,{\rm M}_\odot$ to a few $10^6\,{\rm M}_\odot$, where (mostly heavy) seeds are. Around $z \sim 2-3$, during the peak of the star formation rate and AGN activity in the universe (Madau and Dickinson, 2014; Merloni, 2016), LISA can detect the inspiral of IMBHs of about $10^3\,{\rm M}_\odot$, the full signal of BBHs up to $10^7\,{\rm M}_\odot$, and the merger and

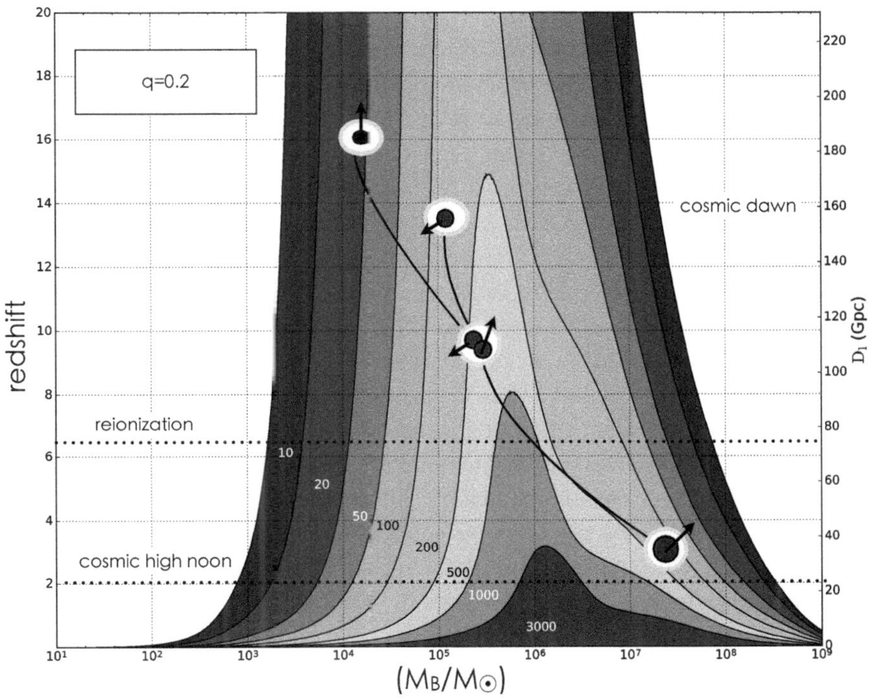

Fig. 5. The $z - M_{\rm B}$ plane, where z is the redshift and $M_{\rm B}$ the rest-frame BBH total mass in units of $M_\odot$. The luminosity distance D_l is plotted on the right of the y-axis. Lines of constant (sky-polarization-and-inclination averaged) SNR are drawn for spinning (non-precessing) BBHs, with mass ratio $q = 0.2$. The spins are aligned with the orbital angular momentum and have equal magnitude $\chi_{\rm spin} = 0.9$. The SNR is computed using the waveform models of Santamaría *et al.* (2010). Dotted lines refer to the redshifts of: *cosmic high noon* ($z \sim 2$, corresponding to the peak of the averaged star formation rate and AGN activity); *cosmic reionization* at about $z \sim 6$, corresponding to the end of the epoch when interstellar hydrogen transited from being neutral to being ionized); and *cosmic dawn* (between $20 < z < 9$, where the first luminous objects start to form). We inserted a BH evolutionary track to show that BHs inevitably cross the LISA bandwidth. Black dots represent BHs, the arrows the spin vectors **S**. BHs are embedded in galactic halos (white circles) and experience episodes of accretion (black lines) and coalescences inside merged halos, which increase the mass and modify the spin. The selected track illustrates the formation of a heavy BH active during cosmic high noon.

ringdown signal of eccentric low-redshift BBHs of $10^8\,M_\odot$, complementing observations from future EM surveys (Nandra *et al.*, 2013).

BHs cross the LISA bandwidth as mergers are inevitable in our hierarchical universe [Amaro-Seoane *et al.* (2013)]. BBH coalescences pinpoint places where galaxy mergers occur. Thus, LISA provides a new and unique way to explore galaxy clustering. This is illustrated in Figure 5 where we depicted two heavy seeds hosted in their dark matter halos, merging around $z \sim 10$. Both seeds increase their mass before coalescing, fed by gas. We allowed the new BH to grow by accretion to deposit a MBH of a few $10^7\,\mathrm{M}_\odot$ detectable as AGN during cosmic high noon.

In virtue of the extremely high SNR of most of the events, the BH parameters will be extracted with exquisite precision (Klein *et al.*, 2016). For the loud sources with SNR$>$ 60, individual redshifted masses can be measured with an error of 0.1% on both components. Even more importantly, the spins of the BHs can be determined to an absolute uncertainty down to 0.01 (Klein *et al.*, 2016), in the loud sources. Information on the astrophysical evolution of massive BHs is encoded in the statistical properties of the observed population. As first illustrated in Sesana *et al.* (2007) observations of multiple BBH coalescences can be combined to learn about their formation path and cosmic evolution. In particular the mass distribution of the ensemble of observed events encodes information on the nature of the first seeds, whereas the spin distribution constrains the primary mode of accretion (whether chaotic or coherent) (Berti and Volonteri, 2008).

The distinctive high SNR of BBH mergers allows for BH spectroscopy, i.e. the direct measure of several frequencies and damping times associated to the quasi-normal modes present in the ringdown signal of the new-born MBH (Berti *et al.*, 2016). This will enable direct precision tests of the "no-hair" theorem in the dynamical sector. Violations of GR prediction indicate new physics, the presence of exotic dark objects and indirect quantum gravity effects (Berti *et al.*, 2015; Cardoso and Pani, 2017).

7. Formation channels of binary black holes and time delays

Detecting seed BHs at high redshifts as GW sources poses a stringent requirement: their existence in binaries in a *GW-hard* state, i.e. in binaries with semi-major axis and eccentricity such that the coalescence time $t_{\rm coal}$ given by equation (3) is a fraction of the cosmic time at the redshift of formation, or equivalently a fraction of the then Hubble time. (GW-hard binaries driven by the sole GW emission are the ones described in § 4.) This level of hardness can be computed posing $t_{\rm coal}(a_\circ, e_\circ;\, z) = t_{\rm Hubble}(z)$. The equality provides the values of the semi-major axis $a_\circ$ and eccentricity $e_\circ$ that a binary should have to be GW-hard, at some stage of its evolution.

A binary is in a *GW-soft* state if its coalescence time at formation is longer than the Hubble time. Some never coalesce, others transits to a GW-hard state. The transition from a GW-soft to a GW-hard state is possible and regulated by dissipative processes of non-GW origin. Hydrodynamical or/and stellar torques acting on the binary control this transition and their strength and effectiveness depend on the environment in which the binary is embedded (Colpi, 2014).

In Nature, binaries are expected to form in the GW-soft state, and this leads to the concept of *time delay*, representing the time lapse between the time of formation of the binary $t_{\rm form}$ at a given redshift z and the time of coalescence as measured in the source frame. Long time delays make it possible to detect sBH which formed at a much earlier epoch, when the host, now incorporated in a larger galaxy, was metal-poor. This is the case of GW150914 (Schneider *et al.*, 2017). But detecting seeds at high redshifts requires binaries to form GW-hard.

Little is known about how seed BHs pair to form GW-hard binaries in the high-z universe. They may form in-situ or via halo-halo mergers. Heavy seeds in GW-hard binaries can form in-situ, i.e. within a halo, if the progenitor is a single supermassive star rotating differentially and subjected to the $m = 2$ mode instability. Hydrodynamical simulations, which incorporate the dynamical space-time evolution, show that an $m = 2$ perturbation nested inside the flow can grow, leading to the formation of a binary of two highly spinning heavy seeds which swiftly merge in a short time (Reisswig *et al.*, 2013). Light seeds forming inside massive ($> 10^7\,{\rm M}_\odot$) dense star clusters of low but not-null metallicity (Devecchi *et al.*, 2012), might pair and form GW-hard binaries through the dynamical channel as they gradually harden via transient three-body or four-body encounters, leading to frequent substitutions of the sBH/IMBH companion, large amplitude oscillations of the eccentricity, and finally to coalescence (Mandel *et al.*, 2008; Haster *et al.*, 2016). Intermediate mass ratio inspirals (so called IMRIs) involving a binary system with a sBH and an IMBH may form in Nature and be detectable as GW sources at relatively low z. These events however may not represent the mergers of light seed BHs in the high redshift universe, and only deep observations with LISA will let us distinguish among possible pathways.

An alternative avenue of formation of heavy seeds in binaries is via halo-halo mergers whose dynamics is influenced by the evolving cosmological background. At present, cosmological simulations do not have the force resolution to track in full detail the dynamics of the seed BHs embedded in their halos (from 10^3 parsecs scale down to a few AU), nor their accretion and associated feedback, which are implemented via sub-grid prescriptions at scales of several parsecs. There exists a sub-grid advection scheme in which BHs are artificially dragged toward the halo center and then forced to merge promptly as the two halos merge. Recent simulations by Tremmel *et al.* (2015), which include fine-tuned dynamics to account for unresolved dynamical friction, find that MBHs from merging dwarf galaxies can spend significant time away from the centre of the remnant galaxy. In future, improving the modeling of seed/IMBH/MBH orbital decay will help in making predictions on the growth, detectability, and merger rates of BHs over a wide mass spectrum and redshifts.

Dedicated, higher resolution numerical simulations of merging galaxies (not in a full cosmological context) indicate that in gas-rich galaxy-galaxy merger the MBHs sink to the center-most regions of the new galaxy to form a Keplerian binary (when the mass in stars and gas inside the BH orbit becomes comparable to $M_{\rm B}$), if the two galaxies have comparable mass. In unequal mass mergers, sinking of the BHs is effective if they are surrounded by a massive stellar cusp that formed during the merger. This process enhances the BH effective mass, and stellar dynamical friction becomes more efficient in braking the BH orbits (Callegari *et al.*, 2009). Due to the richness and variety of galaxy encounters, there is the possibility that in gas-rich galactic discs one or both MBHs are scattered away from the disc plane in their dynamical interaction off massive clouds. Alternatively, one MBH can be trapped in a massive cloud, and this leads to a broadening of the sinking times, which vary from a few hundreds of Myrs to a few Gyrs. The BH binary that forms then hardens via scattering off of individual stars which extract orbital energy and angular momentum (the so called sling-shot mechanism). This process is rapid as long as there is a sufficiently large reservoir of stars on loss cone orbit, in a tri-axial remnant. Processes reminiscent of Type II planet migration (Haiman *et al.*, 2009) can further harden the binary on scales below the mpc if the BHs are surrounded by a circum-binary gas disc. Gravitational gas-dynamical torques are able to taxi the BHs into the GW-driven regime. At present there is only one simulation capable of tracking the full dynamical history of a 10^8 $M_\odot$ BH binary from the kpc scale down to merger proper (Khan *et al.*, 2016). (For a review on the BH dynamics in minor and major galaxy mergers we refer to Colpi (2014).)

8. Coalescence rates of binary black holes in LISA

The expected rate of BBH coalescences is weakly constrained by current EM observations of galaxy-galaxy mergers with MBHs at their center. Estimating the rate is extremely difficult as it depends on a variety of processes difficult to model, among of which are: (i) the mass spectrum of the seeds and their initial occupation fraction in galactic halos; (ii) their BH mass and spin accretion history; (iii) the type of galaxy hosts in which the BHs inhabit; (iv) the dynamics of pairing and hardening of the BHs in the new galaxy halo, depending on the gas and stellar content of the progenitor halos/galaxies and on their orbits in a cosmological context; (v) the extent of GW recoils which may decrease the occupation fraction of BHs in galaxies that can further merge.

Figure 6 shows the rates expected from three models associated to different seeding mechanisms described in the caption (Klein *et al.*, 2016).

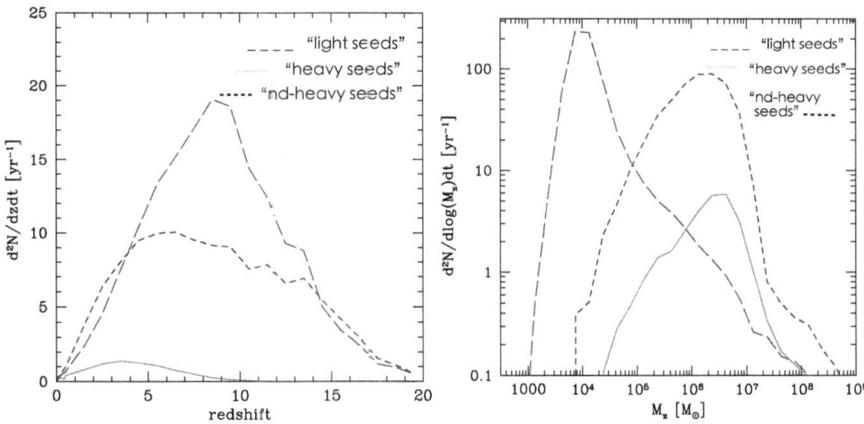

Fig. 6. Predicted BBH coalescence rates per unit redshift (left panel) and per unit binary mass (observer-frame) $M_z = (1+z)M_{\rm B}$ (right panel) for three seed models, from Klein et al. (Klein *et al.*, 2016). The semi-analytical model used follows the evolution of baryonic structures along a dark-matter merger tree produced by an extended Press-Schechter formalism, modified to reproduce results of N-body simulations[Klein *et al.* (2016)]. In the light-seed scenario, seeds are drawn from a log-normal distribution for the progenitor stars centered around $300\,{\rm M}_\odot$ with a rms of 0.2 dex and the exclusion region between $140 - 260\,{\rm M}_\odot$ corresponding to the onset of pair instabilities. Seed BHs carry a mass equal to 2/3 of the mass of the progenitor star. The seeds are implanted in rare massive halos collapsing from the 3.5σ peaks of the primordial density field and are allowed to accrete at twice the Eddington limit. In the heavy-seed scenario, seed BHs carry masses of $10^5\,{\rm M}_\odot$ at form before redshift $z \sim 15$. Both light and heavy seed models account for delays between the formation of the binary in a halo-halo merger and the time of its merging. GW driven coalescence acts on very small galactic scales and dissipative mechanisms driven by the interaction of stars or gas with the binary set the conditions for its coalescence on timescales less than the current age of the universe. Delays have been distributed over a range that varies according to the gas or/and stellar content [rapid (long) when gas (stars) is (are) present, i.e. of order 10^8 yr (10^9 yr)] (see Antonini *et al.* (2015) for details). The figure illustrates how sensitive the merger rates to the physical models are for the seeds and delay time distributions. Courtesy of A. Klein (Klein *et al.*, 2016).

The difference in the merger rates among the models is due to the different mass functions and occupation numbers of the light seeds compared to the heavy seeds, and to the strength of the dynamical interactions that the seeds experience with stars and gas in order to pair on the smallest scales where GWs drive the inspiral. The three models bracket the possible range of coalescence rates. Cosmological simulations of the galaxy assembly anchored to estimates of the local galaxy-merger-rate predict a few to few-hundred coalescences per year for the LISA events.

9. Massive black holes in the near universe and the role of EMRI's LISA sources

Dwarf galaxies are expected to experience relatively quiet merger histories, and thus potentially host the least-massive BHs of the universe today. Extending the local $M_\bullet - \sigma$ correlation and ancillary relations as the $M_\bullet - M_*$ one to low mass galaxies leads to a larger scatter of the data, and a richness in the host galaxy's morphology and kinematics (Kormendy and Ho, 2013). Thus, even in the zero-redshift universe, the knowledge of the low-end of the SMBH mass distribution, and of the occupation fraction in galaxies is incomplete. Often in low mass galaxies, a nuclear star cluster is in place, and a central MBH/IMBH may or may not cohabit the cluster.

A systematic search for the EM signature of an AGN in dwarf galaxies has been carried in Reines *et al.* (2013), who showed interestingly that (even bulge-less) dwarf galaxies host central BHs with median virial masses of $10^5\,\mathrm{M}_\odot$. Whether this mass is close to the seed mass is not predictable. Over the age of the universe, even in the quiet environment of a dwarf, one cannot exclude a secular growth due to stellar tidal captures and disruption events (Stone *et al.*, 2017). This may lead to the growth of the BH of up to a "minimal" mass $M_{\rm min}^{z=0} \sim 10^5\,\mathrm{M}_\odot$, irrespective of the initial mass, thus erasing information on the seed formation mechanisms (Alexander, 2017). EM observations of nearby galactic nuclei will bring further insight into this process of mass saturation at a minimal level.

LISA will also contribute to increment knowledge on the mass distribution of MBH in nearby galaxies in the range between a few $10^4\,\mathrm{M}_\odot$ up to a few $10^6\,\mathrm{M}_\odot$. Exquisite probes are the so called Extreme Mass Ratio Inspirals (EMRIs) (Gair *et al.*, 2017). An EMRI is a extreme mass ratio binary comprising a MBH and a sBH moving on a highly relativistic, mildly eccentric (0-0.2) orbit. The sBH performs about $10^4 - 10^5$ cycles before crossing the horizon of the large BH, and its orbit displays extreme forms of in-plane and off-plane relativistic precession. The waveform encodes information on the masses of the two BHs, the eccentricity of the orbit at plunge and the spin of the large BH to such an extent that it will make it possible to measure deviations from the quadrupole moment of the central object up to a precision of one part in $10^{-4} - 10^{-3}$ (Gair *et al.*, 2017). This shall enable us to test whether the central object is a MBH or a horizonless object as a boson star, and measure the redshifted mass of the sBH with a precision of one part in 10^{-5}. EMRI rates are uncertain and range between 0 and 100 yr^{-1} (Alexander and Bar-Or, 2017). Loss cone dynamics at the gravitational sphere of influence of the massive MBH suggests that an EMRI is produced out of $\sim$ 100 direct plunges, driven by angular momentum changes over the two-body relaxation times. The EMRI's dynamics is extremely sensitive to the stellar distribution around the MBH, so that detecting EMRIs not only will test the nature of the central object but thanks to the accurate mass measurement it can probe the sBH and SMBH mass distribution in Milky Way-like galaxies as well as dwarfs, out to a redshift $z \sim 2$. The EMRI signal is just detectable above the bucket of the LISA observatory sensitivity, where there are also the signals from the coalescing seed BHs.

10. BH archaeology

Following the first direct detection of GWs by the LIGO Scientific Collaboration and Virgo Collaboration, and the success of the LISA Pathfinder mission, we can firmly claim that we are heading into an era of discoveries with the opening of a new window on the universe: *the GW window.* There is a *fil rouge* connecting sBHs and SMBHs. SMBHs, already in place at redshifts as large as $z \sim 7$, may rise from a unique building block, the BHs of stellar origin, or from many avenues, e.g. from a population of IMBHs still elusive to EM observations. A possibility not envisaged in this chapter is that a fraction of BHs in the mass window between $25 - 100\,\mathrm{M}_{\odot}$ are of primordial origin (Carr *et al.*, 2017). The discovery of primordial BHs with mass below the maximum mass of a neutron star, would also be ground-breaking as it would indicate a formation path of cosmological (non-stellar) origin taking us closer to the Big Bang. LISA shall have the capability of detecting the early-forming heavy seeds and track their growth across all cosmic ages, surveying the low-mass tail of the SMBH mass spectrum. Ground based interferometers in their advanced configurations and the third generation of detectors will push observations of sBHs close to the epoch of cosmic reionization, unveiling the high-mass end of sBHs. With this new field of BH archaeology, we will be able to discover if BHs over a wide interval of masses, from about $10\,M_\odot$ up to $10^{10}\,M_\odot$ are the universal outcome of un-halted gravitational collapse, in Nature.

Overall, we can conclude that gravitational waves may provide the perhaps most promising pathway to determine the BH mass function over a large range of redshifts, and it is important to extend current facilities to also probe GWs from black hole mergers in the intermediate or supermassive range. These investigations can and should be complemented with observations of classic EM radiation. Chapter 14 will thus conclude the book with future prospects from EM observations.

Chapter 14

Prospects for detecting the first black holes with the next generation of telescopes

Mark Dijkstra

Institute of Theoretical Astrophysics, University of Oslo, P.O. Box 1029 Blindern, NO-0315 Oslo, Norway

mark.dijkstra@astro.uio.no

This chapter describes the prospects for detecting the first black holes in our Universe, with a specific focus on instruments/telescopes that will come online within the next decade, including e.g. the SKA, WFIRST, EUCLID, JWST, and large ground-based facilities such as E-ELTs, GMT and TMT. This chapter focusses on: (*i*) the indirect detectability of the first generations of stellar mass black holes through their imprint on various Cosmic Background Radiation fields including the 21-cm and X-ray backgrounds (*ii*) direct detectability of line and continuum emission by more massive black holes ($M_{\rm BH} \sim 10^4 - 10^6 M_{\odot}$) that formed via channels other than 'ordinary' stellar evolution.

1. Introduction

Understanding the formation of the first stars, black holes and galaxies in our Universe represents one of the key challenges in cosmology. This represents a well-defined problem with initial conditions that are constrained well by the observed intensity fluctuations in the cosmic microwave background (CMB). However, understanding the physical properties of the first stars and black holes requires following the evolution of density perturbations deep into the non-linear regime, where a number of physical processes — e.g. (*i*) the effects of turbulence on sub-grid scales, which can affect hydrodynamics on resolved scales (Latif *et al.*, 2013a), (*ii*) the presence of weak seed magnetic fields (Sethi *et al.*, 2010; Van Borm and Spaans, 2013; Latif *et al.*, 2014d), (*iii*) the precise nature of dark matter (cold vs. warm vs. self-interacting), (*iv*) streaming velocities between dark matter and baryons (Tseliakhovich and Hirata, 2010; Maio *et al.*, 2011b; O'Leary and

McQuinn, 2012; Latif *et al.*, 2014b), and (v) feedback effects that become relevant with the first onset of nuclear fusion (as in 'quasi-stars', see e.g. Ref Begelman *et al.* (2008)) — can strongly affect the late stages of gravitational collapse (see e.g. Volonteri (2010); Haiman (2013); Latif and Ferrara (2016) for reviews on the formation of the first black holes and Bromm and Larson (2004); Greif (2015) for the formation of the first stars). With little to no guidance by observations, the physical properties of the first stars and black holes are not well known.

The goal of this chapter is to discuss the prospects for improving our observational constraints, and to focus on the prospects for detecting the first black holes that formed in the Universe. I will focus mainly[a] on the detectability of (i) stellar mass black holes, which formed after at the end of the life-time of the first massive stars, and (ii) more massive $M_{\rm BH} \sim 10^4 - 10^6 M_\odot$ that can form in the young Universe through a variety of channels. This discussion focusses on the detectability for the most recent and the next generation of instruments and telescopes. The outline of this chapter is as follows: I will discuss the prospects for the detectability of stellar mass black holes in § 2 and § 3. I then discuss the detectability of more massive black holes in § 4, before I summarise in § 5.

Abbreviation	Full name
BH	black hole
DCBH	direct collapse black hole
CMB	cosmic microwave background
IGM	intergalactic medium
WF	Wouthuysen-Field
XRB	X-ray background
HMXB	high-mass X-ray binary
SFR	star formation rate
SFG	star-forming galaxy
NIRB	near infrared background
OBG	obese black hole galaxy

[a]I will not discuss the detectability of the first primordial black holes that may form in the very early Universe (see e.g. Refs. Zel'dovich and Novikov (1966); Carr and Hawking (1974)). This chapter will in particular be dedicated to electromagnetic observations, while observations through gravitational wave observatories have been discussed in chapter 13. For observational constraints on primordial black holes, see e.g. Ref Mack and Wesley (2008).

2. Prospects for direct detection of the first stellar mass black holes

Observations indicate that the electro-magnetic spectrum associated with accretion onto a stellar mass black hole varies strongly with the rate at which the black hole is accreting. For the highest accretion rate — $\dot{m} \gtrsim 10^{-2}\dot{m}_{\rm edd}$, where[b] $\dot{m}_{\rm edd} \equiv L_{\rm edd}/c^2$ — the accreting black hole is in its high-soft state, during which the spectrum can be decomposed into two components[c]: (*i*) at low energies the standard Shakura–Sunyaev disk model (see e.g. Refs Shakura and Sunyaev (1973); Shimura and Takahara (1995)) applies, which can be approximated as $f_\nu \propto \nu^{1/3}$ out to $h_{\rm p}\nu \lesssim kT_{\rm max} \sim 1 m_{\rm BH}^{-1/4}$ keV, where $m_{\rm BH}$ denotes the black hole mass (in $M_\odot$); (*ii*) a power-law component $f_\nu \propto \nu^{-\alpha_X}$ which extends out to $\sim$ a few to a few hundreds keV (after which it cuts off exponentially, see e.g. Refs Sazonov *et al.* (2004); Feng and Soria (2011)), with α_X having a broad distribution covering $0 \lesssim \alpha_X \lesssim 2$, and peaked at $\alpha_X \sim 1$ (e.g. Ref Feng and Soria (2011)). This power-law component is thought to originate in a hot corona of gas surrounding the black hole which Compton scatters UV photons emitted by the disk to higher energies. The fraction of the total bolometric luminosity that is in either component of the spectrum varies: the power-law component can contain $\sim$ a few to a few tens of per cent of the total bolometric luminosity (e.g. Ref Gilfanov and Merloni (2014)). In other words, the disk component typically contains a fraction of the bolometric luminosity which is of order unity. For a black hole of mass $m_{\rm BH}$ accreting at its Eddington limit, the flux density at an observed frequency $\nu_{\rm obs}$ can be estimated from $f_\nu \sim (1+z)L_{\rm bol}/[4\pi d_{\rm L}^2(z)\nu_{\rm max}]$ (where $d_{\rm L}(z)$ denotes the luminosity distance to redshift z). Plugging in some numerical values yields

$$f_\nu \approx 2\times 10^{-36}\left(\frac{m_{\rm BH}}{100M_\odot}\right)^{5/4}\left(\frac{h_{\rm p}\nu_{\rm obs}}{10\ {\rm eV}}\right)^{1/3}\left(\frac{1+z}{10}\right)^{-1.5}\frac{\rm erg}{\rm s\ cm^2\ Hz}, \quad (1)$$

for $h_{\rm p}\nu_{\rm obs} \lesssim kT_{\rm max}/(1+z)$, where $d_{\rm L}^2(z) \propto (1+z)^{2.5}$ at $z \gg 1$ (approximately). The $m_{\rm BH}^{5/4}$-dependence arises because $\nu_{\rm max} \propto m_{\rm BH}^{-1/4}$. This flux density corresponds to an AB-magnitude of

$$m_{\rm AB} \sim 41 - 3.1\log\left(\frac{m_{\rm BH}}{100M_\odot}\right) - 0.83\log\left(\frac{\nu}{10\ {\rm eV}}\right) + 3.8\log\left(\frac{1+z}{10}\right), \quad (2)$$

[b]For a radiative efficiency η, the maximum mass accretion rate equals $\dot{m} = L_{\rm edd}/(\eta c^2)$.
[c]There is a third 'reflection' component, which consists of high-energy photons originating in the hot corona, but which scattered (either by free electrons, or fluorescently by ions) off the accretion disk. We follow Yue *et al.* (2013) and assume this component to be sub-dominant to the other two.

for $h_p\nu_{obs} \lesssim kT_{max}/(1+z)$. For comparison, NIRCAM on the soon-to-be launched James Webb Space Telescope (JWST) can detect a point source with $S/N = 10$ in after an integration of $t = 10^4$ with an AB-magnitude of ~ 29 in the observed wavelength range $\lambda = 1 - 5\mu$m.[d] Larger ground-based telescopes (such as E-ELT) may go ~ 3 magnitudes deeper, but this is not nearly enough to detect continuum emission from the first black holes accreting at Eddington luminosity.

3. Prospects for indirect detection of the first stellar mass black holes

3.1. *Detecting the first BH through 21-cm*

The prospects for *indirect* detection of the first black holes are much better. X-rays produced in the inner disk or hot corona can penetrate deeply into the neutral intergalactic medium where a significant fraction of their energy is converted into heat.[e] This re-heating of the IGM has a direct observational implication for the detectability of the 21-cm hyperfine transition of atomic hydrogen in the ground state (e.g. Ref Mirabel *et al.* (2011)). Since this line provides such a potentially powerful probe of the high-z Universe (and the first black holes), we summarise the basic physics that regulates its detectability in a bit more detail below.

The visibility of HI gas in its 21-cm line — with frequency $\nu_{21} \sim 1.4$ GHz — is expressed as a *differential* brightness temperature with respect to the background CMB, $\delta T_b(\nu)$ (Furlanetto *et al.*, 2006; Morales and Wyithe, 2010)

$$\delta T_b(\nu) \approx 9x_{HI}(1+\delta)(1+z)^{1/2}\left(1 - \frac{T_{CMB}(z)}{T_S}\right)\left[\frac{H(z)(1+z)}{dv_{||}/dr}\right] \text{ mK}, \quad (3)$$

where $\delta + 1 \equiv \rho/\bar{\rho}$ denotes the overdensity of the gas, $z(= [\nu_{21}/\nu] - 1)$ the redshift of the gas, $T_{CMB}(z) = 2.73(1+z)$ K denotes the temperature of the CMB, the factor in square brackets contains the line-of-sight velocity

[d] https://jwst.stsci.edu/science-planning/proposal-planning-toolbox/sensitivity-overview
[e] In a fully neutral intergalactic medium (IGM), photons with $E \gtrsim 50$ eV deposit $\sim 1/3$ of their energy as heat, $\sim 1/3$ goes into ionization, and the remaining $\sim 1/3$ goes into collisional excitation of transitions in H and He/He$^+$ (e.g. Refs Shull and van Steenberg (1985); Valdes *et al.* (2010)). As the ionization fraction rises above $\sim 10\%$ however, the fraction of photon energy that goes into heat rapidly approaches unity.

gradient $dv_{||}/dr$. The term T_S denotes the spin (or excitation) temperature of the 21-cm transition, which quantifies the number densities of hydrogen atoms in each of the hyperfine transitions, and which is discussed in more detail below (see Eq (4)). The numerical prefactor of 9 mK in Eq (3) applies for gas undergoing Hubble expansion. For slower expansion rates, we increase the number of hydrogen atoms within a fixed velocity (and therefore frequency) range, which enhances the brightness temperature.

Equation (3) states that when $T_{\rm CMB}(z) < T_S$, $\delta T_b(\nu) > 0$, and vice versa. This means that we see HI in absorption [emission] when $T_S < T_{\rm CMB}(z)$ [$T_S > T_{\rm CMB}(z)$]. The spin temperature thus plays a key role in setting the 21-cm signal, and we briefly discuss the physical processes that determine T_s below. Equation (3) highlights why so much effort is going into trying to detect the 21-cm line: having a 3D map of $\delta T_b(\nu)$ provides us with direct constraints on the 3D density [$\delta(\mathbf{r})$], and line-of-sight velocity field [$v_{||}(\mathbf{r})$] of baryons in our Universe. It therefore provides us with a powerful cosmological and astrophysical probe of the high-redshift Universe.

The spin temperature T_S is set by (*i*) collisions, which drive $T_S \rightarrow T_{\rm gas}$, (*ii*) absorption by CMB photons, which drives $T_S \rightarrow T_{\rm CMB}$ (and thus that $\delta T_b \rightarrow 0$, see Equation (3)), and (*iii*) Ly α scattering. Ly α scattering mixes the two hyperfine levels, which drive $T_S \rightarrow T_{\rm gas}$, an effect which is known as the 'Wouthuysen–Field effect' (WF effect). Quantitatively, it has been shown that (Madau *et al*, 1997; Furlanetto *et al.*, 2006; Morales and Wyithe, 2010)

$$\frac{1}{T_S} = \frac{T_{\rm CMB}^{-1} + x_c T_{\rm gas}^{-1} + x_\alpha T_{\rm gas}^{-1}}{1 + x_c + x_\alpha}, \tag{4}$$

where $x_c = \frac{C_{21}T_*}{A_{21}T_{\rm gas}}$ denotes the collisional coupling coefficient, in which $k_B T_*$ denotes the energy difference between the hyperfine levels, C_{21} denotes the collisional de-excitation rate coefficient, and $A_{21} = 2.85 \times 10^{-15}$ s^{-1} denotes the Einstein-A coefficient for the 21-cm transition. Furthermore, $x_\alpha = \frac{4P_\alpha}{27A_{21}T_{\rm gas}}$, in which P_α denotes the Ly α scattering rate.[f] Equation (4) implies that the spin temperature T_s is a weighted average of the gas and CMB temperature. The *left panel* of Figure 1 shows the universally (or globally) averaged temperatures of the gas (*red solid line*) and the CMB (*blue solid line*), and the corresponding 'global' 21-cm signature in the *Right panel*. We discuss these below (this discussion follows Ref Dijkstra (2017)):

[f]We have adopted the common assumption here that the Ly α color temperature $T_\alpha = T_{\rm gas}$.

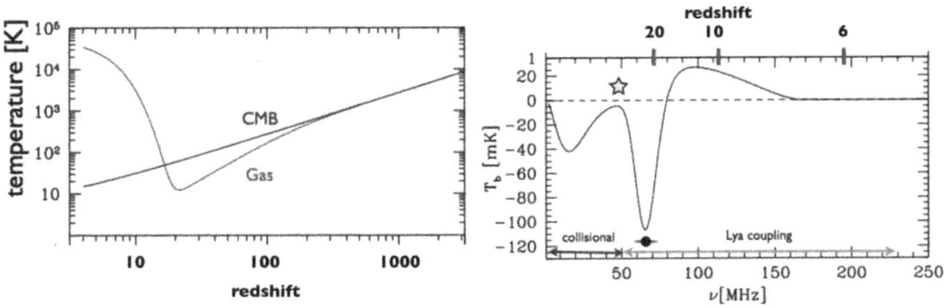

Fig. 1. *Left panel:* Volume-averaged redshift evolution of $T_{\rm gas}$ and the $T_{\rm CMB}$. Right panel: The 'Global' 21-signal, represents the sky-averaged 21-cm brightness temperature $\delta T_{\rm b}(\nu)$. See the main text for a description of each of these curves. The global 21-cm signal constrains when the first stars, black holes, and galaxies formed and depends on their spectra. *Credit: adapted from slides created by J. Pritchard (based on Refs Pritchard and Loeb (2010, 2012)).*

- Adiabatic expansion causes $T_{\rm CMB} \propto (1+z)$ at all z. At $z \gtrsim 100$, $T_{\rm gas} \sim T_{\rm CMB}$ as a result of the interaction between CMB photons and the small residual fraction of free electrons that exist following recombination. When $T_{\rm CMB} = T_{\rm gas}$, we must have $T_{\rm s} = T_{\rm CMB}$, and therefore that $\delta T_{\rm b}(\nu) = 0$ mK, which corresponds to the high-z limit in the *right panel*.
- At $z \lesssim 100$, electron scattering can no longer couple the CMB and gas temperatures, and the (non-relativistic) baryons adiabatically cool as $T_{\rm gas} \propto (1+z)^2$. Because $T_{\rm gas} < T_{\rm CMB}$, we must have that $T_{\rm s} < T_{\rm CMB}$ and we expect to see the 21-cm line in absorption. When $T_{\rm gas}$ first decouples from $T_{\rm CMB}$ the gas densities are high enough for collisions to keep $T_{\rm s}$ locked to $T_{\rm gas}$. However, at $z \sim 70$ ($\nu \sim 20$ MHz) collisions can no longer couple $T_{\rm s}$ to $T_{\rm gas}$, and $T_{\rm s}$ crawls back to $T_{\rm CMB}$, which reduces $\delta T_{\rm b}(\nu)$ (at $\nu \sim 20-50$ MHz, i.e. $z \sim 70-30$).
- The first stars, galaxies, and accreting black holes emitted UV photons in the range $E = 10.2 - 13.6$ eV. These photons travel freely through the neutral IGM, until they redshift into one of the Lyman series resonances, at which point a radiative cascade can produce Ly α. The formation of the first stars thus generates a Ly α background, which initiates the WF-coupling, which forces $T_{\rm s} \rightarrow T_{\rm gas}$. The onset of Ly α scattering — and thus the WF coupling — causes $\delta T_{\rm b}(\nu)$ to drop sharply at $\nu \gtrsim 50$ MHz ($z \lesssim 30$).

- At some point the X-rays produced by accreting black holes easily penetrate deep into the cold, neutral IGM, and reheat the gas. The *left panel* thus has T_{gas} increase at $z \sim 20$, which corresponds to the onset of X-ray heating. In the *right panel* this onset occurs a bit earlier. This difference reflects that the redshift of all the features (minima and maxima) in $\delta T_b(\nu)$ are model dependent, and not well-known (more on this below). The onset of X-ray heating (combined with increasingly efficient WF coupling to the build-up of the Ly α background) enhances $\delta T_b(\nu)$ until it becomes positive when $T_{gas} > T_{CMB}$.
- Finally, $\delta T_b(\nu)$ reaches yet another maximum, which reflects that neutral, X-ray heated gas is reionized away by the ionizing UV-photons emitted by star forming galaxies and quasars. When reionization is complete, there is no diffuse intergalactic neutral hydrogen left, and $\delta T_b(\nu) \rightarrow 0$.

In detail, the onset and redshift evolution of Ly α coupling, X-ray heating, and reionization depend on the redshift evolution of the number densities of galaxies, and their spectral characteristics. All these are uncertain, and it is not possible to make robust predictions for the precise shape of the global 21-cm signature. Instead, one of the main challenges for observational cosmology is to measure the global 21-cm signal, and from this constrain the abundances and characteristic of first generations of galaxies in our Universe. Detecting the global 21-cm is challenging, but especially the deep absorption trough that is expected to exist just prior to the onset of X-ray heating — *which is tightly linked to the formation of the first black holes in our Universe* — at $\nu \sim 70$ MHz is something that may be detectable because of its characteristic spectral shape. Constraints on the global 21-cm signal already rule out scenarios in which the Universe goes from fully neutral at fully ionized over $\Delta z \lesssim 0.06$ at $6 < z < 13$ (see Ref Bowman and Rogers (2010)). While such rapid reionization scenarios are not considered physically plausible, it is remarkable that global 21-cm experiments already have the sensitivity to detect (rapid) changes in the ionization state of the IGM. Encouraged by this success, a number of other experiments aim to detect the global 21-cm signal, including e.g. EDGES 2 (Monsalve *et al.*, 2017), LEDA,[g] BIGHORNS (Sokolowski *et al.*, 2015), and SARAS 2 (Singh *et al.*, 2017).

[g]http://www.tauceti.caltech.edu/leda/

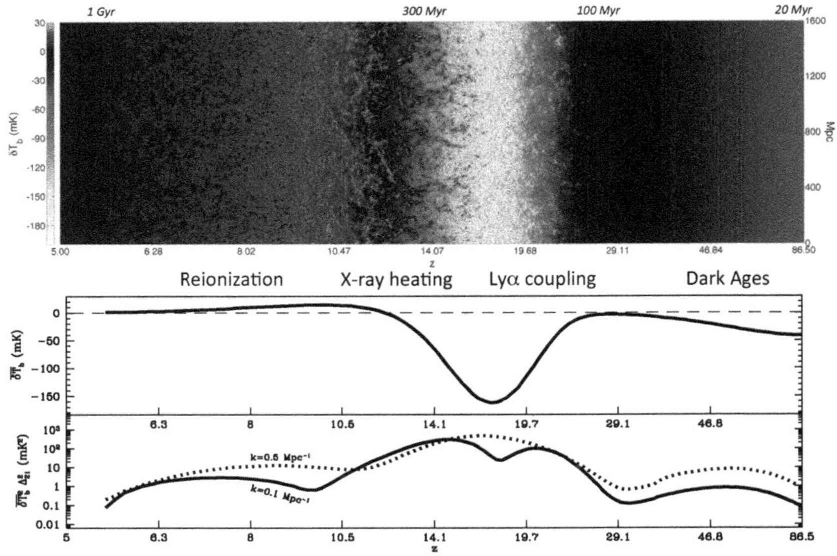

Fig. 2. This figure illustrates that the spatial fluctuations of the 21-cm background provide additional constraints on the first black holes in our Universe (see text). *Credit: Adapted from Figure 4 of Ref Mesinger* et al. *(2016), 'The Evolution Of 21 cm Structure (EOS): public, large-scale simulations of Cosmic Dawn and reionization', MNRAS, 459, 2342M. Reproduced by permission of Oxford University Press/on behalf of the RAS.*

Figure 2 illustrates that there is more information on the first black holes encoded within 21-cm observations: while the global 21-cm signal assigns a single, volume averaged, brightness temperature to each redshift, the *top panel* shows $\delta T_{\rm b}(\nu)$ through a slice of the Universe, which shows that at fixed z there exist spatial fluctuations in $\delta T_{\rm b}(\nu)$. The *bottom panel* shows that the redshift dependence of the *variance* $\delta T_{\rm b}^2$ depends on scale (here represented with wavenumber k). The 3D power-spectrum of the 21-cm fluctuations (which corresponds to $\sim \delta T_{\rm b}^2(k)$) therefore contains additional constraints on the first black holes in the Universe (Pritchard and Furlanetto, 2007; Mesinger *et al.*, 2013; Fialkov *et al.*, 2014; Mesinger *et al.*, 2016).

There is a number of 'first generation' (e.g. PAPER (Parsons *et al.*, 2010), LOFAR (van Haarlem *et al.*, 2013), MWA (Tingay *et al.*, 2013)) and future (e.g. SKA,[h] HERA (DeBoer *et al.*, 2017)) low frequency interferometers that aim to detect fluctuations in the 21-cm background on the scales shown in Figure 2. Mesinger *et al.* (2016) finds that SKA1-low and HERA should be able to detect the epochs of reionization and X-ray heating at a signal-to-noise level of $\sim$ hundreds. With optimistic assumptions on the foregrounds, they also find that all first generations of instruments could detect these epochs at $S/N \gtrsim 3$. In fact, upper limits of the 21-cm power spectrum obtained by PAPER (Pober *et al.*, 2015) already rule out some extreme models, in which the neutral IGM is not heated at all by X-rays even when reionization is well underway. Even though these extreme models are not really physical, it is extremely encouraging that observations have started placing constraints on the temperature of the low-density, neutral intergalactic medium in the very young Universe.

3.2. *Contribution of the early generations of accreting BHs to cosmic backgrounds*

As discussed previously, X-rays emitted by the first black holes play an important role in shaping the thermal history of our Universe. A fraction of these X-rays may be observable today: the neutral IGM is optically thin to photons with energies $E \gtrsim E_{\rm max} = 1.8[(1+z)/15)]^{0.5} x_{\rm HI}^{1/3}$ keV (Dijkstra *et al.*, 2004), and these photons would redshift without absorption and would be observed as a present-day soft X-ray background (soft XRB). It is indeed possible that early generations of accreting black holes contributed significantly to the *unresolved* component of the XRBs, as also discussed in Chapter 12:

- The unresolved soft XRB (here, observed energies $E = 0.5-2$ keV) can rule out certain models in which reionization is dominated by accreting black holes (Dijkstra *et al.*, 2004; Salvaterra *et al.*, 2005; McQuinn, 2012), though quantitatively these constraints depend on the spectral slope $\alpha_{\rm X}$ characterizing the X-ray emission from the hot corona (Dijkstra *et al.*, 2004; McQuinn, 2012).
- Dijkstra *et al.* (2012) (hereafter D12) applied the locally inferred relation between a galaxy's star formation rate and X-ray luminosity

[h]http://skatelescope.org/

(e.g. Lehmer *et al.* (2010); Mineo *et al.* (2012)),[i] and showed that star forming galaxies that are too faint to be detected as individual X-ray sources can fully account for the observed unresolved soft XRB (where 'soft' refers to the observed range $E = 1 - 2$ keV). Interestingly, there are observational indications that the total X-ray luminosity from HMXBs per unit SFR increases with redshift (Basu-Zych *et al.*, 2013), which implies that star forming galaxies could contribute even more.[j]

- A more recent analysis, Cappelluti *et al.* (2017) (which we referred to as C17) shows that there are indications that the spectrum of the unresolved component of the Cosmic X-ray Background (XRB) is harder and more obscured than the collective spectrum of resolved sources. C17 propose a scenario in which a high-z ($z \gtrsim 6-7$) population of direct collapse black holes (DCBHs: black holes that form following the collapse of a pristine gas cloud directly into a massive black hole with no intermediate star formation (Bromm and Loeb, 2003; Volonteri, 2010; Haiman, 2013; Latif and Ferrara, 2016, see also chapter 5)) fully account for the remaining unresolved XRB. This proposal is motivated by theoretical modelling which showed that high-z DCBHs can — under certain assumptions — explain the amplitude of the observed fluctuations in the source extracted Near Infrared Background (NIRB) and its cross-correlation with

---

[i]This correlation — $L_{\rm X} = 2.6 \times 10^{39} \times$ SFR erg s^{-1}, where the star formation rate (SFR) is measured in units of $M_\odot$ yr^{-1} and X-ray luminosity is measured in the energy-range $E = 0.5 - 8.0$ keV [Mineo *et al.* (2012)] — is driven by High-Mass X-ray Binaries (HMXBs). HMXBs are binaries in which a compact object (neutron star or black hole) accretes gas that is stripping off a massive O or B binary companion star. Because these stars are short-lived HMXBs closely trace ongoing star formation.

[j]Specifically, D12 computed that star forming galaxies (SFGs) with SFR> $10^{-2} M_\odot$ yr^{-1} whose X-ray flux falls below the detection limit of the 2 Ms Chandra Deep Field North (Alexander *et al.*, 2003), contribute $\sim 2 - 3 \times 10^{-13}$ erg s^{-1} cm^{-2} deg^{-2} in the 1-2 keV band. In a more recent analysis, Cappelluti *et al.* (2017) (hereafter C17) constrain the contribution of SFGs that are detected in the CANDELS catalogues of the 10 Ms Chandra Deep Fields to the XRB to be $\sim 3 \times 10^{-14}$ erg s^{-1} cm^{-2} deg^{-2}, which is a factor of $\sim 7 - 10$ times lower. There are several reasons for this difference, all of which relate to the fact that C17 considered a subset of SFGs that were included in the D12 analysis: (*i*) the 10 Ms Chandra observations allow for *direct* detection of more SFGs, which can therefore not contribute to the *unresolved* XRB; (*ii*) D12 include all star forming galaxies with SFR> $10^{-2} M_\odot$ yr^{-1}, which corresponds to $M_{\rm UV} \sim -13$ (not corrected for dust), most of which would be too faint to make it into the CANDELS catalogues; (*iii*) in D12 the largest contribution from star forming galaxies to the XRB comes comes from sources at $z < 2$. C17 specifically notes that it is possible for faint sources at these redshifts to contribute significantly to the XRB.

the XRB (Yue *et al.*, 2013). C17 further show that if the unresolved XRB is indeed fully generated by a population of high-z Compton thick accreting black holes, then the observational constraints on the mass density of black holes require that the Compton thick gas that is surrounding the accreting black holes must have gas metallicities[k] $Z \lesssim 10^{-3} Z_{\odot}$.

This discussion highlights that it is possible that early generations of accreting black holes contribute significantly to the unresolved cosmic X-ray backgrounds. However, the nature of the black holes varies greatly between different models, ranging from HMXB powering X-ray emission in undetected ultra-faint $z < 2$ star forming galaxies to $z \gg 6$ newly formed direct collapse black holes. It is possible to distinguish between these two models with future observations as we discuss in the next section.

4. Prospects for detection of the first massive black holes

As already mentioned, the first black holes may have been much more massive, with masses possibly as large as $M_{\rm BH} \sim 10^6 M_{\odot}$ (see Volonteri (2010); Haiman (2013); Latif and Ferrara (2016) for a review of a variety of mechanisms that may give rise to such massive black hole seed masses). These massive objects are more easy to detect. In § 4.1 we compare estimates of the continuum flux density, in § 4.2 we estimate fluxes in some lines, and in § 4.3 we look at observational signatures associated with the formation of these black holes.

4.1. *Direct detection of the continuum emission from first massive BHs*

The *red-dashed line* in Figure 3 shows an 'unprocessed' (by HI surrounding the black hole, including intergalactic HI) model spectrum for a DCBH at $z \sim 9$ from Natarajan *et al.* (2017) for a number of assumptions: (*i*) the total X-ray flux from the accretion disk equals that coming from the hot corona; (*ii*) the spectrum coming from the corona is a power-law with slope $\alpha_{\rm X} = 1$; (*iii*) the exponential cut-off of the spectrum occurs at high ($\gg 10$ keV) energies; (*iv*) the DCBH has grown to $M_{\rm BH} = 1.5 \times 10^6 M_{\odot}$.

[k]This is because lower metallicity gas is more transparent to high energy X-rays. As a result, a *fixed* observed flux in the XRB translates to a *higher* accretion rate in enriched gas (Natarajan *et al.*, 2017), which translates to a larger mass density in accreting black holes.

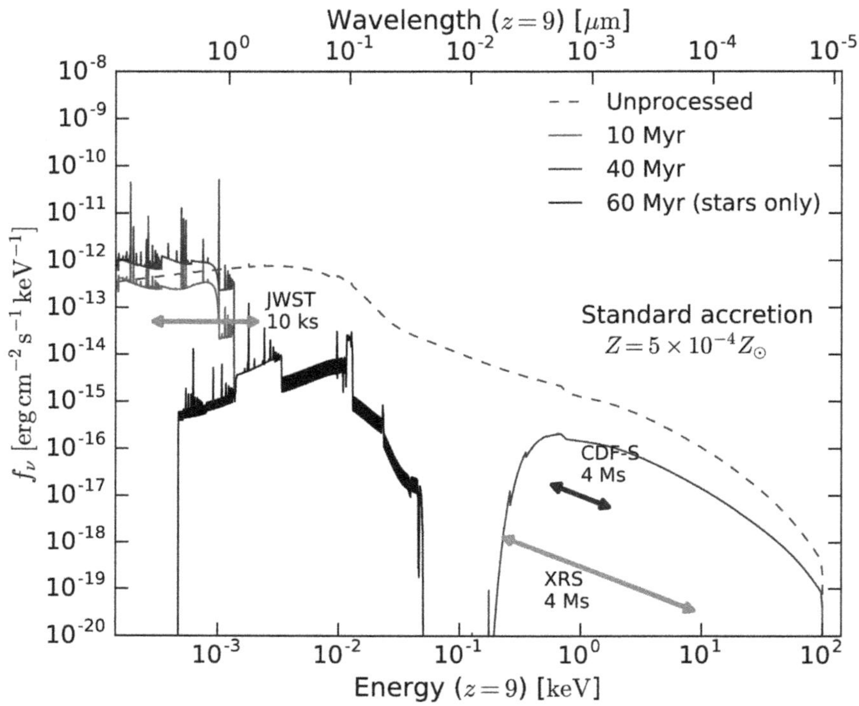

Fig. 3. Model spectra of massive ($M_{\rm BH} \sim 10^4 - 10^6 M_\odot$) accreting black hole. The 'unprocessed' spectrum consists of a disk + hot coronal component. Significant reprocessing of ionizing radiation can occur by atomic hydrogen which is theoretically expected to be surrounding the accreting black hole. Photo-ionization and subsequent recombinating gives rise to strong nebular lines (Ly α, H α, He1640) which can be detected with future (and existing) telescopes & instruments. *Credit: from Figure 1 of Natarajan* et al. *(2017). ©AAS. Reproduced with permission.*

The unprocessed flux density peaks at $h_{\rm p}\nu_{\rm obs} \sim kT_{\rm max}/(1+z) \sim 10$ eV at $f_\nu \sim 10^{-12.5}$ erg s^{-1} cm^{-2} keV^{-1}, which corresponds to $m_{\rm AB} \sim 26.2$. This agrees well with our estimate given in Eq (2), which gives $m_{\rm AB} \sim 27$ for $m_{\rm AB} = 1.5 \times 10^6 M_\odot$.

The *colored solid lines* show the spectrum after it has been processed through a large column density of HI gas (here $N_{\rm HI} \gtrsim 10^{24}$ cm^{-2}). This hydrogen gas absorbs most radiation blueward of the hydrogen ionization threshold at $h_{\rm p}\nu_{\rm obs} \gtrsim 13.6/(1+z) = 1.4$ eV. Because the photoionization cross-section decreases $\propto \nu^{-3}$, the gas becomes increasingly transparent at higher energies, which causes the processes spectrum to rise at $h_{\rm p}\nu_{\rm obs} \gtrsim 0.2$ keV. Lines of different colors represent the accreting black hole at different times — and therefore black hole mass. Ionizing photons that are absorbed in the neutral gas are reprocessed into recombination lines (and continuum), which we will discuss in § 4.2.

Figure 3 also indicates sensitivity of some existing and future observations: the *green line* shows the expected sensitivity of the James Webb Space Telescope in the near-IR ($\lambda \sim 1 - 5\mu$m) for a 10^4 s integration, which corresponds to $f_\nu \sim 10^{-13.3}$ erg s^{-1} cm^{-2} keV^{-1} (i.e. $m_{\rm AB} \sim 28$, which agrees with the sensitivity estimate given in § 2). Continuum magnitudes of $m_{\rm AB} \sim 26 - 27$ in the near-IR are also within reach of WFIRST (Wide Field Infrared Survey Telescope), which is a space-based NIR observatory whose 5σ detection threshold for point sources is expected to reach $26.5 - 27.0$ at $\lambda \sim 1.0 - 1.7$ μm. Deep fields with EUCLID (Laureijs *et al.*, 2011) are expected to reach similar depth over ~ 40 deg^2. Finally, in 'deep drilling fields' LSST can reach $m_{\rm AB} \sim 28$ though this is limited to $\lambda \lesssim 1\mu$m. Also shown are the sensitivity in the (*i*) existing 4Ms Chandra-Deep Field South, and (*ii*) the proposed 'Lynx' X-ray telescope[l] which would have a sensitivity to point sources reaching 10^{-19} erg s^{-1} cm^{-2} — corresponding to a luminosity of $L_X \sim 10^{41}$ erg s^{-1} at $z = 10$ — in a 4Ms observation.[m] It is clear that continuum emission from accretion onto the first massive black holes can be easily detected with future, and even existing telescopes (though in detail this depends on α_X, and where this power-law spectrum cuts off). The more pressing question is therefore how we would be able to identify these objects.

Newly formed massive black holes are thought to rapidly acquire an accompanying stellar component, mostly from mergers with the nearby star forming halos which provided the required Lyman-Werner flux (Dijkstra *et al.*, 2008; Agarwal *et al.*, 2012; Regan *et al.*, 2017) to enable direct collapse of a primordial gas cloud into a massive black hole (Agarwal *et al.*, 2016a; Hartwig *et al.*, 2016a). The mass of the black hole is initially large compared to its accompanying stellar mass, and these galaxies — at least initially — do not obey the locally observed relation that the central black hole mass of

[l]https://wwwastro.msfc.nasa.gov/lynx/

[m]Another X-ray future telescope is ATHENA (http://www.the-athena-x-ray-observatory.eu/mission.html), which will reach a sensitivity of 10^{-17} erg s^{-1} cm^{-2}, which can also detect the continuum of the more massive seeds ($M_{\rm BH} \sim 10^6 M_\odot$), which depends on the slope of the power-law continuum spectrum of the corona.

a galaxy corresponds to $\sim 0.1\%$ of its stellar mass (in the bulge). This initial stage — during which the black hole is overly massive — has been referred to as the 'Obese Black Hole Galaxy' stage (OBG stage, see Agarwal *et al.* (2013)). Natarajan *et al.* (2017) provide several observational diagnostics which may help us identify galaxies in their OBG stage, including: (*i*) a large bolometric luminosity ($L_{\rm bol} \gtrsim 10^{44}$ erg s^{-1}, which corresponds to the Eddington luminosity of a black hole of mass $M_{\rm BH} \sim 10^6 M_\odot$), (*ii*) a large ratio of X-rays to optical flux, and (*iii*) particular colors in the JWST bands. Additional ways to identify galaxies hosting newly formed massive black holes may be through line emission, as we discuss next.

4.2. *Direct detection of lines from the first massive BHs*

Reprocessing of ionizing radiation in the hydrogen gas surrounding the accreting black hole gives rise to strong recombination lines. High-energy UV ($E \gtrsim 55.4$ eV) and X-ray photons can doubly ionize Helium. Recombining doubly ionized Helium produces strong He1640 line emission. The strongest line that is produced however, is H Ly α, which can account for as much as $\sim 40\%$ of the bolometric luminosity of the accreting black hole (see e.g. Dijkstra *et al.* (2016a)).

The recombination lines contain large fluxes, but condensed into narrow spectral ranges, which can make them more easily detectable. For example, a Ly α line with a luminosity of $L_\alpha \sim 10^{43} - 10^{44}$ erg s^{-1} can be detected out to $z \sim 7$ with existing large telescopes such as VLT, Keck & Subaru. At higher redshifts, the Ly α line redshifts into the NIR where the line is better accessible from space, as atmospheric OH lines contaminate ground-based observations. NIRSPEC on JWST has the sensitivity to detect line fluxes of $\sim 1 - 2 \times 10^{-18}$ erg s^{-1} cm^{-2} at $\lambda \sim 1 - 5\mu$m, which corresponds to (apparent) Ly α luminosities of $L_\alpha \sim 10^{42}$ erg s^{-1} at $z \sim 10$. JWST should therefore be able to access Ly α line emission even from the lower mass 'massive' seeds ($M_{\rm BH} \sim 10^4 M_\odot$). For comparison, for 'moderately extended sources', the 7σ detection threshold for emission lines in WFIRST is $\sim 1 - 1.5 \times 10^{-16}$ erg s^{-1} cm^{-2} at $\lambda \sim 1 - 2\mu$m (see Fig 1 of Gehrels *et al.* (2015)), which corresponds to (apparent) Ly α luminosities of $L_\alpha \sim 10^{44}$ erg s^{-1} at $z \sim 10$. EUCLIDs deep, slitless spectroscopic surveys are expected to go similarly deep to $\sim 10^{-16}$ erg s^{-1} cm^{-2} at comparable wavelengths (see e.g. Laureijs *et al.* (2011)). WFIRST & EUCLID may thus be able to directly detect line emission from the most massive black hole seeds. What is more important is that EUCLID & WFIRST will provide interesting targets which can then be followed up spectroscopically with e.g. JWST and the next generation of ground-based facilities such as E-ELT. The E-ELT is expected to be able to detect similar line fluxes to $\lambda \lesssim 2.5\mu$m, by suppressing the narrow atmospheric OH lines. Because JWST and future ground-based facilities are expected to operate deeper into the IR, they can access other redshifted hydrogen lines such as the Hα line is expected to be a factor of ~ 8 times weaker when powered by recombination.

There has been some speculation that we may have already detected such a newly formed massive black hole: CR7 (Cosmos-Redshift 7, Sobral *et al.* (2015)) is one of the most luminous ($L_\alpha \sim 10^{44}$ erg s^{-1}) Ly α emitting galaxies in the Universe, and at $z \sim 6.6$. In addition, the Ly α line is unusually strong, with a rest-frame equivalent width of EW$_\alpha \sim 200$ Å. The strength of this line is especially surprising given that the Ly α forest[n] is opaque to Ly α emission. In addition, CR7 appears to have a strong He1640 line, which is expected if CR7 harbors a source of hard ionizing radiation, such as an accreting black hole. CR7 further consists of several components, one of which consists of an older, passive galaxy. This has fueled speculation that CR7 may harbor a black hole that formed via direct collapse,[o] as the older galaxy would have been a luminous source of Lyman–Werner flux radiation — a key ingredient for direct collapse — when it was forming stars in its past (Agarwal *et al.*, 2016a; Hartwig *et al.*, 2016a).

The extreme physical conditions of the HI gas surrounding the accreting black holes (high density, metal & H_2-free environments) may give rise to interesting Ly α radiative transfer effects, which can give rise to unusual line ratios (this needs to be explored in future work, also see § 4.3). What

[n] The precise opacity of the IGM to radiation that emerges from a galaxy close to the Ly α resonance is strongly frequency-dependent. How much flux is transmitted through the IGM at $z \sim 6.6$ depends on the spectral shape and shift of the Ly α line as it emerges from the galaxy (see e.g. [Dijkstra (2014)] for an extended review).

[o] Other explanations for CR7 have been put forward. Strong Ly α, He1640 line emission, combined with the blue continuum of one of the three components of CR7 has also been associated with Population III star formation (see e.g. Sobral *et al.* (2015); Visbal *et al.* (2016)). However, recent theoretical studies indicate that the required mass of Population III mass that is required to explain CR7 exceeds the maximum mass that is considered theoretically reasonable (Visbal *et al.*, 2017; Xu *et al.*, 2016a). Recent observations by Bowler *et al.* (2017) imply that the strength of the He1640 is weaker, and that [OIII] may have been detected, which would rule out the Population III scenario (but not the DCBH-scenario, see Pacucci *et al.* (2017c)).

is in this sense peculiar about CR7 is that the spectral shape of the Ly α line is similar to that of low-redshift strong Ly α emitting galaxies, in which the Ly α spectral line shape is affected by (likely stellar feedback-driven) outflows. This is important for two reasons: (*i*) the interstellar medium of CR7 is likely shaped by processes similar to that in the local Universe. This is consistent with a scenario in which the dark matter halo hosting a direct collapse black hole rapidly merges with the star forming halo (as we described when introducing the OBG stage). This merging process likely re-shapes the physical conditions of the interstellar medium in a way that more closely resembles that of 'ordinary' galaxies (*ii*) it implies that the accreting black hole is *not* surrounded by a large column density of HI gas, as is thought to be the case for newly formed DCBHs. This is because large column densities of HI gas require Ly α photons to scatter frequently before escaping, which broadens the width of the spectral line. The fact that the observed Ly α line is narrow limits the allowed HI column densities through which Ly α photons are scattering to be $\sim 10^{19} - 10^{20}$ cm^{-2} (see Dijkstra *et al.* (2016a)).

4.3. *Detection of the formation of the first massive BHs*

Inayoshi *et al.* (2016) discuss observational signatures of a particular formation channel of massive ($M_{\rm BH} \sim 10^5 M_\odot$) black holes via Hyper-Eddington accretion. In this scenario, the mass accretion rate onto a central black hole (which could be stellar mass and formed from a Pop. III star, or more massive and formed differently) exceeds $\dot{M}_{\rm BH} \gtrsim 5000 L_{\rm edd}/c^2$. The transition to hyper-Eddington happens when the radius HII region created by hot gas accreting onto the black hole becomes smaller than the Bondi radius of the black hole. Under these conditions, photons are trapped in the accretion flow and carried into the black hole together with the gas. The overall bolometric luminosity of the accreting black hole is the Eddington luminosity. The most massive of these black holes ($M_{\rm BH} \sim 10^6 M_\odot$) could also be accompanied with a Ly α luminosity reaching $\sim 10^{42} - 10^{44}$ erg s^{-1}. Given the lower gas temperatures involved in this mechanism, the He1640 line should be absent. The spectral shape of the Ly α line could distinguish it from other models, though this requires Ly α radiative transfer modelling which needs to be explored in future work. Inayoshi *et al.* (2016) further argue that this Ly α signature likely disappears in the presence of small amounts of dust, which is interesting as it would suggest it is uniquely associated with the *first* black holes that form this way.

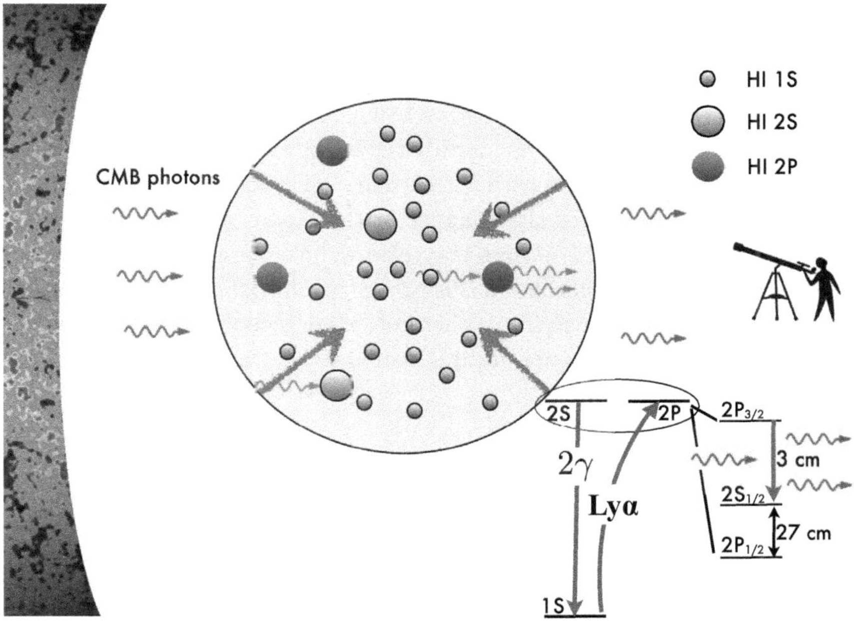

Fig. 4. The physical conditions inside a gas cloud that collapses directly into a massive black hole are optimal for trapping Ly α radiation. Trapped Ly α 'pump' the $2p$-level of HI, and can give rise to stimulated fine-structure emission at a rest-frame wavelength of $\lambda = 3.04$ cm. Stimulated 3-cm emission would be a smoking-gun signature of direct collapse. *Credit: from Figure 1 of Dijkstra* et al. *(2016b).* ©*AAS. Reproduced with permission.*

Another unique signature that is associated with the formation of direct collapse black holes was discussed in Dijkstra *et al.* (2016b): the physical conditions that enable direct collapse: (*i*) the collapse of pristine gas, (*ii*) the (virtual) absence of H_2, (*iii*) the large column densities of HI gas, and (*iv*) suppressed fragmentation, provide optimal conditions for Ly α trapping, which can cause Ly α photons to scatter $\gtrsim 10^8 - 10^9$ times. Under these conditions the $2p$ level (which is the only level of atomic Hydrogen excited by Ly α absorption) can be overpopulated with respect to the $2s$ level. Spin-orbit coupling and the Lamb-shift causes the $2p_{3/2}$ ($2p_{1/2}$) level[p] to lie

[p]We adopt the notation nL_J, where n is the principle quantum number, L denotes the electron's *orbital* angular momentum, and J denotes the *total* (orbital + spin) quantum number.

above (below) the $2s$ level, and gives rise to a fine-structure line with rest-frame wavelength $\lambda = 3.04$ cm ($\lambda = 27$ cm). Overpopulation of the $2p$ level can therefore give rise to stimulated 3-cm emission [Field and Partridge (1961)]. Dijkstra *et al.* (2016b) showed that gas clouds directly collapsing into a massive black holes could amplify the background CMB (see Figure 4) at a level that is detectable with SKA1-mid. This signal is only detectable during a small fraction of the collapse of the cloud, and beaming effects which occur in saturated masers may also reduce the detectability of the signal. However, the 3-cm line has a broad peculiar spectral shape, which makes it easy to distinguish from other lines. If detected, this would provide a true smoking-gun signature of direct collapse.

5. Conclusions & outlook

The main goal of this chapter is summarised in Figure 5: we have presented a brief discussion of the prospects for detecting the first generations of black holes in our Universe. To summarise:

- The first stellar mass black holes can only be detected indirectly: they leave a unique imprint on the 21-cm background in several ways: (*i*) the presence and/or depth of the global 21-cm absorption trough is directly linked to the emergence of the first black holes and their spectral properties; (*ii*) X-ray heating by the first black holes directly affects the spatial fluctuations of the 21-cm signal in ways that can be detected at $\sim$ hundreds-σ level by future low-frequency arrays such that SKA1-low and HERA. Early generations of stellar mass black holes may have contributed significantly to the unresolved soft-XRB and the near-IR background.

- If the first black holes were more massive, as has been argued theoretically, then this greatly improves the prospects for detecting them. Their continuum emission is well within reach of future facilities operating in the near-IR such as WFIRST, JWST, large ground-based facilities (E-ELT, GMT and TMT), while their X-ray emission is within reach of the deepest X-ray observations that have been performed with Chandra. Future X-ray telescopes (ATHENA, Lynx) should be able to detect X-ray emission from individual high-z massive black holes. It is theoretically possible

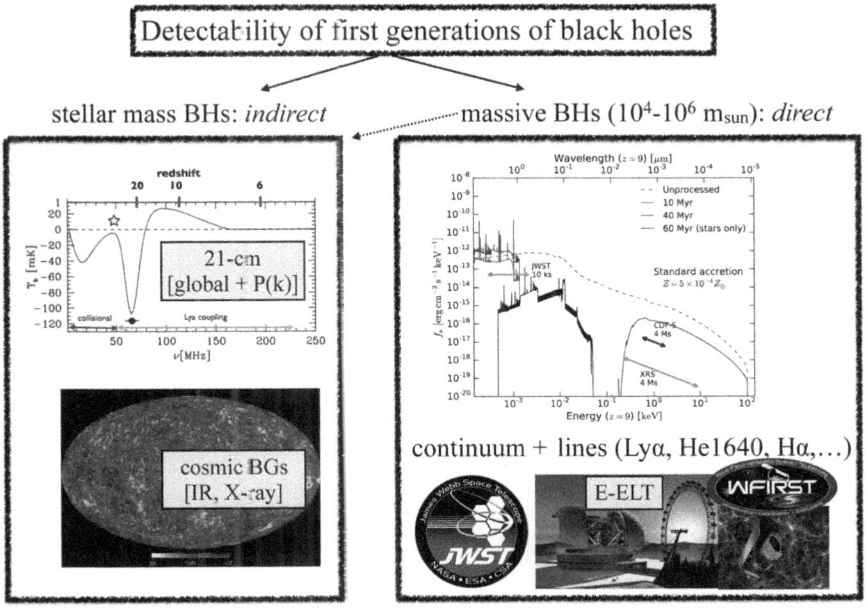

Fig. 5. Schematic representation of this chapter: we have discussed (*i*) the *indirect* detectability of the first generations of stellar mass black holes through their impact on various cosmic radiation backgrounds, ranging from the 21-cm background (*top-left*) to the IR and X-ray backgrounds (*lower left*), (*ii*) the direct detectability of continuum and line emission from the first more massive black hole seeds with a range of future space and ground-based telescopes such as WFIRST, EUCLID, JWST, and E-ELT (*right*).

to identify these black holes among other 'normal' high-z galaxies through a combination of the ratio of their optical-to-X-ray emission, and through their broad-band colors. Recombination lines provide another promising way of both finding and identifying the most distant, massive newly formed black holes, though quantitative predictions require further radiative transfer modelling.

In short: the next generation of telescopes will clearly have the capabilities to detect (indirectly and/or directly) electromagnetic radiation from the first generations of black holes. These detections (and/or non-detections) will provide the first observational constraints on the formation of the first baryonic structures in our Universe, which will represent a milestone for observational cosmology.

Acknowledgements

I would like to thank Marat Gilfanov, Felix Mirabel, and Philippe Laurent for helpful correspondence, Andrei Mesinger for helpful correspondence and for permission to use Figure 2, and Priya Natarajan for permission to use Figure 3. I would like to thank Greg Salvesen and Mike McCourt for helpful discussions, and the physics department at UCSB for their kindly hosting me while I wrote this chapter.

References

Abbott, B. P., Abbott, R., Abbott, T. D., Abernathy, M. R., Acernese, F., Ackley, K., Adams, C., Adams, T., Addesso, P., Adhikari, R. X. and et al. (2016a). GW150914: The advanced LIGO detectors in the era of first discoveries, *Physical Review Letters* **116**, 13, 131103, doi:10.1103/PhysRevLett.116.131103.

Abbott, B. P., Abbott, R., Abbott, T. D., Abernathy, M. R., Acernese, F., Ackley, K., Adams, C., Adams, T., Addesso, P., Adhikari, R. X. and et al. (2016b). Observation of gravitational waves from a binary black hole merger, *Physical Review Letters* **116**, 6, 061102, doi:10.1103/PhysRevLett.116.061102.

Abbott, B. P., Abbott, R., Abbott, T. D., Abernathy, M. R., Acernese, F., Ackley, K., Adams, C., Adams, T., Addesso, P., Adhikari, R. X. and et al. (2016c). Prospects for observing and localizing gravitational-wave transients with advanced LIGO and advanced virgo, *Living Reviews in Relativity* **19**, doi:10.1007/lrr-2016-1.

Abbott, B. P., Abbott, R., Abbott, T. D., Abernathy, M. R., Ackley, K., Adams, C., Addesso, P., Adhikari, R. X., Adya, V. B., Affeldt, C. and et al. (2017a). Exploring the sensitivity of next generation gravitational wave detectors, *Classical and Quantum Gravity* **34**, 4, 044001, doi:10.1088/1361-6382/aa51f4.

Abbott, B. P., Abbott, R., Abbott, T. D., Acernese, F., Ackley, K., Adams, C., Adams, T., Addesso, P., Adhikari, R. X., Adya, V. B. and et al. (2017b). GW170104: Observation of a 50-solar-mass binary black hole coalescence at redshift 0.2, *Physical Review Letters* **118**, 22, 221101, doi:10.1103/PhysRevLett.118.221101.

Abbott, B. P., Abbott, R., Abbott, T. D., Acernese, F., Ackley, K., Adams, C., Adams, T., Addesso, P., Adhikari, R. X., Adya, V. B. and et al. (2017c). GW170608: Observation of a 19 solar-mass binary black hole coalescence, *ApJ* **851**, L35, doi:10.3847/2041-8213/aa9f0c.

Abbott, B. P., Abbott, R., Abbott, T. D., Acernese, F., Ackley, K., Adams, C., Adams, T., Addesso, P., Adhikari, R. X., Adya, V. B. and et al. (2017d). GW170814: A three-detector observation of gravitational waves from a binary black hole coalescence, *Physical Review Letters* **119**, 14, 141101, doi:10.1103/PhysRevLett.119.141101.

Abbott, B. P., Abbott, R., Abbott, T. D., Acernese, F., Ackley, K., Adams, C., Adams, T., Addesso, P., Adhikari, R. X., Adya, V. B. and et al. (2017e). GW170817: Observation of gravitational waves from a binary neutron star inspiral, *Physical Review Letters* **119**, 16, 161101, doi:10.1103/PhysRevLett.119.161101.

Abel, T., Anninos, P., Zhang, Y. and Norman, M. L. (1997). Modeling primordial gas in numerical cosmology, *New Astronomy* **2**, pp. 181–207, doi:10.1016/S1384-1076(97)00010-9.

Abel, T., Bryan, G. L. and Norman, M. L. (2002). The formation of the first star in the universe, *Science* **295**, pp. 93–98, doi:10.1126/science.295.5552.93.

Abel, T., Wise, J. H. and Bryan, G. L. (2007). The H II region of a primordial star, *ApJ* **659**, pp. L87–L90, doi:10.1086/516820.

Abramowicz, M. A., Czerny, B., Lasota, J. P. and Szuszkiewicz, E. (1988). Slim accretion disks, *ApJ* **332**, pp. 646–658, doi:10.1086/166683.

Abramowitz, M. and Stegun, I. A. (1972). *Handbook of Mathematical Functions* (Handbook of Mathematical Functions, New York: Dover, 1972).

Agarwal, B., Cullen, F., Khochfar, S., Klessen, R. S., Glover, S. C. O. and Johnson, J. (2017a). Effects of binary stellar populations on direct collapse black hole formation, *MNRAS* **468**, pp. L82–L86, doi:10.1093/mnrasl/slx028.

Agarwal, B., Dalla Vecchia, C., Johnson, J. L., Khochfar, S. and Paardekooper, J.-P. (2014). The First Billion Years project: birthplaces of direct collapse black holes, *MNRAS* **443**, pp. 648–657, doi:10.1093/mnras/stu1112.

Agarwal, B., Davis, A. J., Khochfar, S., Natarajan, P. and Dunlop, J. S. (2013). Unravelling obese black holes in the first galaxies, *MNRAS* **432**, pp. 3438–3444, doi:10.1093/mnras/stt696.

Agarwal, B., Johnson, J. L., Khochfar, S., Pellegrini, E., Rydberg, C.-E., Klessen, R. S. and Oesch, P. (2017b). Metallicity evolution of direct collapse black hole hosts: CR7 as a case study, *ArXiv e-prints:1702.00407*.

Agarwal, B., Johnson, J. L., Zackrisson, E., Labbe, I., van den Bosch, F. C., Natarajan, P. and Khochfar, S. (2016a). Detecting direct collapse black holes: making the case for CR7, *MNRAS* **460**, pp. 4003–4010, doi:10.1093/mnras/stw1173.

Agarwal, B. and Khochfar, S. (2015). Revised rate coefficients for H_2 and H^- destruction by realistic stellar spectra, *MNRAS* **446**, pp. 160–168, doi:10.1093/mnras/stu1973.

Agarwal, B., Khochfar, S., Johnson, J. L., Neistein, E., Dalla Vecchia, C. and Livio, M. (2012). Ubiquitous seeding of supermassive black holes by direct collapse, *MNRAS* **425**, pp. 2854–2871, doi:10.1111/j.1365-2966.2012.21651.x.

Agarwal, B., Regan, J., Klessen, R. S., Downes, T. P. and Zackrisson, E. (2017c). An analytic resolution to the competition between Lyman-Werner radiation and metal winds in direct collapse black hole hosts, *ArXiv e-prints*.

Agarwal, B., Smith, B., Glover, S., Natarajan, P. and Khochfar, S. (2016b). New constraints on direct collapse black hole formation in the early Universe, *MNRAS* **459**, pp. 4209–4217, doi:10.1093/mnras/stw929.

Ahn, K., Shapiro, P. R., Iliev, I. T., Mellema, G. and Pen, U.-L. (2009). The inhomogeneous background of H_2-dissociating radiation during cosmic reionization, *ApJ* **695**, pp. 1430–1445, doi:10.1088/0004-637X/695/2/1430.

Aird, J., Coil, A. L., Georgakakis, A., Nandra, K., Barro, G. and Pérez-González, P. G. (2015). The evolution of the X-ray luminosity functions of unabsorbed and absorbed AGNs out to z $\sim$ 5, *MNRAS* **451**, pp. 1892–1927, doi:10.1093/mnras/stv1062.

Alcock, C., Allsman, R. A., Alves, D. R., Axelrod, T. S., Becker, A. C., Bennett, D. P., Cook, K. H., Dalal, N., Drake, A. J., Freeman, K. C., Geha, M. and Griest, K. et al. (2000). The MACHO project: Microlensing results from 5.7 years of large magellanic cloud observations, *ApJ* **542**, pp. 281–307, doi:10.1086/309512.

Alexander, D. M., Bauer, F. E., Brandt, W. N., Schneider, D. P., Hornschemeier, A. E. and Vignali, C. et al. (2003). The Chandra Deep Field north survey. XIII. 2 Ms Point-Source Catalogs, *AJ* **126**, pp. 539–574, doi:10.1086/376473.

Alexander, D. M. and Hickox, R. C. (2012). What drives the growth of black holes? *New Astronomy Reviews* **56**, pp. 93–121, doi:10.1016/j.newar.2011.11.003.

Alexander, T. (2017). Stellar dynamics and stellar phenomena near a massive black hole, *ARA&A* **55**, pp. 17–57, doi:10.1146/annurev-astro-091916-055306.

Alexander, T. and Bar-Or, B. (2017). A universal minimal mass scale for present-day central black holes, *Nature Astronomy* **1**, 0147, doi:10.1038/s41550-017-0147.

Alexander, T. and Natarajan, P. (2014). Rapid growth of seed black holes in the early Universe by supra-exponential accretion, *Science* **345**, pp. 1330–1333, doi:10.1126/science.1251053.

Alvarez, M. A., Bromm, V. and Shapiro, P. R. (2006). The H II region of the first star, *ApJ* **639**, pp. 621–632, doi:10.1086/499578.

Alvarez, M. A., Wise, J. H. and Abel, T. (2009). Accretion onto the first stellar-mass black holes, *ApJ* **701**, pp. L133–L137, doi:10.1088/0004-637X/701/2/L133.

Amaro-Seoane, P., Aoudia, S., Babak, S., Binétruy, P., Berti, E., Bohé, A., Caprini, C., Colpi, M., Cornish, N. J., Danzmann, K., Dufaux, J.-F., Gair, J. and Hinder, I. et al. (2013). eLISA: Astrophysics and cosmology in the millihertz regime, *GW Notes, Vol. 6, p. 4-110* **6**, pp. 4–110.

Amaro-Seoane, P., Audley, H., Babak, S., Baker, J., Barausse, E., Bender, P., Berti, E., Binetruy, P., Wass, P., Weber, W., Ziemer, J. and Zweifel, P. (2017). Laser Interferometer Space Antenna, *ArXiv e-prints 1702.00786.*

Antonini, F., Barausse, E. and Silk, J. (2015). The coevolution of nuclear star clusters, massive black holes, and their host galaxies, *ApJ* **812**, 72, doi:10.1088/0004-637X/812/1/72.

Antonov, V. A. (1962). *Solution of the problem of stability of stellar system Emden's density law and the spherical distribution of velocities* (Vestnik Leningradskogo Universiteta, Leningrad).

Aoki, W., Tominaga, N., Beers, T. C., Honda, S. and Lee, Y. S. (2014). A chemical signature of first-generation very massive stars, *Science* **345**, pp. 912–915, doi:10.1126/science.1252633.

Arfken, G. B. and Weber, H. J. (2005). *Mathematical methods for physicists 6th ed.* (Materials and Manufacturing Processes).

Asplund, M., Grevesse, N., Sauval, A. J. and Scott, P. (2009). The chemical composition of the Sun, *ARA&A* **47**, pp. 481–522, doi:10.1146/annurev.astro.46.060407.145222.

Athanassoula, E. (1992). The existence and shapes of dust lanes in galactic bars, *MNRAS* **259**, pp. 345–364, doi:10.1093/mnras/259.2.345.

Aykutalp, A. and Wise, J. (2017). Co-evolution of central direct collapse black holes and stellar populations in the early Universe, in *American Astronomical Society Meeting Abstracts*, *American Astronomical Society Meeting Abstracts*, Vol. 229, p. 406.05.

Aykutalp, A., Wise, J. H., Spaans, M. and Meijerink, R. (2014). Songlines from direct collapse seed black holes: effects of X-rays on black hole growth and stellar populations, *ApJ* **797**, 139, doi:10.1088/0004-637X/797/2/139.

Bañados, E., Connor, T., Stern, D., Mulchaey, J., Fan, X., Decarli, R., Farina, E. P., Mazzucchelli, C., Venemans, B. P., Walter, F., Wang, F. and Yang, J. (2018a). Chandra X-Rays from the Redshift 7.54 Quasar ULAS J1342+0928, *ApJ* **856**, L25, doi:10.3847/2041-8213/aab61e.

Bañados, E., Venemans, B. P., Decarli, R., Farina, E. P., Mazzucchelli, C., Walter, F., Fan, X., Stern, D., Schlafly, E. and Chambers, K. C. et al. (2016). The Pan-STARRS1 Distant z 5.6 Quasar Survey: More than 100 quasars within the first Gyr of the Universe, *ApJS* **227**, 11, doi:10.3847/0067-0049/227/1/11.

Bañados, E., Venemans, B. P., Mazzucchelli, C., Farina, E. P., Walter, F., Wang, F. and Decarli, R. et al. (2018b). An 800-million-solar-mass black hole in a significantly neutral Universe at a redshift of 7.5, *Nature* **553**, pp. 473–476, doi:10.1038/nature25180.

Bañados, E., Venemans, B. P., Morganson, E., Decarli, R., Walter, F., Chambers, K. C., Rix, H.-W., Farina, E. P., Fan, X., Jiang, L. and McGreer, I. et al. (2014). Discovery of Eight z ˜ 6 Quasars from Pan-STARRS1, *AJ* **148**, 14, doi:10.1088/0004-6256/148/1/14.

Baade, W. and Zwicky, F. (1934). On Super-novae, *Proceedings of the National Academy of Science* **20**, pp. 254–259, doi:10.1073/pnas.20.5.254.

Babak, S., Gair, J., Sesana, A., Barausse, E., Sopuerta, C. F., Berry, C. P. L., Berti, E., Amaro-Seoane, P., Petiteau, A. and Klein, A. (2017). Science with the space-based interferometer LISA. V. Extreme mass-ratio inspirals, *Phys. Rev. D* **95**, 10, 103012, doi:10.1103/PhysRevD.95.103012.

Bachetti, M., Harrison, F. A., Walton, D. J., Grefenstette, B. W., Chakrabarty, D., Fürst, F., Barret, D., Beloborodov, A., Boggs, S. E. and Christensen, F. E. et al. (2014). An ultraluminous X-ray source powered by an accreting neutron star, *Nature* **514**, pp. 202–204, doi:10.1038/nature13791.

Bade, N., Komossa, S. and Dahlem, M. (1996). Detection of an extremely soft X-ray outburst in the HII-like nucleus of NGC 5905. *A&A* **309**, pp. L35–L38.

Balbus, S. A. and Hawley, J. F. (1991). A powerful local shear instability in weakly magnetized disks. I - Linear analysis. II - Nonlinear evolution, *ApJ* **376**, pp. 214–233, doi:10.1086/170270.

Balbus, S. A. and Hawley, J. F. (1998). Instability, turbulence, and enhanced transport in accretion disks, *Reviews of Modern Physics* **70**, pp. 1–53, doi: 10.1103/RevModPhys.70.1.

Baldassare, V. F., Reines, A. E., Gallo, E. and Greene, J. E. (2015). A 50,000 solar mass black hole in the nucleus of RGG 118, *ApJ* **809**, L14, doi: 10.1088/2041-8205/809/1/L14.

Ball, W. H., Tout, C. A., Żytkow, A. N. and Eldridge, J. J. (2011). The structure and evolution of quasi-stars, *MNRAS* **414**, pp. 2751–2762, doi:10.1111/j.1365-2966.2011.18591.x.

Banerjee, R. and Jedamzik, K. (2003). Are cluster magnetic fields primordial? *Physical Review Letters* **91**, 25, 251301, doi:10.1103/PhysRevLett.91.251301.

Barausse, E., Morozova, V. and Rezzolla, L. (2012). On the mass radiated by coalescing black hole binaries, *ApJ* **758**, 63, doi:10.1088/0004-637X/758/1/63.

Bardeen, J. M. (1970). Kerr metric black holes, *Nature* **226**, pp. 64–65, doi: 10.1038/226064a0.

Bardeen, J. M., Bond, J. R., Kaiser, N. and Szalay, A. S. (1986). The statistics of peaks of Gaussian random fields, *ApJ* **304**, pp. 15–61, doi:10.1086/164143.

Bardeen, J. M. and Petterson, J. A. (1975). The lense-thirring effect and accretion disks around Kerr black holes, *ApJ* **195**, p. L65, doi:10.1086/181711.

Bardeen, J. M., Press, W. H. and Teukolsky, S. A. (1972). Rotating black holes: locally nonrotating frames, energy extraction, and scalar synchrotron radiation, *ApJ* **178**, pp. 347–370, doi:10.1086/151796.

Barack, L., Cardoso, V., Nissanke, S., Sotiriou, T. P., Askar, A., Belczynski, C., Bertone, G., Bon, E., Blas, D., Brito, R., Bulik, T., Burrage, C., Byrnes, C. T., Caprini, C., Chernyakova, M. *et al.* (2018). Black holes, gravitational waves and fundamental physics: a roadmap, ArXiv e-prints 1806.05195.

Barkana, R. and Loeb, A. (2001). In the beginning: the first sources of light and the reionization of the universe, *Phys. Rep.* **349**, pp. 125–238, doi: 10.1016/S0370-1573(01)00019-9.

Barnes, J. E. (2002). Formation of gas discs in merging galaxies, *MNRAS* **333**, pp. 481–494, doi:10.1046/j.1365-8711.2002.05335.x.

Barnes, J. E. and Hernquist, L. (1996). Transformations of galaxies. II. Gasdynamics in merging disk galaxies, *ApJ* **471**, p. 115, doi:10.1086/177957.

Barrow, K. S. S., Wise, J. H., Norman, M. L., O'Shea, B. W. and Xu, H. (2017). First light: exploring the spectra of high-redshift galaxies in the Renaissance Simulations, *MNRAS* **469**, pp. 4863–4878, doi:10.1093/mnras/stx1181.

Bartelmann, M., Fabis, F., Berg, D., Kozlikin, E., Lilow, R. and Viermann, C. (2014). A microscopic, non-equilibrium, statistical field theory for cosmic structure formation, *ArXiv e-prints:1411.0806*.

Basu-Zych, A. R., Lehmer, B. D., Hornschemeier, A. E., Bouwens, R. J., Fragos, T., Oesch, P. A., Belczynski, K., Brandt, W. N., Kalogera, V., Luo, B., Miller, N., Mullaney, J. R., Tzanavaris, P., Xue, Y. and Zezas, A. (2013). The X-ray star formation story as told by Lyman Break Galaxies in the 4 Ms CDF-S, *ApJ* **762**, 45, doi:10.1088/0004-637X/762/1/45.

Bate, M. R., Bonnell, I. A. and Price, N. M. (1995). Modelling accretion in protobinary systems, *MNRAS* **277**, pp. 362–376.

Baumann, D. and Peiris, H. V. (2008). Cosmological inflation: theory and observations, *ArXiv e-prints 0810.3022*.

Becerra, F., Greif, T. H., Springel, V. and Hernquist, L. E. (2015). Formation of massive protostars in atomic cooling haloes, *MNRAS* **446**, pp. 2380–2393, doi:10.1093/mnras/stu2284.

Beck, R., Brandenburg, A., Moss, D., Shukurov, A. and Sokoloff, D. (1996). Galactic magnetism: recent developments and perspectives, *ARA&A* **34**, pp. 155–206, doi:10.1146/annurev.astro.34.1.155.

Beers, T. C. and Christlieb, N. (2005). The discovery and analysis of very metal-poor stars in the galaxy, *ARA&A* **43**, pp. 531–580, doi:10.1146/annurev.astro.42.053102.134057.

Begelman, M. C. (1979). Can a spherically accreting black hole radiate very near the Eddington limit, *MNRAS* **187**, pp. 237–251, doi:10.1093/mnras/187.2.237.

Begelman, M. C. (2010). Evolution of supermassive stars as a pathway to black hole formation, *MNRAS* **402**, pp. 673–681, doi:10.1111/j.1365-2966.2009.15916.x.

Begelman, M. C. (2012). Radiatively inefficient accretion: breezes, winds and hyperaccretion, *MNRAS* **420**, pp. 2912–2923, doi:10.1111/j.1365-2966.2011.20071.x.

Begelman, M. C., Blandford, R. D. and Rees, M. J. (1980). Massive black hole binaries in active galactic nuclei, *Nature* **287**, pp. 307–309, doi:10.1038/287307a0.

Begelman, M. C., Blandford, R. D. and Rees, M. J. (1984). Theory of extragalactic radio sources, *Reviews of Modern Physics* **56**, pp. 255–351, doi:10.1103/RevModPhys.56.255.

Begelman, M. C. and Rees, M. J. (1978). The fate of dense stellar systems, *MNRAS* **185**, pp. 847–860, doi:10.1093/mnras/185.4.847.

Begelman, M. C., Rossi, E. M. and Armitage, P. J. (2008). Quasi-stars: accreting black holes inside massive envelopes, *MNRAS* **387**, pp. 1649–1659, doi:10.1111/j.1365-2966.2008.13344.x.

Begelman, M. C. and Shlosman, I. (2009). Angular momentum transfer and lack of fragmentation in self-gravitating accretion flows, *ApJ* **702**, pp. L5–L8, doi:10.1088/0004-637X/702/1/L5.

Begelman, M. C., Volonteri, M. and Rees, M. J. (2006). Formation of supermassive black holes by direct collapse in pre-galactic haloes, *MNRAS* **370**, pp. 289–298, doi:10.1111/j.1365-2966.2006.10467.x.

Behrend, R. and Maeder, A. (2001). Formation of massive stars by growing accretion rate, *A&A* **373**, pp. 190–198, doi:10.1051/0004-6361:20010585.

Belczynski, K., Dominik, M., Bulik, T., O'Shaughnessy, R., Fryer, C. and Holz, D. E. (2010). The effect of metallicity on the detection prospects for gravitational waves, *ApJ* **715**, pp. L138–L141, doi:10.1088/2041-8205/715/2/L138.

Belczynski, K., Heger, A., Gladysz, W., Ruiter, A. J., Woosley, S., Wiktorowicz, G., Chen, H.-Y., Bulik, T., O'Shaughnessy, R., Holz, D. E., Fryer, C. L. and Berti, E. (2016a). The effect of pair-instability mass loss on black-hole mergers, *A&A* **594**, A97, doi:10.1051/0004-6361/201628980.

Belczynski, K., Holz, D. E., Bulik, T. and O'Shaughnessy, R. (2016b). The first gravitational-wave source from the isolated evolution of two stars in the 40-100 solar mass range, *Nature* **534**, pp. 512–515, doi:10.1038/nature18322.

Belczynski, K., Ryu, T., Perna, R., Berti, E., Tanaka, T. L. and Bulik, T. (2017). On the likelihood of detecting gravitational waves from Population III compact object binaries, *MNRAS* **471**, pp. 4702–4721, doi:10.1093/mnras/stx1759.

Bellovary, J., Volonteri, M., Governato, F., Shen, S., Quinn, T. and Wadsley, J. (2011). The first massive black hole seeds and their hosts, *ApJ* **742**, 13, doi:10.1088/0004-637X/742/1/13.

Bellovary, J. M., Mac Low, M.-M., McKernan, B. and Ford, K. E. S. (2016). Migration traps in disks around supermassive black holes, *ApJ* **819**, L17, doi:10.3847/2041-8205/819/2/L17.

Bennett, D. P., Becker, A. C., Quinn, J. L., Tomaney, A. B., Alcock, C., Allsman, R. A., Alves, D. R., Axelrod, T. S., Calitz, J. J., Cook, K. H. and Drake, A. J. et al. (2002). Gravitational microlensing events due to stellar-mass black holes, *ApJ* **579**, pp. 639–659, doi:10.1086/342225.

Benson, A. J. (2010). Galaxy formation theory, *Phys. Rep.* **495**, pp. 33–86, doi:10.1016/j.physrep.2010.06.001.

Bernal, J. L., Raccanelli, A., Verde, L. and Silk, J. (2018). Signatures of primordial black holes as seeds of supermassive black holes, *JCAP* **5**, 17, doi:10.1088/1475-7516/2018/05/017.

Bernardeau, F., Colombi, S., Gaztañaga, E. and Scoccimarro, R. (2002). Large-scale structure of the Universe and cosmological perturbation theory, *Phys. Rep.* **367**, pp. 1–3.

Bernardeau, F., Crocce, M. and Scoccimarro, R. (2008). Multipoint propagators in cosmological gravitational instability, *Phys. Rev. D* **78**, 10, pp. 103521–+, doi:10.1103/PhysRevD.78.103521.

Berti, E., Barausse, E., Cardoso, V., Gualtieri, L., Pani, P., Sperhake, U., Stein, L. C., Wex, N., Yagi, K., Baker, T., Burgess, C. P., Coelho, F. S., Doneva, D., De Felice, A. Ferreira, P. G., Freire, P. C. C. and Healy, J. et al. (2015). Testing general relativity with present and future astrophysical observations, *Classical and Quantum Gravity* **32**, 24, 243001, doi:10.1088/0264-9381/32/24/243001.

Berti, E., Sesana, A., Barausse, E., Cardoso, V. and Belczynski, K. (2016). Spectroscopy of Kerr black holes with earth- and space-based interferometers, *Physical Review Letters* **117**, 10, 101102, doi:10.1103/PhysRevLett.117.101102.

Berti, E. and Volonteri, M. (2008). Cosmological black hole spin evolution by Mergers and Accretion, *ApJ* **684**, 822-828, doi:10.1086/590379.

Bertone, G., Hooper, D. and Silk, J. (2005). Particle dark matter: evidence, candidates and constraints, *Phys. Rep.* **405**, pp. 279–390, doi:10.1016/j.physrep.2004.08.031.

Bieri, R., Dubois, Y., Rosdahl, J., Wagner, A., Silk, J. and Mamon, G. A. (2017). Outflows driven by quasars in high-redshift galaxies with radiation hydrodynamics, *MNRAS* **464**, pp. 1854–1873, doi:10.1093/mnras/stw2380.

Biermann, L. (1950). Über den Ursprung der Magnetfelder auf Sternen und im interstellaren Raum (miteinem Anhang von A. Schlüter), *Zeitschrift Naturforschung Teil A* **5**, p. 65.

Binney, J. and Tremaine, S. (1987). *Galactic dynamics* (Princeton University Press).

Binney, J. and Tremaine, S. (2008). *Galactic Dynamics: Second Edition* (Princeton: Princeton University Press).

Black, J. H. (2006). Chemistry and cosmology, *Faraday Discussions* **133**, p. 27, doi:10.1039/b516837f.

Blanchet, L. (2014). Gravitational radiation from post-Newtonian sources and inspiralling compact binaries, *Living Reviews in Relativity* **17**, doi:10.12942/lrr-2014-2.

Blandford, R. D. and McKee, C. F. (1982). Reverberation mapping of the emission line regions of Seyfert galaxies and quasars, *ApJ* **255**, pp. 419–439, doi:10.1086/159843.

Blandford, R. D. and Payne, D. G. (1982). Hydromagnetic flows from accretion discs and the production of radio jets, *MNRAS* **199**, pp. 883–903, doi:10.1093/mnras/199.4.883.

Blandford, R. D. and Znajek, R. L. (1977). Electromagnetic extraction of energy from Kerr black holes, *MNRAS* **179**, pp. 433–456, doi:10.1093/mnras/179.3.433.

Blecha, L., Loeb, A. and Narayan, R. (2013). Double-peaked narrow-line signatures of dual supermassive black holes in galaxy merger simulations, *MNRAS* **429**, pp. 2594–2616, doi:10.1093/mnras/sts533.

Blecha, L., Snyder, G. F., Satyapal, S. and Ellison, S. L. (2018). The power of infrared AGN selection in mergers: a theoretical study, *MNRAS* **478**, pp. 3056–3071, doi:10.1093/mnras/sty1274.

Bleuler, A. and Teyssier, R. (2014). Towards a more realistic sink particle algorithm for the RAMSES CODE, *MNRAS* **445**, pp. 4015–4036, doi:10.1093/mnras/stu2005.

Bloom, J. S., Giannios, D., Metzger, B. D., Cenko, S. B., Perley, D. A., Butler, N. R., Tanvir, N. R., Levan, A. J., O'Brien, P. T. and Strubbe, L. E. et al. (2011). A possible relativistic jetted outburst from a massive black hole fed by a tidally disrupted star, *Science* **333**, p. 203, doi:10.1126/science.1207150.

Blumenthal, G. R., Faber, S. M., Flores, R. and Primack, J. R. (1986). Contraction of dark matter galactic halos due to baryonic infall, *ApJ* **301**, pp. 27–34, doi:10.1086/163867.

Boekholt, T. C. N., Schleicher, D. R. G., Fellhauer, M., Klessen, R. S., Reinoso, B., Stutz, A. M and Haemmerlé, L. (2018). Formation of massive seed black holes via collisions and accretion, *MNRAS* **476**, pp. 366–380, doi: 10.1093/mnras/sty208.

Boerner, G. (2003). *The early universe: facts and fiction* (The early universe: facts and fiction, 4th ed. By G. Boerner. Astronomy and astrophysics library. Berlin: Springer, 2003.).

Bond, J. R., Cole, S., Efstathiou, G. and Kaiser, N. (1991). Excursion set mass functions for hierarchical Gaussian fluctuations, *ApJ* **379**, pp. 440–460, doi:10.1086/170520.

Bondi, H. (1952). On spherically symmetrical accretion, *MNRAS* **112**, p. 195, doi:10.1093/mnras/112.2.195.

Bondi, H. and Hoyle, F. (1944). On the mechanism of accretion by stars, *MNRAS* **104**, p. 273, doi:10.1093/mnras/104.5.273.

Bonetti, M., Haardt, F., Sesana, A. and Barausse, E. (2017). Post-Newtonian evolution of massive black hole triplets in galactic nuclei – II. Survey of the parameter space, *ArXiv e-prints 1709.06088*.

Bonetti, M., Perego, A., Capelo, P. R., Dotti, M. and Miller, M. C. (2018). r-process nucleosynthesis in the early Universe through fast mergers of compact binaries in triple systems, *PASA* **35**, e017, doi:10.1017/pasa.2018.11.

Bonifacio, P., Caffau, E., Spite, M., Limongi, M., Chieffi, A., Klessen, R. S., François, P., Molaro, P., Ludwig, H.-G., Zaggia, S. and Spite, F. et al. (2015). TOPoS . II. On the bimodality of carbon abundance in CEMP stars Implications on the early chemical evolution of galaxies, *A&A* **579**, A28, doi:10.1051/0004-6361/201425266.

Bonnell, I. A., Bate, M. R., Clarke, C. J. and Pringle, J. E. (2001). Competitive accretion in embedded stellar clusters, *MNRAS* **323**, pp. 785–794, doi: 10.1046/j.1365-8711.2001.04270.x.

Booth, C. and Schaye, J. (2009). Cosmological simulations of the growth of supermassive black holes and feedback from active galactic nuclei: method and tests, *MNRAS* **398**, 53.

Bouchet, F. R., Juszkiewicz, R., Colombi, S. and Pellat, R. (1992). Weakly nonlinear gravitational instability for arbitrary Omega, *ApJ* **394**, pp. L5–L8, doi:10.1086/186459.

Bouwens, R. J., Illingworth, G. D., Oesch, P. A., Labbé, I., Trenti, M., van Dokkum, P., Franx, M., Stiavelli, M., Carollo, C. M., Magee, D. and Gonzalez, V. (2011). Ultraviolet luminosity functions from 132 z ˜ 7 and z ˜ 8 Lyman-break galaxies in the ultra-deep HUDF09 and wide-area early release science WFC3/IR observations, *ApJ* **737**, 90, doi: 10.1088/0004-637X/737/2/90.

Bovino, S., Grassi, T., Latif, M. A. and Schleicher, D. R. G. (2013a). Impact of an accurate modelling of primordial chemistry in high-resolution studies, *MNRAS* **434**, pp. L36–L40, doi:10.1093/mnrasl/slt072.

Bovino, S., Grassi, T., Schleicher, D. R. G. and Banerjee, R. (2016). The Formation of the Primitive Star SDSS J102915+172927: Effect of the dust mass and the grain-size distribution, *ApJ* **832**, 154, doi:10.3847/0004-637X/832/2/154.

Bovino, S., Latif, M. A., Grassi, T. and Schleicher, D. R. G. (2014a). Dark-matter halo mergers as a fertile environment for low-mass Population III star formation, *MNRAS* **441**, pp. 2181–2187, doi:10.1093/mnras/stu714.

Bovino, S., Schleicher, D. R. G. and Grassi, T. (2014b). Primordial star formation: relative impact of H_2 three-body rates and initial conditions, *A&A* **561**, A13, doi:10.1051/0004-6361/201322387.

Bovino, S., Schleicher, D. R. G. and Schober, J. (2013b). Turbulent magnetic field amplification from the smallest to the largest magnetic Prandtl numbers, *New Journal of Physics* **15**, 1, 013055, doi:10.1088/1367-2630/15/1/013055.

Bowler, R. A. A., McLure, R. J., Dunlop, J. S., McLeod, D. J., Stanway, E. R., Eldridge, J. J. and Jarvis, M. J. (2017). No evidence for Population III stars or a Direct Collapse Black Hole in the z = 6.6 Lyman-α emitter 'CR7', *MNRAS* **469**, pp. 448–458, doi:10.1093/mnras/stx839.

Bowman, J. D. and Rogers, A. E. E. (2010). A lower limit of $\Delta z > 0.06$ for the duration of the reionization epoch, *Nature* **468**, pp. 796–798, doi:10.1038/nature09601.

Boyer, R. H. and Lindquist, R. W. (1967). Maximal analytic extension of the Kerr metric, *Journal of Mathematical Physics* **8**, pp. 265–281, doi:10.1063/1.1705193.

Brandenburg, A. and Subramanian, K. (2005). Astrophysical magnetic fields and nonlinear dynamo theory, *Phys. Rep.* **417**, pp. 1–209, doi:10.1016/j.physrep.2005.06.005.

Brandt, W. N. and Alexander, D. M. (2015). Cosmic X-ray surveys of distant active galaxies. The demographics, physics, and ecology of growing supermassive black holes, *A&ARv* **23**, 1, doi:10.1007/s00159-014-0081-z.

Bridle, A. H. and Perley, R. A. (1984). Extragalactic radio jets, *ARA&A* **22**, pp. 319–358, doi:10.1146/annurev.aa.22.090184.001535.

Bromm, V. (2013). Formation of the first stars, *Reports on Progress in Physics* **76**, 11, 112901, doi:10.1088/0034-4885/76/11/112901.

Bromm, V., Coppi, P. S. and Larson, R. B. (2002). The formation of the first stars. I. the primordial star-forming cloud, *ApJ* **564**, pp. 23–51, doi:10.1086/323947.

Bromm, V. and Larson, R. B. (2004). The first stars, *ARA&A* **42**, pp. 79–118, doi:10.1146/annurev.astro.42.053102.134034.

Bromm, V. and Loeb, A. (2003). Formation of the first supermassive black holes, *ApJ* **596**, pp. 34–46, doi:10.1086/377529.

Bromm, V. and Yoshida, N. (2011). The first galaxies, *ARA&A* **49**, pp. 373–407, doi:10.1146/annurev-astro-081710-102608.

Bruhns, H., Kreckel, H., Miller, K. A., Urbain, X. and Savin, D. W. (2010). Absolute energy-resolved measurements of the $H^-+H\rightarrow H_2+e^-$ associative detachment reaction using a merged-beam apparatus, *Phys. Rev. A* **82**, 4, 042708, doi:10.1103/PhysRevA.82.042708.

Buchert, T. (1989). A class of solutions in Newtonian cosmology and the pancake theory, *A&A* **223**, pp. 9–24.

Buchert, T. (1992). Lagrangian theory of gravitational instability of Friedman-Lemaitre cosmologies and the 'Zel'dovich approximation', *MNRAS* **254**, pp. 729–737.

Buchert, T. (1994). Lagrangian theory of gravitational instability of Friedman-Lemaitre cosmologies — a generic third-order model for nonlinear clustering, *MNRAS* **267**, pp. 811–+.

Buchert, T. and Götz, G. (1987). A class of solutions for self-gravitating dust in Newtonian gravity, *Journal of Mathematical Physics* **28**, pp. 2714–2719.

Buonanno, A. and Damour, T. (1999). Effective one-body approach to general relativistic two-body dynamics, *Phys. Rev. D* **59**, 8, 084006, doi:10.1103/PhysRevD.59.084006.

Buonanno, A. and Sathyaprakash, B. S. (2014). Sources of gravitational waves: theory and observations, *ArXiv e-prints 1410.7832.*

Caffau, E., Bonifacio, P., François, P., Sbordone, L., Monaco, L., Spite, M., Spite, F., Ludwig, H.-G., Cayrel, R., Zaggia, S., Hammer, F., Randich, S., Molaro, P. and Hill, V. (2011a). An extremely primitive star in the Galactic halo, *Nature* **477**, pp. 67–69, doi:10.1038/nature10377.

Caffau, E., Bonifacio, P., François, P., Spite, M., Spite, F., Zaggia, S., Ludwig, H.-G., Steffen, M., Mashonkina, L., Monaco, L. and Sbordone, L. et al. (2012). A primordial star in the heart of the Lion, *A&A* **542**, A51, doi:10.1051/0004-6361/201118744.

Caffau, E., Bonifacio, P., Sbordone, L., François, P., Monaco, L., Spite, M., Plez, B., Cayrel, R., Christlieb, N., Clark, P., Glover, S. and Klessen, R. et al. (2013). TOPoS. I. Survey design and analysis of the first sample, *A&A* **560**, A71, doi:10.1051/0004-6361/201322488.

Caffau, E., Ludwig, H.-G., Steffen, M., Freytag, B. and Bonifacio, P. (2011b). Solar chemical abundances determined with a CO5BOLD 3D model atmosphere, *Sol. Phys.* **268**, pp. 255–269, doi:10.1007/s11207-010-9541-4.

Callegari, S., Kazantzidis, S., Mayer, L., Colpi, M., Bellovary, J. M., Quinn, T. and Wadsley, J. (2011). Growing massive black hole pairs in minor mergers of disk galaxies, *ApJ* **729**, 85, doi:10.1088/0004-637X/729/2/85.

Callegari, S., Mayer, L., Kazantzidis, S., Colpi, M., Governato, F., Quinn, T. and Wadsley, J. (2009). Pairing of supermassive black holes in unequal-mass galaxy mergers, *ApJ* **696**, pp. L89–L92, doi:10.1088/0004-637X/696/1/L89.

Campanelli, M., Lousto, C. O., Zlochower, Y. and Merritt, D. (2007). Maximum gravitational recoil, *Physical Review Letters* **98**, 23, 231102, doi:10.1103/PhysRevLett.98.231102.

Capelo, P. R. and Dotti, M. (2017). Shocks and angular momentum flips: a different path to feeding the nuclear regions of merging galaxies, *MNRAS* **465**, pp. 2643–2653, doi:10.1093/mnras/stw2872.

Capelo, P. R., Dotti, M., Volonteri, M., Mayer, L., Bellovary, J. M. and Shen, S. (2017). A survey of dual active galactic nuclei in simulations of galaxy mergers: frequency and properties, *MNRAS* **469**, pp. 4437–4454, doi:10.1093/mnras/stx1067.

Capelo, P. R., Volonteri, M., Dotti, M., Bellovary, J. M., Mayer, L. and Governato, F. (2015). Growth and activity of black holes in galaxy mergers with varying mass ratios, *MNRAS* **447**, pp. 2123–2143, doi:10.1093/mnras/stu2500.

Cappelluti, N., Li, Y., Ricarte, A., Agarwal, B., Allevato, V., Tasnim Ananna, T., Ajello, M., Civano, F., Comastri, A., Elvis, M. and Finoguenov, A. et al. (2017). The Chandra COSMOS legacy survey: energy spectrum of the cosmic X-ray background and constraints on undetected populations, *ApJ* **837**, 19, doi:10.3847/1538-4357/aa5ea4.

Cardoso, V. and Pani, P. (2017). Tests for the existence of black holes through gravitational wave echoes, *Nature Astronomy* **1**, pp. 586–591, doi:10.1038/s41550-017-0225-y.

Carnall, A. C., Shanks, T., Chehade, B., Fumagalli, M., Rauch, M., Irwin, M. J., Gonzalez-Solares, E., Findlay, J. R. and Metcalfe, N. (2015). Two bright z 6 quasars from VST ATLAS and a new method of optical plus mid-infrared colour selection, *MNRAS* **451**, pp. L16–L20, doi:10.1093/mnrasl/slv057.

Carr, B., Kühnel, F. and Sandstad, M. (2016a). Primordial black holes as dark matter, *Phys. Rev. D* **94**, 8, 083504, doi:10.1103/PhysRevD.94.083504.

Carr, B., Raidal, M., Tenkanen, T., Vaskonen, V. and Veermäe, H. (2017). Primordial black hole constraints for extended mass functions, *Phys. Rev. D* **96**, 2, 023514, doi:10.1103/PhysRevD.96.023514.

Carr, B. J. (2005). Primordial black holes recent developments, in P. Chen, E. Bloom, G. Madejski and V. Patrosian (eds.), *22nd Texas Symposium on Relativistic Astrophysics*, pp. 89–100.

Carr, B. J. and Hawking, S. W. (1974). Black holes in the early Universe, *MNRAS* **168**, pp. 399–416, doi:10.1093/mnras/168.2.399.

Carr, B. J., Kohri, K., Sendouda, Y. and Yokoyama, J. (2016b). Constraints on primordial black holes from the Galactic gamma-ray background, *Phys. Rev. D* **94**, 4, 044029, doi:10.1103/PhysRevD.94.044029.

Catelan, P. (1995). Lagrangian dynamics in non-flat universes and non-linear gravitational evolution, *MNRAS* **276**, pp. 115–124.

Catelan, P., Lucchin, F., Matarrese, S. and Moscardini, L. (1995). Eulerian perturbation theory in non-flat universes: second-order approximation, *MNRAS* **276**, pp. 39–56.

Catelan, P. and Theuns, T. (1996). Evolution of the angular momentum of protogalaxies from tidal torques: Zel'dovich approximation, *MNRAS* **282**, pp. 436–454.

Chabrier, G. (2003). Galactic stellar and substellar initial mass function, *PASP* **115**, pp. 763–795, doi:10.1086/376392.

Chan, C. and Heger, A. (2017). Combined Nucleosynthetic Yields of Multiple First Stars, in S. Kubono, T. Kajino, S. Nishimura, T. Isobe, S. Nagataki, T. Shima and Y. Takeda (eds.), *14th International Symposium on Nuclei in the Cosmos (NIC2016)*, p. 020209, doi:10.7566/JPSCP.14.020209.

Chandrasekhar, S. (1931). The maximum mass of ideal white dwarfs, *ApJ* **74**, p. 81, doi:10.1086/143324.

Chandrasekhar, S. (1943). Dynamical friction. I. general considerations: the coefficient of dynamical friction. *ApJ* **97**, p. 255, doi:10.1086/144517.

Chandrasekhar, S. (1951a). The fluctuations of density in isotropic turbulence, *Proceedings of the Royal Society of London Series A* **210**, pp. 18–25, doi: 10.1098/rspa.1951.0227.

Chandrasekhar, S. (1951b). The gravitational instability of an infinite homogeneous turbulent medium, *Proceedings of the Royal Society of London Series A* **210**, pp. 26–29, doi:10.1098/rspa.1951.0228.

Chandrasekhar, S. (1964). The dynamical instability of gaseous masses approaching the schwarzschild limit in general relativity. *ApJ* **140**, p. 417, doi: 10.1086/147938.

Chehade, B., Carnall, A. C., Shanks, T., Diener, C., Fumagalli, M., Findlay, J. R., Metcalfe, N., Hennawi, J., Leibler, C., Murphy, D. N. A., Prochaska, J. X., Irwin, M. J. and Gonzalez-Solares, E. (2018). Two more, bright, z 6 quasars from VST ATLAS and WISE, *MNRAS* **478**, pp. 1649–1659, doi:10.1093/mnras/sty690.

Chen, P., Wise, J. H., Norman, M. L., Xu, H. and O'Shea, B. W. (2014). Scaling relations for galaxies prior to reionization, *ApJ* **795**, 144, doi: 10.1088/0004-637X/795/2/144.

Cheng, T.-P. (2005). *Relativity, gravitation and cosmology. A basic introduction* (Oxford University Press).

Chevallier, M. and Polarski, D. (2001). Accelerating universes with scaling dark matter, *International Journal of Modern Physics D* **10**, pp. 213–223, doi: 10.1142/S0218271801000822.

Chiaki, G., Nozawa, T. and Yoshida, N. (2013a). Growth of dust grains in a low-metallicity gas and its effect on the cloud fragmentation, *ApJ* **765**, L3, doi:10.1088/2041-8205/765/1/L3.

Chiaki, G., Schneider, R., Nozawa, T., Omukai, K., Limongi, M., Yoshida, N. and Chieffi, A. (2014). Dust grain growth and the formation of the extremely primitive star SDSS J102915+172927, *MNRAS* **439**, pp. 3121–3127, doi: 10.1093/mnras/stu178.

Chiaki, G., Yoshida, N. and Hirano, S. (2016). Gravitational collapse and the thermal evolution of low-metallicity gas clouds in the early Universe, *MNRAS* **463**, pp. 2781–2798, doi:10.1093/mnras/stw2120.

Chiaki, G., Yoshida, N. and Kitayama, T. (2013b). Low-mass star formation triggered by early supernova explosions, *ApJ* **762**, 50, doi:10.1088/0004-637X/762/1/50.

Choi, J.-H., Shlosman, I. and Begelman, M. C. (2013). Supermassive black hole formation at high redshifts via direct collapse: Physical Processes in the Early Stage, *ApJ* **774**, 149, doi:10.1088/0004-637X/774/2/149.

Choi, J.-H., Shlosman, I. and Begelman, M. C. (2015). Supermassive black hole formation at high redshifts via direct collapse in a cosmological context, *MNRAS* **450**, pp. 4411–4423, doi:10.1093/mnras/stv694.

Chon, S., Hirano, S., Hosokawa, T. and Yoshida, N. (2016). Cosmological Sismulations of early black hole formation: halo mergers, tidal disruption, and the conditions for direct collapse, *ApJ* **832**, 134, doi:10.3847/0004-637X/832/2/134.

Chon, S. and Latif, M. A. (2017). The impact of ionizing radiation on the formation of a supermassive star in the early Universe, *MNRAS*.

Chuzhoy, L., Kuhlen, M. and Shapiro, P. R. (2007). Feedback from first radiation sources: H^- photodissociation, *ApJ* **665**, pp. L85–L88, doi:10.1086/521438.

Ciardi, B., Ferrara, A. and Abel, T. (2000). Intergalactic H_2 photodissociation and the soft ultraviolet background produced by Population III objects, *ApJ* **533**, pp. 594–600, doi:10.1086/308673.

Clark, P. C., Glover, S. C. O. and Klessen, R. S. (2008). The First stellar cluster, *ApJ* **672**, 757-764, doi:10.1086/524187.

Clark, P. C., Glover, S. C. O., Klessen, R. S. and Bromm, V. (2011a). Gravitational fragmentation in turbulent primordial gas and the initial mass function of Population III stars, *ApJ* **727**, 110, doi:10.1088/0004-637X/727/2/110.

Clark, P. C., Glover, S. C. O., Smith, R. J., Greif, T. H., Klessen, R. S. and Bromm, V. (2011b). The formation and fragmentation of disks around primordial protostars, *Science* **331**, pp. 1040–, doi:10.1126/science.1198027.

Cohn, H. (1980). Late core collapse in star clusters and the gravothermal instability, *ApJ* **242**, pp. 765–771, doi:10.1086/158511.

Colpi, M. (2014). Massive binary black holes in galactic Nnuclei and their path to coalescence, *Space Sci. Rev.* **183**, pp. 189–221, doi:10.1007/s11214-014-0067-1.

Colpi, M. and Sesana, A. (2017). *Gravitational wave sources in the era of multi-band gravitational wave astronomy* (World Scientific Publishing Co), pp. 43–140, doi:10.1142/9789813141766_0002.

Commerçon, B., Hennebelle, P. and Henning, T. (2011). Collapse of massive magnetized dense cores using radiation magnetohydrodynamics: early fragmentation inhibition, *ApJ* **742**, L9, doi:10.1088/2041-8205/742/1/L9.

Cooke, R. J. and Madau, P. (2014). Carbon-enhanced Metal-poor stars: relics from the dark ages, *ApJ* **791**, 116, doi:10.1088/0004-637X/791/2/116.

Coppola, C. M., D'Introno, R., Galli, D., Tennyson, J. and Longo, S. (2012). Non-equilibrium H_2 formation in the early Universe: energy exchanges, rate coefficients, and spectral distortions, *ApJS* **199**, 16, doi:10.1088/0067-0049/199/1/16.

Coppola, C. M., Diomede, P., Longo, S. and Capitelli, M. (2011). H_2 and HD Direct Photodissociation in the chemistry of the primordial universe, *ApJ* **727**, 37, doi:10.1088/0004-637X/727/1/37.

Coppola, C. M., Galli, D., Palla, F., Longo, S. and Chluba, J. (2013). Non-thermal photons and H_2 formation in the early Universe, *MNRAS* **434**, pp. 114–122, doi:10.1093/mnras/stt1007.

Cowie, L. L., Barger, A. J. and Hasinger, G. (2012). The faintest X-ray sources from z = 0 TO 8, *ApJ* **748**, 50, doi:10.1088/0004-637X/748/1/50.

Cowie, L. L., Ostriker, J. P. and Stark, A. A. (1978). Time-dependent spherically symmetric accretion onto compact X-ray sources, *ApJ* **226**, pp. 1041–1062, doi:10.1086/156682.

Cox, T. J., Jonsson, P., Somerville, R. S., Primack, J. R. and Dekel, A. (2008). The effect of galaxy mass ratio on merger-driven starbursts, *MNRAS* **384**, pp. 386–409, doi:10.1111/j.1365-2966.2007.12730.x.

Crittenden, R. G., Natarajan, P., Pen, U.-L. and Theuns, T. (2001). Spin-induced galaxy Aalignments and their implications for weak-lensing measurements, *ApJ* **559**, pp. 552–571, doi 10.1086/322370.

Crocce, M. and Scoccimarro, R. (2006a). Memory of initial conditions in gravitational clustering, *Phys. Rev. D* **73**, 6, pp. 063520–+, doi:10.1103/PhysRevD.73.063520.

Crocce, M. and Scoccimarro, R. (2006b). Renormalized cosmological perturbation theory, *Phys. Rev. D* **73**, 6, pp. 063519–+, doi:10.1103/PhysRevD.73.063519.

Croton, D. J., Springel, V., White, S. D. M., De Lucia, G., Frenk, C. S., Gao, L., Jenkins, A., Kauffmann, G., Navarro, J. F. and Yoshida, N. (2006). The many lives of active galactic nuclei: cooling flows, black holes and the luminosities and colours of galaxies, *MNRAS* **365**, pp. 11–28, doi:10.1111/j.1365-2966.2005.09675.x.

Dale, J. E. and Davies, M. B. (2006). Collisions and close encounters involving massive main-sequence stars, *MNRAS* **366**, pp. 1424–1436, doi:10.1111/j.1365-2966.2005.09937.x.

Davies, M. B., Miller, M. C. and Bellovary, J. M. (2011). Supermassive black hole formation via gas accretion in nuclear stellar clusters, *ApJ* **740**, L42, doi:10.1088/2041-8205/740/2/L42.

de Bennassuti, M., Schneider, R., Valiante, R. and Salvadori, S. (2014). Decoding the stellar fossils of the dusty Milky Way progenitors, *MNRAS* **445**, pp. 3039–3054, doi:10.1093/mnras/stu1962.

de Jong, T. (1972). The density of H2 molecules in dark interstellar clouds, *A&A* **20**, p. 263.

De Rosa, G., Venemans, B. P., Decarli, R., Gennaro, M., Simcoe, R. A., Dietrich, M., Peterson, B. M. and Walter, F. et al. (2014). Black hole mass estimates and emission-line properties of a sample of redshift z > 6.5 Quasars, *ApJ* **790**, 145, doi:10.1088/0004-637X/790/2/145.

DeBoer, D. R., Parsons, A. R., Aguirre, J. E., Alexander, P., Ali, Z. S., Beardsley, A. P., Bernardi, G., Bowman, J. D., Bradley, R. F., Carilli, C. L., Cheng, C. and et al. (2017). Hydrogen Epoch of Reionization Array (HERA), *PASP* **129**, 4, p. 045001, doi:10.1088/1538-3873/129/974/045001.

Dekel, A. and Birnboim, Y. (2006). Galaxy bimodality due to cold flows and shock heating, *MNRAS* **368**, pp. 2–20, doi:10.1111/j.1365-2966.2006.10145.x.

Dekel, A. and Rees, M. J. (1987). Physical mechanisms for biased galaxy formation, *Nature* **326**, pp. 455–462, doi:10.1038/326455a0.

Devecchi, B. and Volonteri, M. (2009). Formation of the First Nuclear Clusters and Massive Black Holes at High Redshift, *ApJ* **694**, pp. 302–313, doi:10.1088/0004-637X/694/1/302.

Devecchi, B., Volonteri, M., Rossi, E. M., Colpi, M. and Portegies Zwart, S. (2012). High-redshift formation and evolution of central massive objects - II. The census of BH seeds, *MNRAS* **421**, pp. 1465–1475, doi: 10.1111/j.1365-2966.2012.20406.x.

Di Matteo, T., Khandai, N., DeGraf, C., Feng, Y., Croft, R. A. C., Lopez, J. and Springel, V. (2012). Cold flows and the first quasars, *ApJ* **745**, L29, doi:10.1088/2041-8205/745/2/L29.

Di Matteo, T., Springel, V. and Hernquist, L. (2005). Energy input from quasars regulates the growth and activity of black holes and their host galaxies, *Nature* **433**, pp. 604–607, doi:10.1038/nature03335.

Dib, S., Piau, L., Mohanty, S. and Braine, J. (2011). Star formation efficiency as a function of metallicity: from star clusters to galaxies, *MNRAS* **415**, pp. 3439–3454, doi:10.1111/j.1365-2966.2011.18966.x.

Dijkstra, M. (2014). Lyα Emitting galaxies as a probe of reionisation, *PASA* **31**, e040, doi:10.1017/pasa.2014.33.

Dijkstra, M. (2017). Saas-Fee lecture notes: physics of Lyman Alpha radiative transfer, *ArXiv e-prints 1704.03416*.

Dijkstra, M., Ferrara, A. and Mesinger, A. (2014). Feedback-regulated supermassive black hole seed formation, *MNRAS* **442**, pp. 2036–2047, doi: 10.1093/mnras/stu1007.

Dijkstra, M., Gilfanov, M., Loeb, A. and Sunyaev, R. (2012). Constraints on the redshift evolution of the L_X-SFR relation from the cosmic X-ray backgrounds, *MNRAS* **421**, pp. 213–223, doi:10.1111/j.1365-2966.2011.20292.x.

Dijkstra, M., Gronke, M. and Sobral, D. (2016a). Lyα signatures from direct collapse black holes, *ApJ* **823**, 74, doi:10.3847/0004-637X/823/2/74.

Dijkstra, M., Haiman, Z. and Loeb, A. (2004). A limit from the X-ray background on the contribution of quasars to reionization, *ApJ* **613**, pp. 646–654, doi: 10.1086/422167.

Dijkstra, M., Haiman, Z., Mesinger, A. and Wyithe, J. S. B. (2008). Fluctuations in the high-redshift Lyman-Werner background: close halo pairs as the origin of supermassive black holes, *MNRAS* **391**, pp. 1961–1972, doi:10. 1111/j.1365-2966.2008.14031.x.

Dijkstra, M., Sethi, S. and Loeb, A. (2016b). 3-cm Fine Structure Masers: a unique signature of supermassive black hole formation via direct Ccollapse in the early Universe, *ApJ* **820**, 10, doi:10.3847/0004-637X/820/1/10.

Dimopoulos, S. and Landsberg, G. (2001). Black holes at the Large Hadron Collider, *Physical Review Letters* **87**, 16, 161602, doi:10.1103/PhysRevLett.87. 161602.

Dopcke, G., Glover, S. C. O., Clark, P. C. and Klessen, R. S. (2013). On the initial mass function of low-metallicity stars: the importance of dust Cooling, *ApJ* **766**, 103, doi:10.1088/0004-637X/766/2/103.

Dotan, C., Rossi, E. M. and Shaviv, N. J. (2011). A lower limit on the halo mass to form supermassive black holes, *MNRAS* **417**, pp. 3035–3046, doi: 10.1111/j.1365-2966.2011.19461.x.

Dotti, M., Colpi, M., Haardt, F. and Mayer, L. (2007). Supermassive black hole binaries in gaseous and stellar circumnuclear discs: orbital dynamics and gas accretion, *MNRAS* **379**, pp. 956–962, doi:10.1111/j.1365-2966.2007.12010.x.

Dotti, M., Colpi, M., Pallini, S., Perego, A. and Volonteri, M. (2013). On the orientation and magnitude of the black hole spin in galactic nuclei, *ApJ* **762**, 68, doi:10.1088/0004-637X/762/2/68.

Draine, B. T. and Bertoldi, F. (1996). Structure of stationary photodissociation Fronts, *ApJ* **468**, p. 269, doi:10.1086/177689.

Droste, J. (1917). The field of a single centre in Einstein's theory of gravitation, and the motion of a particle in that field, *Koninklijke Nederlandse Akademie van Wetenschappen Proceedings Series B Physical Sciences* **19**, pp. 197–215.

Dubois, Y., Devriendt, J., Slyz, A. and Teyssier, R. (2012). Self-regulated growth of supermassive black holes by a dual jet-heating active galactic nucleus feedback mechanism: methods, tests and implications for cosmological simulations, *MNRAS* **420**, pp. 2662–2683, doi:10.1111/j.1365-2966.2011.20236.x.

Dubois, Y., Pichon, C., Welker, C., Le Borgne, D., Devriendt, J., Laigle, C., Codis, S., Pogosyan, D., Arnouts, S. and Benabed, K. et al. (2014a). Dancing in the dark: galactic properties trace spin swings along the cosmic web, *MNRAS* **444**, pp. 1453–1468, doi:10.1093/mnras/stu1227.

Dubois, Y., Volonteri, M. and Silk, J. (2014b). Black hole evolution — III. Statistical properties of mass growth and spin evolution using large-scale hydrodynamical cosmological simulations, *MNRAS* **440**, pp. 1590–1606, doi:10.1093/mnras/stu373.

Durrer, R. (2008). *The Cosmic Microwave Background* (Cambridge University Press).

Eddington, A. S. (1916). On the radiative equilibrium of the stars, *MNRAS* **77**, pp. 16–35, doi:10.1093/mnras/77.1.16.

Ehlers, J. and Buchert, T. (1997). Newtonian cosmology in Lagrangian formulation: foundations and perturbation theory, *General Relativity and Gravitation* **29**, pp. 733–764.

Einstein, A. (1915). Die Feldgleichungen der Gravitation, *Sitzungsberichte der Königlich Preußischen Akademie der Wissenschaften (Berlin), Seite 844-847.*.

Einstein, A. (1916). Die Grundlage der allgemeinen Relativitätstheorie, *Annalen der Physik* **354**, pp. 769–822, doi:10.1002/andp.19163540702.

Einstein, A. (1939). On a stationary system with spherical symmetry consisting of many gravitating masses, *Annals of Mathematics* **40**, p. 922.

Eisenstein, D. J. and Hu, W. (1998). Baryonic features in the matter transfer function, *ApJ* **496**, pp. 605–+, doi:10.1086/305424.

Eisenstein, D. J. and Hu, W. (1999). Power spectra for cold dark matter and its variants, *ApJ* **511**, pp. 5–15, doi:10.1086/306640.

Eisenstein, D. J. and Loeb, A. (1995). Origin of quasar progenitors from the collapse of low-spin cosmological perturbations, *ApJ* **443**, pp. 11–17, doi:10.1086/175498.

eLISA Consortium, Amaro Secane, P., Aoudia, S., Audley, H., Auger, G., Babak, S., Baker, J., Barausse, E., Barke, S., Bassan, M., Beckmann, V., Benacquista, M., Bender, P. L., Berti, E. and Binétruy, P. et al. (2013). The gravitational Universe, *ArXiv e-prints 1305.5720.*

Elmegreen, B. G. (2002). Star formation from galaxies to globules, *ApJ* **577**, pp. 206–220, doi:10.1086/342177.

Espinoza, P., Selman, F. J. and Melnick, J. (2009). The massive star initial mass function of the Arches cluster, *A&A* **501**, pp. 563–583, doi: 10.1051/0004-6361/20078597.

Evans, C. R. and Kochanek, C. S. (1989). The tidal disruption of a star by a massive black hole, *ApJ* **346**, pp. L13–L16, doi:10.1086/185567.

Fabian, A. C. (2012). Observational Evidence of Active Galactic Nuclei Feedback, *ARA&A* **50**, pp. 455–489, doi:10.1146/annurev-astro-081811-125521.

Fabian, A. C., Celotti, A., Blundell, K. M., Kassim, N. E. and Perley, R. A. (2002). The properties of the X-ray holes in the intracluster medium of the Perseus cluster, *MNRAS* **331**, pp. 369–375, doi:10.1046/j.1365-8711.2002. 05182.x.

Fabian, A. C., Pringle, J. E. and Rees, M. J. (1975). Tidal capture formation of binary systems and X-ray sources in globular clusters, *MNRAS* **172**, pp. 15p–18p, doi:10.1093/mnras/172.1.15P.

Fan, X., Carilli, C. L. and Keating, B. (2006a). Observational constraints on cosmic reionization, *ARA&A* **44**, pp. 415–462, doi:10.1146/annurev.astro. 44.051905.092514.

Fan, X., Hennawi, J. F., Richards, G. T., Strauss, M. A., Schneider, D. P., Donley, J. L., Young, J. E., Annis, J., Lin, H., Lampeitl, H. and Lupton, R. H. et al. (2004) .

Fan, X., Narayanan, V. K., Lupton, R. H., Strauss, M. A., Knapp, G. R., Becker, R. H., White, R. L., Pentericci, L., Leggett, S. K., Haiman, Z., Gunn, J. E. and Ivezić, Ž. et al. (2001). A survey of z5.8 quasars in the Sloan Digital Sky survey. I. discovery of three new quasars and the spatial density of luminous quasars at z˜6, *AJ* **122**, pp. 2833–2849, doi:10.1086/324111.

Fan, X., Narayanan, V. K., Strauss, M. A., White, R. L., Becker, R. H., Pentericci, L. and Rix, H.-W. (2002). Evolution of the ionizing background and the epoch of reionization from the spectra of z˜6 quasars, *AJ* **123**, pp. 1247–1257, doi:10.1086/339030.

Fan, X., Strauss, M. A., Becker, R. H., White, R. L., Gunn, J. E., Knapp, G. R., Richards, G. T., Schneider, D. P., Brinkmann, J. and Fukugita, M. (2006b). Constraining the evolution of the ionizing background and the epoch of reionization with z˜6 quasars. II. A Sample of 19 Quasars, *AJ* **132**, pp. 117–136, doi:10.1086/504836.

Fan, X., Strauss, M. A., Richards, G. T., Hennawi, J. F., Becker, R. H., White, R. L. and Diamond-Stanic, A. M. (2006c). A survey of z ¿ 5.7 quasars in the sloan digital Sky survey. IV. discovery of seven additional quasars, *AJ* **131**, pp. 1203–1209, doi:10.1086/500296.

Fan, X., Strauss, M. A., Schneider, D. P., Becker, R. H., White, R. L., Haiman, Z. and Gregg, M. (2003). A survey of z 5.7 quasars in the sloan digital

Sky survey. II. discovery of three additional quasars at z6, *AJ* **125**, pp. 1649–1659, doi:10.1086/368246.

Fan, X., White, R. L., Davis, M., Becker, R. H., Strauss, M. A., Haiman, Z., Schneider, D. P., Gregg, M. D., Gunn, J. E., Knapp, G. R. and Lupton, R. H. et al. (2000). The discovery of a luminous Z=5.80 quasar from the sloan digital Sky survey, *AJ* **120**, pp. 1167–1174, doi:10.1086/301534.

Fanali, R., Dotti, M., Fiacconi, D. and Haardt, F. (2015). Bar formation as driver of gas inflows in isolated disc galaxies, *MNRAS* **454**, pp. 3641–3652, doi: 10.1093/mnras/stv2247.

Farrell, S. A., Webb, N. A., Barret, D., Godet, O. and Rodrigues, J. M. (2009). An intermediate-mass black hole of over 500 solar masses in the galaxy ESO243-49, *Nature* **460**, pp. 73–75, doi:10.1038/nature08083.

Federrath, C., Chabrier, G., Schober, J., Banerjee, R., Klessen, R. S. and Schleicher, D. R. G. (2011a). Mach number dependence of turbulent magnetic field amplification: solenoidal versus compressive flows, *Physical Review Letters* **107**, 11, 114504, doi:10.1103/PhysRevLett.107.114504.

Federrath, C. and Klessen, R. S. (2012). The star formation rate of turbulent magnetized clouds: comparing theory, simulations, and observations, *ApJ* **761**, 156, doi:10.1088/0004-637X/761/2/156.

Federrath, C., Roman-Duval, J., Klessen, R. S., Schmidt, W. and Mac Low, M.-M. (2010). Comparing the statistics of interstellar turbulence in simulations and observations. Solenoidal versus compressive turbulence forcing, *A&A* **512**, A81, doi:10.1051/0004-6361/200912437.

Federrath, C., Schober, J., Bovino, S. and Schleicher, D. R. G. (2014). The turbulent dynamo in highly compressible supersonic plasmas, *ApJ* **797**, L19, doi:10.1088/2041-8205/797/2/L19.

Federrath, C., Sur, S., Schleicher, D. R. G., Banerjee, R. and Klessen, R. S. (2011b). A new Jeans resolution criterion for (M)HD simulations of self-gravitating gas: application to magnetic field amplification by gravity-driven turbulence, *ApJ* **731**, 62, doi:10.1088/0004-637X/731/1/62.

Feng, H. and Soria, R. (2011). Ultraluminous X-ray sources in the Chandra and XMM-Newton era, *New Astronomy Reviews* **55**, pp. 166–183, doi:10.1016/j.newar.2011.08.002.

Feng, Y., Di Matteo, T., Croft R. and Khandai, N. (2014). High-redshift supermassive black holes: accretion through cold flows, *MNRAS* **440**, pp. 1865–1879, doi:10.1093/mnras/stu432.

Feng, Y., Di-Matteo, T., Croft, R. A., Bird, S., Battaglia, N. and Wilkins, S. (2016). The BlueTides simulation: first galaxies and reionization, *MNRAS* **455**, pp. 2778–2791, doi:10.1093/mnras/stv2484.

Fernandez, R., Bryan, G. L., Haiman, Z. and Li, M. (2014). H_2 suppression with shocking inflows: testing a pathway for supermassive black hole formation, *MNRAS* **439**, pp. 3798–3807, doi:10.1093/mnras/stu230.

Ferrara, A., Haardt, F. and Salvaterra, R. (2013). Can supermassive black hole seeds form in galaxy mergers? *MNRAS* **434**, pp. 2600–2605, doi:10.1093/mnras/stt1350.

Ferrara, A., Salvadori, S., Yue, B. and Schleicher, D. (2014). Initial mass function of intermediate-mass black hole seeds, *MNRAS* **443**, pp. 2410–2425, doi: 10.1093/mnras/stu1280.

Ferrarese, L. and Merritt, D. (2000). A fundamental relation between supermassive black holes and their Host galaxies, *ApJ* **539**, pp. L9–L12, doi: 10.1086/312838.

Ferrière, K., Gillard, W. and Jean, P. (2007). Spatial distribution of interstellar gas in the innermost 3 kpc of our galaxy, *A&A* **467**, pp. 611–627, doi: 10.1051/0004-6361:20066992.

Fiacconi, D., Mayer, L., Madau, P., Lupi, A., Dotti, M. and Haardt, F. (2017a). Young and turbulent: the early life of massive galaxy progenitors, *MNRAS* **467**, pp. 4080–4100, doi:10.1093/mnras/stx335.

Fiacconi, D., Mayer, L., Roškar, R. and Colpi, M. (2013). Massive black hole pairs in clumpy, self-gravitating circumnuclear disks: stochastic orbital decay, *ApJ* **777**, L14, doi:10.1088/2041-8205/777/1/L14.

Fiacconi, D., Pinto, C., Walton, D. J. and Fabian, A. C. (2017b). Constraining the mass of accreting black holes in ultraluminous X-ray sources with ultrafast outflows, *MNRAS* **469**, pp. L99–L103, doi:10.1093/mnrasl/slx065.

Fiacconi, D. and Rossi, E. M. (2016). Bright vigorous winds as signposts of supermassive black hole birth, *MNRAS* **455**, pp. 2–16, doi:10.1093/mnras/stv2237.

Fiacconi, D. and Rossi, E. M. (2017). Light or heavy supermassive black hole seeds: the role of internal rotation in the fate of supermassive stars, *MNRAS* **464**, pp. 2259–2269, doi:10.1093/mnras/stw2505.

Fiacconi, D., Sijacki, D. and Pringle, J. E. (2018). Galactic nuclei evolution with spinning black holes: method and implementation, *MNRAS* **477**, pp. 3807–3835, doi:10.1093/mnras/sty893.

Fialkov, A., Barkana, R., Tseliakhovich, D. and Hirata, C. M. (2012). Impact of the relative motion between the dark matter and baryons on the first stars: semi-analytical modelling, *MNRAS* **424**, pp. 1335–1345, doi:10.1111/j.1365-2966.2012.21318.x.

Fialkov, A., Barkana, R. and Visbal, E. (2014). The observable signature of late heating of the Universe during cosmic reionization, *Nature* **506**, pp. 197–199, doi:10.1038/nature12999.

Field, G. B. and Partridge, R. B. (1961). Stimulated emission of the 3.04-CM fine-structure line of Hydrogen in diffuse nebulae. *ApJ* **134**, p. 959, doi: 10.1086/147222.

Field, G. B., Somerville, W. B. and Dressler, K. (1966). Hydrogen molecules in astronomy, *ARA&A* **4**, p. 207, doi:10.1146/annurev.aa.04.090166.001231.

Finkelstein, S. L., Papovich, C., Salmon, B., Finlator, K., Dickinson, M., Ferguson, H. C., Giavalisco, M., Koekemoer, A. M. and Reddy, N. A. et al. (2012). Candels: The evolution of galaxy rest-frame ultraviolet colors from z = 8 to 4, *ApJ* **756**, 164, doi:10.1088/0004-637X/756/2/164.

Frank, J., King, A. and Raine, D. J. (2002). *Accretion Power in Astrophysics: Third Edition* (Cambridge: Cambridge University Press).

Frebel, A. (2010). Stellar archaeology: exploring the Universe with metal-poor stars, *Astronomische Nachrichten* **331**, pp. 474–488, doi:10.1002/asna.201011362.

Frebel, A., Aoki, W., Christlieb, N., Ando, H., Asplund, M., Barklem, P. S., Beers, T. C., Eriksson, K., Fechner, C., Fujimoto, M. Y., Honda, S. and Kajino, T. et al. (2005). Nucleosynthetic signatures of the first stars, *Nature* **434**, pp. 871–873, doi:10.1038/nature03455.

Frebel, A., Johnson, J. L. and Bromm, V. (2009). The minimum stellar metallicity observable in the Galaxy, *MNRAS* **392**, pp. L50–L54, doi:10.1111/j.1745-3933.2008.00587.x.

Freese, K., Bodenheimer, P., Gondolo, P. and Spolyar, D. (2009). Dark stars: a new study of the first stars in the Universe, *New Journal of Physics* **11**, 10, 105014, doi:10.1088/1367-2630/11/10/105014.

Freese, K., Bodenheimer, P., Spolyar, D. and Gondolo, P. (2008). Stellar structure of dark stars: a first phase of stellar evolution resulting from dark matter annihilation, *ApJ* **685**, L101, doi:10.1086/592685.

Fricke, K. J. (1973). Dynamical Phases of Supermassive Stars, *ApJ* **183**, pp. 941–958, doi:10.1086/152280.

Fukugita, M., Ichikawa, T., Gunn J. E., Doi, M., Shimasaku, K. and Schneider, D. P. (1996). The Sloan Digital Sky survey photometric system, *AJ* **111**, p. 1748, doi:10.1086/117915.

Fuller, G. M., Woosley, S. E. and Weaver, T. A. (1986). The evolution of radiation-dominated stars. I — Nonrotating supermassive stars, *ApJ* **307**, pp. 675–686, doi:10.1086/164452.

Fumagalli, M., O'Meara, J. M. and Prochaska, J. X. (2011). Detection of pristine gas two billion years after the Big bang, *Science* **334**, p. 1245, doi:10.1126/science.1213581.

Furlanetto, S. R., Oh, S. P. and Briggs, F. H. (2006). Cosmology at low frequencies: The 21 cm transition and the high-redshift Universe, *Phys. Rep.* **433**, pp. 181–301, doi:10.1016/j.physrep.2006.08.002.

Furusawa, H., Kosugi, G., Akiyama, M., Takata, T., Sekiguchi, K., Tanaka, I., Iwata, I., Kajisawa, M., Yasuda, N., Doi, M., Ouchi, M. and Simpson, C et al. (2008). The Subaru/XMM-Newton Deep Survey (SXDS). II. optical imaging and photometric catalogs, *ApJS* **176**, 1-18, doi:10.1086/527321.

Gabor, J. M. and Bournaud, F. (2013). Simulations of supermassive black hole growth in high-redshift disc galaxies, *MNRAS* **434**, pp. 606–620, doi:10.1093/mnras/stt1043.

Gabor, J. M., Capelo, P. R., Volonteri, M., Bournaud, F., Bellovary, J., Governato, F. and Quinn, T. (2016). Comparison of black hole growth in galaxy mergers with GASOLINE and RAMSES, *A&A* **592**, A62, doi:10.1051/0004-6361/201527143.

Gair, J. R., Babak, S., Sesana, A., Amaro-Seoane, P., Barausse, E., Berry, C. P. L., Berti, E. and Sopuerta, C. (2017). Prospects for observing extreme-mass-ratio inspirals with LISA, in *Journal of Physics Conference Series, Journal of Physics Conference Series*, Vol. 840, p. 012021, doi:10.1088/1742-6596/840/1/012021.

Gallerani, S., Fan, X., Maiolino, R. and Pacucci, F. (2017). Physical properties of the first quasars, *Publications of the Astronomical Society of Australia* **34**, e022, doi:10.1017/pasa.2017.14.

Galli, D. and Palla, F. (1998). The chemistry of the early Universe, *A&A* **335**, pp. 403–420.

Gammie, C. F. (2001). Nonlinear outcome of gravitational instability in cooling, gaseous disks, *ApJ* **553**, pp. 174–183, doi:10.1086/320631.

García-Bellido, J. (2017). Massive primordial black holes as dark matter and their detection with Gravitational waves, in *Journal of Physics Conference Series, Journal of Physics Conference Series*, Vol. 840, p. 012032, doi:10.1088/1742-6596/840/1/012032.

Gardner, J. P., Mather, J. C., Clampin, M., Doyon, R., Greenhouse, M. A., Hammel, H. B., Hutchings, J. B., Jakobsen, P., Lilly, S. J. and Long, K. S. e. a. (2006). The James Webb Space Telescope, *Space Science Reviews*.

Gebhardt, K., Bender, R., Bower, G., Dressler, A., Faber, S. M., Filippenko, A. V., Green, R., Grillmair, C., Ho, L. C., Kormendy, J., Lauer, T. R. and Magorrian, J. et al. (2000). A relationship between nuclear black hole mass and galaxy velocity dispersion, *ApJ* **539**, pp. L13–L16, doi:10.1086/312840.

Gehrels, N., Spergel, D. and WFIRST SDT Project (2015). Wide-Field InfraRed Survey Telescope (WFIRST) Mission and Synergies with LISA and LIGO-Virgo, in *Journal of Physics Conference Series, Journal of Physics Conference Series*, Vol. 610, p. 012007, doi:10.1088/1742-6596/610/1/012007.

Genzel, R., Eisenhauer, F. and Gillessen, S. (2010). The galactic center massive black hole and nuclear star cluster, *Reviews of Modern Physics* **82**, pp. 3121–3195, doi:10.1103/RevModPhys.82.3121.

Ghez, A. M., Salim, S., Weinberg, N. N., Lu, J. R., Do, T., Dunn, J. K., Matthews, K., Morris, M. R., Yelda, S., Becklin, E. E., Kremenek, T., Milosavljevic, M. and Naiman, J. (2008). Measuring distance and properties of the Milky Way's central supermassive black hole with stellar orbits, *ApJ* **689**, 1044-1062, doi:10.1086/592738.

Giacconi, R., Gursky, H., Paolini, F. R. and Rossi, B. B. (1962). Evidence for X rays from sources outside the Solar system, *Physical Review Letters* **9**, pp. 439–443, doi:10.1103/PhysRevLett.9.439.

Giacobbo, N., Mapelli, M. and Spera, M. (2018). Merging black hole binaries: the effects of progenitor's metallicity, mass-loss rate and Eddington factor, *MNRAS* **474**, pp. 2959–2974, doi:10.1093/mnras/stx2933.

Gibbons, G. W. (1975). Vacuum polarization and the spontaneous loss of charge by black holes, *Communications in Mathematical Physics* **44**, pp. 245–264, doi:10.1007/BF01609829.

Gilfanov, M. and Merloni, A. (2014). Observational appearance of black holes in X-ray binaries and AGN, *Space Sci. Rev.* **183**, pp. 121–148, doi:10.1007/s11214-014-0071-5.

Gillessen, S., Plewa, P. M., Eisenhauer, F., Sari, R., Waisberg, I., Habibi, M., Pfuhl, O., George, E., Dexter, J., von Fellenberg, S., Ott, T. and Genzel, R. (2017). An update on monitoring stellar orbits in the galactic center, *ApJ* **837**, 30, doi:10.3847/1538-4357/aa5c41.

Girichidis, P., Federrath, C., Allison, R., Banerjee, R. and Klessen, R. S. (2012a). Importance of the initial conditions for star formation — III. statistical properties of embedded protostellar clusters, *MNRAS* **420**, pp. 3264–3280, doi:10.1111/j.1365-2966.2011.20250.x.

Girichidis, P., Federrath, C., Banerjee, R. and Klessen, R. S. (2011). Importance of the initial conditions for star formation — I. cloud evolution and morphology, *MNRAS* **413**, pp. 2741–2759, doi:10.1111/j.1365-2966.2011.18348.x.

Girichidis, P., Federrath, C., Banerjee, R. and Klessen, R. S. (2012b). Importance of the initial conditions for star formation — II. fragmentation-induced starvation and accretion shielding, *MNRAS* **420**, pp. 613–626, doi:10.1111/j.1365-2966.2011.20073.x.

Glassgold, A. E., Galli, D. and Padovani, M. (2012). Cosmic-ray and X-ray heating of interstellar clouds and protoplanetary disks, *ApJ* **756**, 157, doi:10.1088/0004-637X/756/2/157.

Glebbeek, E., Gaburov, E., de Mink, S. E., Pols, O. R. and Portegies Zwart, S. F. (2009). The evolution of runaway stellar collision products, *A&A* **497**, pp. 255–264, doi:10.1051/0004-6361/200810425.

Glebbeek, E., Gaburov, E., Portegies Zwart, S. and Pols, O. R. (2013). Structure and evolution of high-mass stellar mergers, *MNRAS* **434**, pp. 3497–3510, doi:10.1093/mnras/stt1268.

Glebbeek, E., Pols, O. R. and Hurley, J. R. (2008). Evolution of stellar collision products in open clusters. I. blue stragglers in N-body models of M 67, *AAP* **488**, pp. 1007–1015, doi:10.1051/0004-6361:200809930.

Glikman, E., Eigenbrod, A., Djorgovski, S. G., Meylan, G., Thompson, D., Mahabal, A. and Courbin, F. (2008). An exploratory search for z gsim 6 quasars in the Ukidss Early Data Release, *AJ* **136**, pp. 954–962, doi:10.1088/0004-6256/136/3/954.

Glover, S. (2005). The formation of the first stars In the Universe, *Space Sci. Rev.* **117**, pp. 445–508, doi:10.1007/s11214-005-5821-y.

Glover, S. (2013). The First Stars, in T. Wiklind, B. Mobasher and V. Bromm (eds.), *The First Galaxies Astrophysics and Space Science Library*, Vol. 396, p. 103, doi:10.1007/978-3-642-32362-1_3.

Glover, S. C. O. (2007). Radiative feedback from ionized gas, *MNRAS* **379**, pp. 1352–1358, doi:10.1111/j.1365-2966.2007.11787.x.

Glover, S. C. O. (2015a). Simulating the formation of massive seed black holes in the early Universe - I. An improved chemical model, *MNRAS* **451**, pp. 2082–2096, doi:10.1093/mnras/stv1059.

Glover, S. C. O. (2015b). Simulating the formation of massive seed black holes in the early Universe - II. Impact of rate coefficient uncertainties, *MNRAS* **453**, pp. 2901–2918, doi:10.1093/mnras/stv1781.

Glover, S. C. O. (2016). Simulating the formation of massive seed black holes in the early Universe III: The influence of X-rays, *ArXiv e-prints 1610.05679.*

Glover, S. C. O. and Abel, T. (2008). Uncertainties in H_2 and HD chemistry and cooling and their role in early structure formation, *MNRAS* **388**, pp. 1627–1651, doi:10.1111/j.1365-2966.2008.13224.x.

Glover, S. C. O. and Brand, P. W. J. L. (2003). Radiative feedback from an early X-ray background, *MNRAS* **340**, pp. 210–226, doi:10.1046/j.1365-8711.2003.06311.x.

Glover, S. C. O. and Clark, P. C. (2014). Molecular cooling in the diffuse interstellar medium, *MNRAS* **437**, pp. 9–20, doi:10.1093/mnras/stt1809.

Glover, S. C. O., Federrath, C., Mac Low, M.-M. and Klessen, R. S. (2010). Modelling CO formation in the turbulent interstellar medium, *MNRAS* **404**, pp. 2–29, doi:10.1111/j.1365-2966.2009.15718.x.

Goldsmith, P. F. and Langer, W. D. (1978). Molecular cooling and thermal balance of dense interstellar clouds, *ApJ* **222**, pp. 881–895, doi:10.1086/156206.

Governato, F., Brook, C., Mayer, L., Brooks, A., Rhee, G., Wadsley, J., Jonsson, P., Willman, B., Stinson, G., Quinn, T. and Madau, P. (2010). Bulgeless dwarf galaxies and dark matter cores from supernova-driven outflows, *Nature* **463**, pp. 203–206, doi:10.1038/nature08640.

Graham, A. W. and Spitler, L. R. (2009). Quantifying the coexistence of massive black holes and dense nuclear star clusters, *MNRAS* **397**, pp. 2148–2162, doi:10.1111/j.1365-2966.2009.15118.x.

Grassi, T., Bovino, S., Haugbølle, T. and Schleicher, D. R. G. (2017). A detailed framework to incorporate dust in hydrodynamical simulations, *MNRAS* **466**, pp. 1259–1274, doi:10.1093/mnras/stw2871.

Grassi, T., Bovino, S., Schleicher, D. R. G., Prieto, J., Seifried, D., Simoncini, E. and Gianturco, F. A. (2014). KROME — a package to embed chemistry in astrophysical simulations, *MNRAS* **439**, pp. 2386–2419, doi:10.1093/mnras/stu114.

Grasso, D. and Rubinstein, H. R. (2001). Magnetic fields in the early Universe, *Phys. Rep.* **348**, pp. 163–266, doi:10.1016/S0370-1573(00)00110-1.

GRAVITY Collaboration (2018). Detection of the gravitational redshift in the orbit of the star S2 near the Galactic centre massive black hole, *A&A* **615**, L15, doi:10.1051/0004-6361/201833718.

Grazian, A., Fontana, A., Santini, P., Dunlop, J. S., Salvato, M., Sommariva, V., Song, M. and Vanzella, E. (2015). The galaxy stellar mass function in the CANDELS, GOODS-South, and HUDF fields, *A&A* **575**, A96, doi:10.1051/0004-6361/201424750.

Greene, J. E. (2012). Low-mass black holes as the remnants of primordial black hole formation, *Nature Communications* **3**, 1304, doi:10.1038/ncomms2314.

Greif, T. H. (2014). The numerical frontier of the high-redshift Universe, *ArXiv e-prints:1410.3482*.

Greif, T. H. (2015). The numerical frontier of the high-redshift Universe, *Computational Astrophysics and Cosmology* **2**, 3, doi:10.1186/s40668-014-0006-2.

Greif, T. H. and Bromm, V. (2006). Two populations of metal-free stars in the early Universe, *MNRAS* **373**, pp. 128–138, doi:10.1111/j.1365-2966.2006.11017.x.

Greif, T. H., Bromm, V., Clark, P. C., Glover, S. C. O., Smith, R. J., Klessen, R. S., Yoshida, N. and Springel, V. (2012). Formation and evolution of primordial protostellar systems, *MNRAS* **424**, pp. 399–415, doi:10.1111/j.1365-2966.2012.21212.x.

Greif, T. H., Johnson, J. L., Klessen, R. S. and Bromm, V. (2008). The first galaxies: assembly, cooling and the onset of turbulence, *MNRAS* **387**, pp. 1021–1036, doi:10.1111/j.1365-2966.2008.13326.x.

Greif, T. H., White, S. D. M., Klessen, R. S. and Springel, V. (2011). The delay of Population III star formation by supersonic streaming velocities, *ApJ* **736**, 147, doi:10.1088/0004-637X/736/2/147.

Grier, C. J., Hall, P. B., Brandt, W. N., Trump, J. R., Shen, Y., Vivek, M., Filiz Ak, N., Chen, Y., Dawson, K. S., Denney, K. D., Green, P. J. and Jiang, L. et al. (2015). The Sloan Digital Sky Survey reverberation Mapping Project: rapid CIV broad absorption line variability, *ApJ* **806**, 111, doi: 10.1088/0004-637X/806/1/111.

Griffen, B. F., Dooley, G. A., Ji, A. P., O'Shea, B. W., Gómez, F. A. and Frebel, A. (2016). Tracing the first stars and galaxies of the Milky Way, *MNRAS* **474**, pp. 443–459, doi:10.1093/mnras/stx2749.

Gültekin, K., Richstone, D. O., Gebhardt, K., Lauer, T. R., Tremaine, S., Aller, M. C., Bender, R., Dressler, A., Faber, S. M., Filippenko, A. V., Green, R., Ho, L. C., Kormendy, J., Magorrian, J., Pinkney, J. and Siopis, C. (2009). The M-sigma and M-L relations in galactic bulges, and determinations of their intrinsic scatter, *ApJ* **698**, pp. 198–221, doi:10.1088/0004-637X/698/1/198.

Guo, Q. and White, S. (2014). Numerical resolution limits on subhalo abundance matching, *MNRAS* **437**, pp. 3228–3235, doi:10.1093/mnras/stt2116.

Habouzit, M., Volonteri, M. and Dubois, Y. (2017). Blossoms from black hole seeds: properties and early growth regulated by supernova feedback, *MNRAS* **468**, pp. 3935–3948, doi:10.1093/mnras/stx666.

Habouzit, M., Volonteri, M., Latif, M., Dubois, Y. and Peirani, S. (2016a). On the number density of 'direct collapse' black hole seeds, *MNRAS* **463**, pp. 529–540, doi:10.1093/mnras/stw1924.

Habouzit, M., Volonteri, M., Latif, M., Nishimichi, T., Peirani, S., Dubois, Y., Mamon, G. A., Silk, J. and Chevallard, J. (2016b). Black hole formation and growth with non-Gaussian primordial density perturbations, *MNRAS* **456**, pp. 1901–1912, doi:10.1093/mnras/stv2740.

Haehnelt, M. G., Natarajan, P. and Rees, M. J. (1998). High-redshift galaxies, their active nuclei and central black holes, *MNRAS* **300**, pp. 817–827, doi: 10.1046/j.1365-8711.1998.01951.x.

Haehnelt, M. G. and Rees, M. J. (1993). The formation of nuclei in newly formed galaxies and the evolution of the quasar population, *MNRAS* **263**, pp. 168–178, doi:10.1093/mnras/263.1.168.

Haemmerlé, L., Eggenberger, P., Meynet, G., Maeder, A. and Charbonnel, C. (2016). Massive star formation by accretion. I. disc accretion, *A&A* **585**, A65, doi:10.1051/0004-6361/201527202.

Haemmerlé, L., Woods, T. E., Klessen, R. S., Heger, A. and Whalen, D. J. (2017a). On the rotation of supermassive stars, *ArXiv e-prints 1711.09916*.

Haemmerlé, L., Woods, T. E., Klessen, R. S., Heger, A. and Whalen, D. J. (2017b). The evolution of supermassive Population III stars, *ArXiv e-prints 1705.09301*.

Haiman, Z. (2013). The formation of the first massive black holes, in T. Wiklind, B. Mobasher and V. Bromm (eds.), *The First Galaxies, Astrophysics and Space Science Library*, Vol. 396, p. 293, doi:10.1007/978-3-642-32362-1_6.

Haiman, Z., Abel, T. and Rees, M. J. (2000). The radiative feedback of the first cosmological objects, *ApJ* **534**, pp. 11–24, doi:10.1086/308723.

Haiman, Z., Kocsis, B. and Menou, K. (2009). The population of viscosity- and gravitational wave-driven supermassive black hole binaries among luminous active galactic nuclei, *ApJ* **700**, pp. 1952–1969, doi:10.1088/0004-637X/700/2/1952.

Haiman, Z., Rees, M. J. and Loeb, A. (1996). H 2 Cooling of primordial gas triggered by UV irradiation, *Astrophysical Journal v.467*.

Haiman, Z., Rees, M. J. and Loeb, A. (1997). Destruction of molecular Hydrogen during cosmological reionization, *ApJ* **476**, pp. 458–463, doi:10.1086/303647.

Hamann, F., Kaplan, K. F., Rodríguez Hidalgo, P., Prochaska, J. X. and Herbert-Fort, S. (2008). Emergence of a quasar outflow, *MNRAS* **391**, pp. L39–L43, doi:10.1111/j.1745-3933.2008.00554.x.

Hartwig, T., Bromm, V., Klessen, R. S. and Glover, S. C. O. (2015a). Constraining the primordial initial mass function with stellar archaeology, *MNRAS* **447**, pp. 3892–3908, doi:10.1093/mnras/stu2740.

Hartwig, T., Clark, P. C., Glover, S. C. O., Klessen, R. S. and Sasaki, M. (2015b). A new approach to determine optically thick H_2 cooling and its effect on primordial star formation, *ApJ* **799**, 114, doi:10.1088/0004-637X/799/2/114.

Hartwig, T., Glover, S. C. O., Klessen, R. S., Latif, M. A. and Volonteri, M. (2015c). How an improved implementation of H_2 self-shielding influences the formation of massive stars and black holes, *MNRAS* **452**, pp. 1233–1244, doi:10.1093/mnras/stv1368.

Hartwig, T., Latif, M. A., Magg, M., Bromm, V., Klessen, R. S., Glover, S. C. O., Whalen, D. J., Pellegrini, E. W. and Volonteri, M. (2016a). Exploring the nature of the Lyman-α emitter CR7, *MNRAS* **462**, pp. 2184–2202, doi:10.1093/mnras/stw1775.

Hartwig, T., Volonteri, M., Bromm, V., Klessen, R. S., Barausse, E., Magg, M. and Stacy, A. (2016b). Gravitational waves from the remnants of the first stars, *MNRAS* **460**, pp. L74–L78, doi:10.1093/mnrasl/slw074.

Haster, C.-J., Antonini, F., Kalogera, V. and Mandel, I. (2016). N-body dynamics of intermediate mass-ratio inspirals in star clusters, *ApJ* **832**, 192, doi:10.3847/0004-637X/832/2/192.

Hawking, S. (1971). Gravitationally collapsed objects of very low mass, *MNRAS* **152**, p. 75, doi:10.1093/mnras/152.1.75.

Hawking, S. W. (1974). Black hole explosions? *Nature* **248**, pp. 30–31, doi:10.1038/248030a0.

Hawking, S. W. (1989). Black holes from cosmic strings, *Physics Letters B* **231**, pp. 237–239, doi:10.1016/0370-2693(89)90206-2.

Hawking, S. W., Moss, I. G. and Stewart, J. M. (1982). Bubble collisions in the very early universe, *Phys. Rev. D* **26**, pp. 2681–2693, doi:10.1103/PhysRevD.26.2681.

Hayward, C. C., Torrey, P., Springel, V., Hernquist, L. and Vogelsberger, M. (2014). Galaxy mergers on a moving mesh: a comparison with smoothed particle hydrodynamics, *MNRAS* **442**, pp. 1992–2016, doi:10.1093/mnras/stu957.

Heckman, T. M. and Best, P. N. (2014). The coevolution of galaxies and supermassive black holes: insights from surveys of the contemporary Universe, *ARA&A* **52**, pp. 589–660, doi:10.1146/annurev-astro-081913-035722.

Heger, A., Fryer, C. L., Woosley, S. E., Langer, N. and Hartmann, D. H. (2003a). How massive single stars end their life, *ApJ* **591**, pp. 288–300, doi:10.1086/375341.

Heger, A. and Woosley, S. E. (2002). The nucleosynthetic signature of Population III, *ApJ* **567**, pp. 532–543, doi:10.1086/338487.

Heger, A. and Woosley, S. E. (2010). Nucleosynthesis and evolution of massive metal-free stars, *ApJ* **724**, pp. 341–373, doi:10.1088/0004-637X/724/1/341.

Heger, A., Woosley, S. E., Fryer, C. L. and Langer, N. (2003b). Massive star evolution through the ages, in W. Hillebrandt and B. Leibundgut (eds.), *From Twilight to Highlight: The Physics of Supernovae*, p. 3, doi:10.1007/10828549_1.

Heggie, D. C. (1975). Binary evolution in stellar dynamics, *MNRAS* **173**, pp. 729–787, doi:10.1093/mnras/173.3.729.

Hilbert, D. (1917). Die Grundlagen der Physik (Zweite Mitteilung.), *Nachr. Ges. Wiss. Göttingen, Math. Phys. Kl.* **53**.

Hills, J. G. (1975). Possible power source of Seyfert galaxies and QSOs, *Nature* **254**, pp. 295–298, doi:10.1038/254295a0.

Hills, J. G. (1988). Hyper-velocity and tidal stars from binaries disrupted by a massive Galactic black hole, *Nature* **331**, pp. 687–689, doi:10.1038/331687a0.

Hills, J. G. and Day, C. A. (1976). Stellar collisions in globular clusters, *Astrophys. Lett.* **17**, p. 87.

Hirano, S., Hosokawa, T., Yoshida, N., Omukai, K. and Yorke, H. W. (2015). Primordial star formation under the influence of far ultraviolet radiation: 1540 cosmological haloes and the stellar mass distribution, *MNRAS* **448**, pp. 568–587, doi:10.1093/mnras/stv044.

Hirano, S., Hosokawa, T., Yoshida, N., Umeda, H., Omukai, K., Chiaki, G. and Yorke, H. W. (2014). One hundred first stars: protostellar evolution and the final masses, *ApJ* **781**, 60, doi:10.1088/0004-637X/781/2/60.

Hirano, S., Umeda, H. and Yoshida, N. (2011). Evolution of primordial stars powered by dark matter annihilation up to the main-sequence stage, *ApJ* **736**, 58, doi:10.1088/0004-637X/736/1/58.

Hirano, S. and Yoshida, N. (2013). Radiative cooling implementations in simulations of primordial star formation, *ApJ* **763**, 52, doi:10.1088/0004-637X/763/1/52.

Hirschmann, M., Khochfar, S., Burkert, A., Naab, T., Genel, S. and Somerville, R. S. (2010). On the evolution of the intrinsic scatter in black hole versus galaxy mass relations, *MNRAS* **407**, pp. 1016–1032, doi:10.1111/j.1365-2966.2010.17006.x.

Ho, L. C. (2008). Nuclear activity in nearby galaxies, *ARA&A* **46**, pp. 475–539, doi:10.1146/annurev.astro.45.051806.110546.

Hobbs, A., Nayakshin, S., Power, C. and King, A. (2011). Feeding supermassive black holes through supersonic turbulence and ballistic accretion, *MNRAS* **413**, pp. 2633–2650, doi:10.1111/j.1365-2966.2011.18333.x.

Hobbs, G., Archibald, A., Arzoumanian, Z., Backer, D., Bailes, M., Bhat, N. D. R., Burgay, M., Burke-Spolaor, S., van Straten, W., Verbiest, J. P. W., Yardley, D. R. B. and You, X. P. (2010). The International Pulsar Timing Array project: using pulsars as a gravitational wave detector, *Classical and Quantum Gravity* **27**, 8, 084013, doi:10.1088/0264-9381/27/8/084013.

Hobson, M. P., Efstathiou, G. P. and Lasenby, A. N. (2006). *General Relativity* (Cambridge University Press), doi:10.2277/0521829518.

Hollenbach, D. and McKee, C. F. (1979). Molecule formation and infrared emission in fast interstellar shocks. I Physical processes, *ApJS* **41**, pp. 555–592, doi:10.1086/190631.

Holmberg, E. (1941). On the clustering tendencies among the nebulae. II. a study of encounters between laboratory models of stellar systems by a new integration procedure. *ApJ* **94**, p. 385, doi:10.1086/144344.

Holzbauer, L. N. and Furlanetto, S. R. (2012). Fluctuations in the high-redshift Lyman-Werner and Lyα radiation backgrounds, *MNRAS* **419**, pp. 718–731, doi:10.1111/j.1365-2966.2011.19752.x.

Hopkins, P. (2013). A general class of Lagrangian smoothed particle hydrodynamics methods and implications for fluid mixing problems, *MNRAS* **428**, 2840.

Hopkins, P. F., Hernquist, L., Cox, T. J., Di Matteo, T., Robertson, B. and Springel, V. (2006). A unified, merger-driven model of the origin of starbursts, quasars, the cosmic X-ray background, supermassive black holes, and galaxy spheroids, *ApJS* **163**, pp. 1–49, doi:10.1086/499298.

Hopkins, P. F. and Quataert, E. (2010). How do massive black holes get their gas? *MNRAS* **407**, pp. 1529–1564, doi:10.1111/j.1365-2966.2010.17064.x.

Hopkins, P. F. and Quataert, E. (2011). An analytic model of angular momentum transport by gravitational torques: from galaxies to massive black holes, *MNRAS*.

Hosokawa, T. (2017). Formation of massive primordial stars controlled by the protostellar evolution , *Mem. Soc. Astron. Italiana* **88**, p. 689.

Hosokawa, T., Hirano, S., Kuiper, R., Yorke, H. W., Omukai, K. and Yoshida, N. (2016). Formation of massive primordial stars: intermittent UV feedback with episodic mass accretion, *ApJ* **824**, 119, doi:10.3847/0004-637X/824/2/119.

Hosokawa, T. and Omukai, K. (2009). Evolution of massive protostars with high accretion rates, *ApJ* **691**, pp. 823–846, doi:10.1088/0004-637X/691/1/823.

Hosokawa, T., Omukai, K. and Yorke, H. W. (2012a). Rapidly accreting supergiant protostars: embryos of supermassive black holes? *ApJ* **756**, 93, doi:10.1088/0004-637X/756/1/93.

Hosokawa, T., Omukai, K., Yoshida, N. and Yorke, H. W. (2011). Protostellar feedback halts the growth of the first stars in the Universe, *Science* **334**, pp. 1250–, doi:10.1126/science.1207433.

Hosokawa, T., Yorke, H. W., Inayoshi, K., Omukai, K. and Yoshida, N. (2013). Formation of primordial supermassive stars by rapid mass accretion, *ApJ* **778**, 178, doi:10.1088/0004-637X/778/2/178.

Hosokawa, T., Yorke, H. W. and Omukai, K. (2010). Evolution of massive protostars via disk accretion, *ApJ* **721**, pp. 478–492, doi:10.1088/0004-637X/721/1/478.

Hosokawa, T., Yoshida, N., Omukai, K. and Yorke, H. W. (2012b). Protostellar feedback and final mass of the second-generation primordial stars, *ApJ* **760**, L37, doi:10.1088/2041-8205/760/2/L37.

Howard, C. S., Pudritz, R. E. and Harris, W. E. (2014). Cluster formation in molecular clouds — I. stellar populations, star formation rates and ionizing radiation, *MNRAS* **438**, pp. 1305–1317, doi:10.1093/mnras/stt2275.

Hoyle, F. and Lyttleton, R. A. (1939). The evolution of the stars, *Proceedings of the Cambridge Philosophical Society* **35**, p. 592, doi:10.1017/S0305004100021368.

Hummel, J. A., Pawlik, A. H., Milosavljević, M. and Bromm, V. (2012). The source density and observability of pair-instability supernovae from the first stars, *ApJ* **755**, 72, doi:10.1088/0004-637X/755/1/72.

Hummel, J. A., Stacy, A. and Bromm, V. (2016). The first stars: formation under cosmic ray feedback, *MNRAS* **460**, pp. 2432–2444, doi:10.1093/mnras/stw1127.

Hummel, J. A., Stacy, A., Jeon, M., Oliveri, A. and Bromm, V. (2015). The first stars: formation under X-ray feedback, *MNRAS* **453**, pp. 4136–4147, doi:10.1093/mnras/stv1902.

Hut, P., McMillan, S., Goodman, J., Mateo, M., Phinney, E. S., Pryor, C., Richer, H. B., Verbunt, F. and Weinberg, M. (1992). Binaries in globular clusters, *PASP* **104**, pp. 981–1034, doi:10.1086/133085.

Iben, I., Jr. (1963). Massive stars in quasi-static equlibrium. *ApJ* **138**, p. 1090, doi:10.1086/147708.

Iliev, I. T., Scannapieco, E., Martel, H. and Shapiro, P. R. (2003). Non-linear clustering during the cosmic Dark Ages and its effect on the 21-cm background from minihaloes, *MNRAS* **341**, pp. 81–90, doi:10.1046/j.1365-8711.2003.06410.x.

Inayoshi, K. and Haiman, Z. (2014). Does disc fragmentation prevent the formation of supermassive stars in protogalaxies? *MNRAS* **445**, pp. 1549–1557, doi:10.1093/mnras/stu1870.

Inayoshi, K. and Haiman, Z. (2016). Is there a maximum mass for black holes in galactic nuclei? *ApJ* **828**, 110, doi:10.3847/0004-637X/828/2/110.

Inayoshi, K., Haiman, Z. and Ostriker, J. P. (2016). Hyper-Eddington accretion flows on to massive black holes, *MNRAS* **459**, pp. 3738–3755, doi:10.1093/mnras/stw836.

Inayoshi, K., Hosokawa, T. and Omukai, K. (2013). Pulsational instability of supergiant protostars: do they grow supermassive by accretion? *MNRAS* **431**, pp. 3036–3044, doi:10.1093/mnras/stt362.

Inayoshi, K. and Omukai, K. (2011). Effect of cosmic ray/X-ray ionization on supermassive black hole formation, *MNRAS* **416**, pp. 2748–2759, doi:10.1111/j.1365-2966.2011.19229.x.

Inayoshi, K., Omukai, K. and Tasker, E. (2014). Formation of an embryonic supermassive star in the first galaxy, *MNRAS* **445**, pp. L109–L113, doi:10.1093/mnrasl/slu151.

Inayoshi, K. and Tanaka, T. L. (2015). The suppression of direct collapse black hole formation by soft X-ray irradiation, *MNRAS* **450**, pp. 4350–4363, doi:10.1093/mnras/stv871.

Iocco, F. (2008). Dark matter capture and annihilation on the first stars: preliminary estimates, *ApJ* **677**, L1, doi:10.1086/587959.

Iocco, F., Bressan, A., Ripamonti, E., Schneider, R., Ferrara, A. and Marigo, P. (2008). Dark matter annihilation effects on the first stars, *MNRAS* **390**, pp. 1655–1669, doi:10.1111/j.1365-2966.2008.13853.x.

Israel, W. (1967). Event horizons in static vacuum space-times, *Physical Review* **164**, pp. 1776–1779, doi:10.1103/PhysRev.164.1776.

Iwamoto, N., Umeda, H., Tominaga, N., Nomoto, K. and Maeda, K. (2005). The first chemical enrichment in the Universe and the formation of hyper metal-poor stars, *Science* **309**, pp. 451–453, doi:10.1126/science.1112997.

Jahnke, K. and Macciò, A. V. (2011). The non-causal origin of the black-hole-galaxy scaling relations, *ApJ* **734**, 92, doi:10.1088/0004-637X/734/2/92.

Jappsen, A., Klessen, R. S., Glover, S. C. O. and Mac Low, M. (2009). Star formation at very low metallicity. IV. fragmentation does not depend on metallicity for cold initial conditions, *ApJ* **696**, pp. 1065–1074, doi:10.1088/0004-637X/696/2/1065.

Jappsen, A.-K., Glover, S. C. O., Klessen, R. S. and Mac Low, M.-M. (2007). Star formation at very low metallicity. II. on the insignificance of metal-line cooling during the early stages of gravitational collapse, *ApJ* **660**, pp. 1332–1343, doi:10.1086/513085.

Jappsen, A.-K., Klessen, R. S., Larson, R. B., Li, Y. and Mac Low, M.-M. (2005). The stellar mass spectrum from non-isothermal gravoturbulent fragmentation, *A&A* **435**, pp. 611–623, doi:10.1051/0004-6361:20042178.

Jasche, J., Ciardi, B. and Enßlin, T. A. (2007). Cosmic rays and the primordial gas, *MNRAS* **380**, pp. 417–429, doi:10.1111/j.1365-2966.2007.12058.x.

Jeans, J. H. (1902). The Stability of a Spherical Nebula, *Philosophical Transactions of the Royal Society of London Series A* **199**, pp. 1–53, doi:10.1098/rsta.1902.0012.

Jeon, M., Pawlik, A. H., Bromm, V. and Milosavljević, M. (2014a). Radiative feedback from high-mass X-ray binaries on the formation of the first galaxies and early reionization, *MNRAS* **440**, pp. 3778–3796, doi:10.1093/mnras/stu444.

Jeon, M., Pawlik, A. H., Bromm, V. and Milosavljević, M. (2014b). Recovery from Population III supernova explosions and the onset of second-generation star formation, *MNRAS* **444**, pp. 3288–3300, doi:10.1093/mnras/stu1980.

Jeon, M., Pawlik, A. H., Greif, T. H., Glover, S. C. O., Bromm, V., Milosavljević, M. and Klessen, R. S. (2012). The first galaxies: assembly with black hole feedback, *ApJ* **754**, 34, doi:10.1088/0004-637X/754/1/34.

Ji, A. P., Frebel, A., Ezzeddine, R. and Casey, A. R. (2016a). Chemical diversity in the ultra-faint dwarf galaxy Tucana II, *ApJ* **832**, L3, doi: 10.3847/2041-8205/832/1/L3.

Ji, A. P., Frebel, A., Simon, J. D. and Chiti, A. (2016b). Complete element abundances of nine stars in the r-process galaxy reticulum II, *ApJ* **830**, 93, doi:10.3847/0004-637X/830/2/93.

Jiang, L., Fan, X., Annis, J., Becker, R. H., White, R. L., Chiu, K., Lin, H., Lupton, R. H., Richards, G. T., Strauss, M. A., Jester, S. and Schneider, D. P. (2008). A survey of z 6 quasars in the Sloan Digital Sky Survey Deep Stripe. I. a flux-limited sample at z AB ¡ 21, *AJ* **135**, pp. 1057–1066, doi: 10.1088/0004-6256/135/3/1057.

Jiang, L., Fan, X., Bian, F., Annis, J., Chiu, K., Jester, S., Lin, H., Lupton, R. H., Richards, G. T., Strauss, M. A., Malanushenko, V., Malanushenko, E. and Schneider, D. P. (2009). A Survey of z 6 quasars in the Sloan Digital Sky Survey Deep Stripe. II. discovery of six quasars at z AB ¿ 21, *AJ* **138**, pp. 305–311, doi:10.1088/0004-6256/138/1/305.

Jiang, L., McGreer, I. D., Fan, X., Strauss, M. A., Bañados, E., Becker, R. H., Bian, F., Farnsworth, K., Shen, Y., Wang, F., Wang, R. and Wang, S. et al. (2016). The final SDSS high-redshift quasar sample of 52 quasars at z5.7, *ApJ* **833**, 222, doi:10.3847/1538-4357/833/2/222.

Jiang, Y.-F., Stone, J. and Davis, S. W. (2017). Super-Eddington accretion disks around supermassive black holes, *ArXiv e-prints:1709.02845*.

Jiang, Y.-F., Stone, J. M. and Davis, S. W. (2014). A Global three-dimensional radiation magneto-hydrodynamic simulation of super-eddington accretion disks, *ApJ* **796**, 106, doi:10.1088/0004-637X/796/2/106.

Jogee, S. (2006). The Fueling and Evolution of AGN: Internal and External Triggers, in D. Alloin (ed.), *Physics of Active Galactic Nuclei at all Scales, Lecture Notes in Physics, Berlin Springer Verlag*, Vol. 693, p. 143, doi: 10.1007/3-540-34621-X_6.

Joggerst, C. C., Almgren, A., Bell, J., Heger, A., Whalen, D. and Woosley, S. E. (2010). The nucleosynthetic imprint of 15-40 M $_{sun}$ primordial supernovae on metal-poor stars, *ApJ* **709**, pp. 11–26, doi:10.1088/0004-637X/709/1/11.

Joggerst, C. C., Woosley, S. E. and Heger, A. (2009). Mixing in Zero- and Solar-Metallicity Supernovae, *ApJ* **693**, pp. 1780–1802, doi:10.1088/0004-637X/693/2/1780.

Johansson, P. H., Burkert, A. and Naab, T. (2009). The evolution of black hole scaling relations in galaxy mergers, *ApJ* **707**, pp. L184–L189, doi:10.1088/0004-637X/707/2/L184.

John, T. L. (1988). Continuous absorption by the negative hydrogen ion reconsidered, *A&A*.

Johnson, J. L. (2010). Population III star clusters in the reionized Universe, *MNRAS* **404**, pp. 1425–1436, doi:10.1111/j.1365-2966.2010.16351.x.

Johnson, J. L. and Bromm, V. (2006). The cooling of shock-compressed primordial gas, *MNRAS* **366**, pp. 247–256, doi:10.1111/j.1365-2966.2005.09846.x.

Johnson, J. L., Dalla Vecchia, C. and Khochfar, S. (2013a). The First Billion Years project: the impact of stellar radiation on the co-evolution of Populations II and III, *MNRAS* **428**, pp. 1857–1872, doi:10.1093/mnras/sts011.

Johnson, J. L. and Dijkstra, M. (2017). Enhanced direct collapse due to Lyman α feedback, *A&A* **601**, A138, doi:10.1051/0004-6361/201630010.

Johnson, J. L., Greif, T. H. and Bromm, V. (2008). Occurrence of metal-free galaxies in the early Universe, *MNRAS* **388**, pp. 26–38, doi:10.1111/j.1365-2966.2008.13381.x.

Johnson, J. L., Khochfar, S., Greif, T. H. and Durier, F. (2011). Accretion on to black holes formed by direct collapse, *MNRAS* **410**, pp. 919–933, doi:10.1111/j.1365-2966.2010.17491.x.

Johnson, J. L., Whalen, D. J., Agarwal, B., Paardekooper, J.-P. and Khochfar, S. (2014). The impact of reionization on the formation of supermassive black hole seeds, *MNRAS* **445**, pp. 686–693, doi:10.1093/mnras/stu1676.

Johnson, J. L., Whalen, D. J., Li, H. and Holz, D. E. (2013b). Supermassive seeds for supermassive black holes, *ApJ* **771**, 116, doi:10.1088/0004-637X/771/2/116.

Kaiser, N., Aussel, H., Burke, B. E., Boesgaard, H., Chambers, K., Chun, M. R., Heasley, J. N., Hodapp, K.-W., Hunt, B., Jedicke, R., Jewitt, D. and Kudritzki, R. et al. (2002). Pan-STARRS: A large synoptic survey telescope array, in J. A. Tyson and S. Wolff (eds.), *Survey and Other Telescope Technologies and Discoveries*, *Proceedings of the SPIE*, Vol. 4836, pp. 154–164, doi:10.1117/12.457365.

Kaiser, N., Burgett, W., Chambers, K., Denneau, L., Heasley, J., Jedicke, R., Magnier, E., Morgan, J., Onaka, P. and Tonry, J. (2010). The Pan-STARRS wide-field optical/NIR imaging survey, in *Ground-based and Airborne Telescopes III*, *Proceedings of the SPIE*, Vol. 7733, p. 77330E, doi:10.1117/12.859188.

Kasen, D., Woosley, S. E. and Heger, A. (2011). Pair instability supernovae: light curves, spectra, and shock breakout, *ApJ* **734**, 102, doi:10.1088/0004-637X/734/2/102.

Kashikawa, N., Ishizaki, Y., Willott, C. J., Onoue, M., Im, M., Furusawa, H., Toshikawa, J., Ishikawa, S., Niino, Y., Shimasaku, K., Ouchi, M. and Hibon, P. (2015). The Subaru high-z quasar survey: discovery of faint z ˜ 6 quasars, *ApJ* **798**, 28, doi:10.1088/0004-637X/798/1/28.

Katz, H., Sijacki, D. and Haehnelt, M. G. (2015). Seeding high-redshift QSOs by collisional runaway in primordial star clusters, *MNRAS* **451**, pp. 2352–2369, doi:10.1093/mnras/stv1048.

Katz, N. (1992). Dissipational galaxy formation. II — Effects of star formation, *ApJ* **391**, pp. 502–517, doi:10.1086/171366.

Katz, N., Weinberg, D. H. and Hernquist, L. (1996). Cosmological simulations with TreeSPH, *ApJS* **105**, p. 19, doi:10.1086/192305.

Kauffmann, G. and White, S. D. M. (1993). The merging history of dark matter haloes in a hierarchical universe. *MNRAS* **261**, doi:10.1093/mnras/261.4.921.

Kazantsev, A. P. (1968). Enhancement of a magnetic field by a conducting fluid, *Soviet Journal of Experimental and Theoretical Physics* **26**, p. 1031.

Keller, S. C., Bessell, M. S., Frebel, A., Casey, A. R., Asplund, M., Jacobson, H. R., Lind, K., Norris, J. E., Yong, D., Heger, A., Magic, Z. and da Costa, G. S. et al. (2014). A single low-energy, iron-poor supernova as the source of metals in the star SMSS J031300.36-670839.3, *Nature* **506**, pp. 463–466, doi:10.1038/nature12990.

Keres, D., Katz, N., Dave, R., Fardal, M. and Weinberg, D. H. (2009). Galaxies in a simulated Lambda CDM universe - II. Observable properties and constraints on feedback, *MNRAS* **396**, pp. 2332–2344, doi:10.1111/j.1365-2966.2009.14924.x.

Kerr, R. P. (1963). Gravitational field of a spinning mass as an example of algebraically special metrics, *Physical Review Letters* **11**, pp. 237–238, doi:10.1103/PhysRevLett.11.237.

Keto, E. (2007). The formation of massive stars: accretion, disks, and the development of hypercompact H II regions, *ApJ* **666**, pp. 976–981, doi:10.1086/520320.

Khan, F. M., Fiacconi, D., Mayer, L., Berczik, P. and Just, A. (2016). Swift coalescence of supermassive black holes in cosmological mergers of massive galaxies, *ApJ* **828**, 73, doi:10.3847/0004-637X/828/2/73.

Khan, F. M., Capelo, P. R., Mayer, L. and Berczik, P. (2018). Dynamical Evolution and Merger Time-scales of LISA Massive Black Hole Binaries in Disk Galaxy Mergers, *ApJ* **868**, 97, doi:10.3847/1538-4357/aae77b.

Khandai, N., Di Matteo, T., Croft, R., Wilkins, S., Feng, Y., Tucker, E., DeGraf, C. and Liu, M.-S. (2015). The MassiveBlack-II simulation: the evolution of haloes and galaxies to z = 0, *MNRAS* **450**, pp. 1349–1374, doi:10.1093/mnras/stv627.

Kim, J.-h., Wise, J. H., Alvarez, M. A. and Abel, T. (2011). Galaxy formation with self-consistently modeled stars and massive black holes. I. feedback-regulated star formation and black hole growth, *ApJ* **738**, 54, doi:10.1088/0004-637X/738/1/54.

King, A. (2016). How big can a black hole grow? *MNRAS* **456**, pp. L109–L112, doi:10.1093/mnrasl/slv186.

King, A. and Pounds, K. (2015). Powerful outflows and feedback from active galactic nuclei, *ARA&A* **53**, pp. 115–154, doi:10.1146/annurev-astro-082214-122316.

King, A. and Pringle, J. (2006). Growing supermassive black holes by chaotic accretion, *MNRAS* **373**, doi:10.1111/j.1745-3933.2006.00249.x.

King, A. R., Lubow, S. H., Ogilvie, G. I. and Pringle, J. E. (2005). Aligning spinning black holes and accretion discs, *MNRAS* **363**, pp. 49–56, doi:10.1111/j.1365-2966.2005.09378.x.

Kippenhahn, R. and Meyer-Hofmeister, E. (1977). On the radii of accreting main sequence stars, *A&A* **54**, pp. 539–542.

Kippenhahn, R., Weigert, A. and Weiss, A. (2012). *Stellar Structure and Evolution* (Springer Verlag Berlin Heidelberg), doi:10.1007/978-3-642-30304-3.

Kirby, E. N., Guo, M., Zhang, A. J., Deng, M., Cohen, J. G., Guhathakurta, P., Shetrone, M. D., Lee, Y. S. and Rizzi, L. (2015). Carbon in red giants in globular clusters and dwarf spheroidal galaxies, *ApJ* **801**, 125, doi:10.1088/0004-637X/801/2/125

Kitayama, T., Yoshida, N., Susa, H. and Umemura, M. (2004). The structure and evolution of early cosmological H II regions, *ApJ* **613**, pp. 631–645, doi:10.1086/423313.

Kızıltan, B., Baumgardt, H. and Loeb, A. (2017). An intermediate-mass black hole in the centre of the globular cluster 47 Tucanae, *Nature* **542**, pp. 203–205, doi:10.1038/nature21361.

Klein, A., Barausse, E., Sesana, A., Petiteau, A., Berti, E., Babak, S., Gair, J., Aoudia, S., Hinder, I., Ohme, F. and Wardell, B. (2016). Science with the space-based interferometer eLISA: supermassive black hole binaries, *Phys. Rev. D* **93**, 2, 024003, doi:10.1103/PhysRevD.93.024003.

Klessen, R. S. and Glover, S. C. O. (2016). Physical processes in the interstellar medium, *Star Formation in Galaxy Evolution: Connecting Numerical Models to Reality, Saas-Fee Advanced Course, Volume 43. ISBN 978-3-662-47889-9. Springer-Verlag Berlin Heidelberg, 2016, p. 85* **43**, p. 85, doi: 10.1007/978-3-662-47890-5_2.

Klessen, R. S., Glover, S. C. O. and Clark, P. C. (2012). On the formation of very metal poor stars: the case of SDSS J1029151+172927, *MNRAS* **421**, pp. 3217–3221, doi:10.1111/j.1365-2966.2012.20544.x.

Koch, A., Feltzing, S., Adén, D. and Matteucci, F. (2013). Neutron-capture element deficiency of the Hercules dwarf spheroidal galaxy, *A&A* **554**, A5, doi:10.1051/0004-6361/201220742.

Kormendy, J. (2013). *Secular Evolution in Disk Galaxies* (Cambridge: Cambridge University Press), p. 1.

Kormendy, J. and Ho, L. C. (2013). Coevolution (Or Not) of supermassive black holes and host galaxies, *ARA&A* **51**, pp. 511–653, doi:10.1146/annurev-astro-082708-101811.

Kormendy, J. and Richstone, D. (1995). Inward bound—the search for supermassive black holes in galactic nuclei, *ARA&A* **33**, p. 581, doi:10.1146/annurev.aa.33.090195.003053.

Koushiappas, S. M., Bullock, J. S. and Dekel, A. (2004). Massive black hole seeds from low angular momentum material, *MNRAS* **354**, pp. 292–304, doi:10.1111/j.1365-2966.2004.08190.x.

Kovetz, E. D., Cholis, I., Breysse, P. C. and Kamionkowski, M. (2017). Black hole mass function from gravitational wave measurements, *Phys. Rev. D* **95**, 10, 103010, doi:10.1103/PhysRevD.95.103010.

Kratter, K. and Lodato, G. (2016). Gravitational instabilities in circumstellar disks, *ARA&A* **54**, pp. 271–311, doi:10.1146/annurev-astro-081915-023307.

Kratter, K. M. and Matzner, C. D. (2006). Fragmentation of massive protostellar discs, *MNRAS* **373**, pp. 1563–1576, doi:10.1111/j.1365-2966.2006.11103.x.

Kratter, K. M., Murray-Clay, R. A. and Youdin, A. N. (2010). The runts of the litter: why planets formed through gravitational instability can only be failed binary stars, *ApJ* **710**, pp. 1375–1386, doi:10.1088/0004-637X/710/2/1375.

Kreckel, H., Bruhns, H., Čížek, M., Glover, S. C. O., Miller, K. A., Urbain, X. and Savin, D. W. (2010). Experimental results for H_2 formation from H^- and H and implications for first star formation, *Science* **329**, pp. 69–, doi:10.1126/science.1187191.

Krolik, J. H. (1998). *Active Galactic Nuclei: From the Central Black Hole to the Galactic Environment* (Princeton: Princeton University Press).

Kroupa, P. (2002). The initial mass function of stars: evidence for uniformity in variable systems, *Science* **295**, pp. 82–91, doi:10.1126/science.1067524.

Krumholz, M. R. (2015). Notes on star formation, *ArXiv e-prints:1511.03457*.

Krumholz, M. R., Klein R. I., McKee, C. F., Offner, S. S. R. and Cunningham, A. J. (2009). The formation of massive star systems by accretion, *Science* **323**, p. 754, doi:10.1126/science.1165857.

Krumholz, M. R., McKee, C. F. and Klein, R. I. (2004). Embedding Lagrangian sink particles in Eulerian grids, *ApJ* **611**, pp. 399–412, doi:10.1086/421935.

Kruskal, M. D. (1960). Maximal extension of Schwarzschild metric, *Physical Review* **119**, pp. 1743–1745, doi:10.1103/PhysRev.119.1743.

Kuiper, R., Klahr, H., Beuther, H. and Henning, T. (2010). Circumventing the radiation pressure barrier in the formation of massive stars via disk accretion, *ApJ* **722**, pp. 1556–1576, doi:10.1088/0004-637X/722/2/1556.

Kuiper, R., Klahr, H., Beuther, H. and Henning, T. (2011). Three-dimensional simulation of massive star formation in the disk accretion scenario, *ApJ* **732**, 20, doi:10.1088/0004-637X/732/1/20.

Kulkarni, G., Hennawi, J. F., Rollinde, E. and Vangioni, E. (2014). Chemical constraints on the contribution of Population III stars to cosmic reionization, *ApJ* **787**, 64, doi:10.1088/0004-637X/787/1/64.

Kulkarni, G., Rollinde, E., Hennawi, J. F. and Vangioni, E. (2013). Chemical enrichment of damped Lyα systems as a direct constraint on Population III star formation, *ApJ* **772**, 93, doi:10.1088/0004-637X/772/2/93.

Lacey, C. and Cole, S. (1993). Merger rates in hierarchical models of galaxy formation, *MNRAS* **262**, pp. 627–649, doi:10.1093/mnras/262.3.627.

Lada, C. J. and Lada, E. A. (2003). Embedded clusters in molecular clouds, *ARA&A* **41**, pp. 57–115, doi:10.1146/annurev.astro.41.011802.094844.

Lai, D. K., Bolte, M., Johnson, J. A., Lucatello, S., Heger, A. and Woosley, S. E. (2008). Detailed abundances for 28 metal-poor stars: stellar relics in the Milky Way, *ApJ* **681**, 1524–1556, doi:10.1086/588811.

Landau, L. D. (1932). To the stars theory, *Phys. Zs. Sowjet., vol.1, p.285, 1932 (English and German)* **1**, p. 285.

Landau, L. D. and Lifshitz, E. M. (1962). *The classical theory of fields* (Oxford: Pergamon Press).

Langlois, D. (2010). Lectures on inflation and cosmological perturbations, *ArXiv e-prints 1001.5259*.

Laplace, P. S. (1796). *Exposition du système du monde*.

Larson, R. B. (1969). Numerical calculations of the dynamics of collapsing protostar, *MNRAS* **145**, p. 271.

Latif, M. A., Bovino, S., Grassi, T., Schleicher, D. R. G. and Spaans, M. (2015). How realistic UV spectra and X-rays suppress the abundance of direct collapse black holes, *MNRAS* **446**, pp. 3163–3177, doi:10.1093/mnras/stu2244.

Latif, M. A., Bovino, S., Van Borm, C., Grassi, T., Schleicher, D. R. G. and Spaans, M. (2014a). A UV flux constraint on the formation of direct collapse black holes, *MNRAS* **443**, pp. 1979–1987, doi:10.1093/mnras/stu1230.

Latif, M. A. and Ferrara, A. (2016). Formation of supermassive black hole seeds, *PASA* **33**, e051, doi:10.1017/pasa.2016.41.

Latif, M. A., Niemeyer, J. C. and Schleicher, D. R. G. (2014b). Impact of baryonic streaming velocities on the formation of supermassive black holes via direct collapse, *MNRAS* **440**, pp. 2969–2975, doi:10.1093/mnras/stu489.

Latif, M. A., Omukai, K., Habouzit, M., Schleicher, D. R. G. and Volonteri, M. (2016a). Impact of dust cooling on direct-collapse black hole formation, *ApJ* **823**, 40, doi:10.3847/0004-637X/823/1/40.

Latif, M. A. and Schleicher, D. R. G. (2015a). Disc fragmentation and the formation of Population III stars, *MNRAS* **449**, pp. 77–87, doi:10.1093/mnras/stu2573.

Latif, M. A. and Schleicher, D. R. G. (2015b). The formation of supermassive black holes in rapidly rotating disks, *A&A* **578**, A118, doi:10.1051/0004-6361/201525855.

Latif, M. A., Schleicher, D. R. G., Bovino, S., Grassi, T. and Spaans, M. (2014c). The formation of massive primordial stars in the presence of moderate UV backgrounds, *ApJ* **792**, 78, doi:10.1088/0004-637X/792/1/78.

Latif, M. A., Schleicher, D. R. G. and Hartwig, T. (2016b). Witnessing the birth of a supermassive protostar, *MNRAS* **458**, pp. 233–241, doi:10.1093/mnras/stw297.

Latif, M. A., Schleicher, D. R. G. and Schmidt, W. (2014d). Magnetic fields during the formation of supermassive black holes, *MNRAS* **440**, pp. 1551–1561, doi:10.1093/mnras/stu357.

Latif, M. A., Schleicher, D. R. G., Schmidt, W. and Niemeyer, J. (2013a). Black hole formation in the early Universe, *MNRAS* **433**, pp. 1607–1618, doi:10.1093/mnras/stt834.

Latif, M. A., Schleicher, D. R. G., Schmidt, W. and Niemeyer, J. (2013b). The small-scale dynamo and the amplification of magnetic fields in massive primordial haloes, *MNRAS* **432**, pp. 668–678, doi:10.1093/mnras/stt503.

Latif, M. A., Schleicher, D. R. G., Schmidt, W. and Niemeyer, J. C. (2013c). The characteristic black hole mass resulting from direct collapse in the early Universe, *MNRAS* **436**, pp. 2989–2996, doi:10.1093/mnras/stt1786.

Latif, M. A., Schleicher, D. R. G., Spaans, M. and Zaroubi, S. (2011a). Lyman alpha emission from the first galaxies: implications of UV backgrounds and the formation of molecules, *A&A* **532**, A66, doi:10.1051/0004-6361/201117144.

Latif, M. A. and Volonteri, M. (2015). Assessing inflow rates in atomic cooling haloes: implications for direct collapse black holes, *MNRAS* **452**, pp. 1026–1044, doi:10.1093/mnras/stv1337.

Latif, M. A., Zaroubi, S. and Spaans, M. (2011b). The impact of Lyman α trapping on the formation of primordial objects, *MNRAS* **411**, pp. 1659–1670, doi:10.1111/j.1365-2966.2010.17796.x.

Lattimer, J. M. and Schramm, D. N. (1974). Black-hole-neutron-star collisions, *ApJ* **192**, pp. L145–L147, doi:10.1086/181612.

Laureijs, R., Amiaux, J., Arduini, S., Auguères, J., Brinchmann, J., Cole, R., Cropper, M., Dabin, C., Duvet, L., Ealet, A. and et al. (2011). Euclid definition Study Report, *ArXiv e-prints:1110.3193*.

Lawrence, A., Warren, S. J., Almaini, O., Edge, A. C., Hambly, N. C., Jameson, R. F., Lucas, P., Casali, M , Adamson, A., Dye, S., Emerson, J. P. and Foucaud, S. et al. (2007). The UKIRT Infrared Deep Sky Survey (UKIDSS), *MNRAS* **379**, pp. 1599–1617, doi:10.1111/j.1365-2966.2007.12040.x.

Lazar, M., Schlickeiser, R., Wielebinski, R. and Poedts, S. (2009). Cosmological effects of weibel-type instabilities, *ApJ* **693**, pp. 1133–1141, doi:10.1088/0004-637X/693/2/1133.

Le Bourlot, J., Pineau des Forêts, G. and Flower, D. R. (1999). The cooling of astrophysical media by H_2, *MNRAS* **305**, pp. 802–810, doi:10.1046/j.1365-8711.1999.02497.x.

Lee, H. and Yoon, S.-C. (2016). On the role of the $\Omega\Gamma$ limit in the formation of Population III massive stars, *ApJ* **820**, 135, doi:10.3847/0004-637X/820/2/135.

Lehmer, B. D., Alexander, D. M., Bauer, F. E., Brandt, W. N., Goulding, A. D., Jenkins, L. P., Ptak, A. and Roberts, T. P. (2010). A Chandra perspective on galaxy-wide X-ray binary emission and its correlation with star formation rate and stellar mass: new results from luminous infrared galaxies, *ApJ* **724**, pp. 559–571, doi:10.1088/0004-637X/724/1/559.

Lehner, L. and Pretorius, F. (2014). Numerical relativity and astrophysics, *ARA&A* **52**, pp. 661–694, doi:10.1146/annurev-astro-081913-040031.

Leitherer, C., Schaerer, D., Goldader, J. D., Delgado, R. M. G., Robert, C., Kune, D. F., de Mello, D. F., Devost, D. and Heckman, T. M. (1999). Starburst99: Synthesis models for galaxies with active star formation, *ApJS* **123**, pp. 3–40, doi:10.1086/313233.

Lense, J. and Thirring, H. (1918). Über den Einfluß der Eigenrotation der Zentralkörper auf die Bewegung der Planeten und Monde nach der Einsteinschen Gravitationstheorie, *Physikalische Zeitschrift* **19**.

Lenzuni, P., Chernoff, D. F. and Salpeter, E. E. (1991). Rosseland and Planck mean opacities of a zero-metallicity gas, *ApJS* **76**, pp. 759–801, doi:10.1086/191580.

Lepp, S. and Shull, J. M. (1983). The kinetic theory of H2 dissociation, *ApJ* **270**, pp. 578–582, doi:10.1086/161149.

Li, Y., Hernquist, L. and Finkbeiner, D. (2006). The infrared properties of galaxies and quasars at z 6, in *American Astronomical Society Meeting Abstracts*, *Bulletin of the American Astronomical Society*, Vol. 38, p. 964.

Li, Y., Hernquist, L., Robertson, B., Cox, T. J., Hopkins, P. F., Springel, V., Gao, L., Di Matteo, T., Zentner, A. R., Jenkins, A. and Yoshida, N. (2007). Formation of z˜6 quasars from hierarchical galaxy mergers, *ApJ* **665**, pp. 187–208, doi:10.1086/519297.

Liddle, A. R. and Lyth D. H. (2000). *Cosmological Inflation and Large-Scale Structure* (Cosmological Inflation and Large-Scale Structure, by Andrew R. Liddle and David H. Lyth, pp. 414. ISBN 052166022X. Cambridge, UK: Cambridge University Press, April 2000.).

Linder, E. V. (2005). Cosmic growth history and expansion history, *Phys. Rev. D* **72**, 4, pp. 043529–+, doi:10.1103/PhysRevD.72.043529.

Linder, E. V. and Jenkins, A. (2003). Cosmic structure growth and dark energy, *MNRAS* **346**, pp. 573–583, doi:10.1046/j.1365-2966.2003.07112.x.

Lipovka, A., Núñez-López, R. and Avila-Reese, V. (2005). The cooling function of HD molecule revisited, *MNRAS* **361**, pp. 850–854, doi:10.1111/j.1365-2966.2005.09226.x.

Lodato, G. (2007). Self-gravitating accretion discs, *Nuovo Cimento Rivista Serie* **30**, p. 293, doi:10.1393/ncr/i2007-10022-x.

Lodato, G. and Natarajan, P. (2006). Supermassive black hole formation during the assembly of pre-galactic discs, *MNRAS* **371**, pp. 1813–1823, doi:10.1111/j.1365-2966.2006.10801.x.

Lodato, G. and Rice, W. (2004). Testing the locality of transport in self-gravitating accretion discs, *MNRAS* **351**, 630.

Loeb, A. (2010). *How Did the First Stars and Galaxies Form? By Abraham Loeb. Princeton University Press, 2010. ISBN: 978-1-4008-3406-8* (Princeton University Press).

Loeb, A. and Furlanetto, S. R. (2013). *The First Galaxies in the Universe* (Princeton University Press).

Loeb, A. and Rasio, F. A. (1994). Collapse of primordial gas clouds and the formation of quasar black holes, *ApJ* **432**, pp. 52–61, doi:10.1086/174548.

Lombardi, J. C., Jr., Rasio, F. A. and Shapiro, S. L. (1996). Collisions of main-sequence stars and the formation of blue stragglers in globular clusters, *ApJ* **468**, p. 797, doi:10.1086/177736.

Longair, M. S. (2008). *Galaxy Formation* (Galaxy Formation, by Malcolm S. Longair Berlin: Springer, 2008. ISBN 978-3-540-73477-2).

Lu, Y., Mo, H. J., Weinberg, M. D. and Katz, N. (2011). A Bayesian approach to the semi-analytic model of galaxy formation: methodology, *MNRAS* **416**, pp. 1949–1964, doi:10.1111/j.1365-2966.2011.19170.x.

Luo, B., Brandt, W. N., Alexander, D. M., Stern, D., Teng, S. H., Arévalo, P., Bauer, F. E., Boggs, S. E., Christensen, F. E., Comastri, A. and Craig, W. W. et al. (2014). Weak hard X-ray emission from broad absorption line quasars: evidence for intrinsic X-ray weakness, *ApJ* **794**, 70, doi:10.1088/0004-637X/794/1/70.

Luo, B., Brandt, W. N., Xue, Y. Q., Alexander, D. M., Brusa, M., Bauer, F. E., Comastri, A., Fabian, A. C., Gilli, R., Lehmer, B. D., Rafferty, D. A., Schneider, D. P. and Vignali, C. (2011). Revealing a population of heavily obscured active galactic nuclei at z 0.5-1 in the Chandra Deep Field-South, *ApJ* **740**, 37, doi:10.1088/0004-637X/740/1/37.

Lupi, A., Colpi, M., Devecchi, B., Galanti, G. and Volonteri, M. (2014). Constraining the high-redshift formation of black hole seeds in nuclear star clusters with gas inflows, *MNRAS* **442**, pp. 3616–3626, doi:10.1093/mnras/stu1120.

Lupi, A., Haardt, F., Dotti, M., Fiacconi, D., Mayer, L. and Madau, P. (2016). Growing massive black holes through supercritical accretion of stellar-mass seeds, *MNRAS* **456**, pp. 2993–3003, doi:10.1093/mnras/stv2877.

Lynden-Bell, D. (1969). Galactic nuclei as collapsed old quasars, *Nature* **223**, pp. 690–694, doi:10.1038/223690a0.

Lynden-Bell, D. and Wood, R. (1968). The gravo-thermal catastrophe in isothermal spheres and the onset of red-giant structure for stellar systems, *MNRAS* **138**, p. 495, doi:10.1093/mnras/138.4.495.

Lyne, A. G., Biggs, J. D , Brinklow, A., McKenna, J. and Ashworth, M. (1988). Discovery of a binary millisecond pulsar in the globular cluster M4, *Nature* **332**, pp. 45–47, doi:10.1038/332045a0.

Lyne, A. G., Brinklow, A., Middleditch, J., Kulkarni, S. R. and Backer, D. C. (1987). The discovery of a millisecond pulsar in the globular cluster M28, *Nature* **328**, pp. 399–401, doi:10.1038/328399a0.

Ma, X., Hopkins, P. F., Garrison-Kimmel, S., Faucher-Giguère, C.-A., Quataert, E., Boylan-Kolchin, M., Hayward, C. C., Feldmann, R. and Kereš, D. (2017). Simulating galaxies in the reionization era with FIRE-2: galaxy scaling relations, stellar mass functions, and luminosity functions, *ArXiv e-prints:1706.06605*.

Mac Low, M.-M. and Klessen, R. S. (2004). Control of star formation by supersonic turbulence, *Reviews of Modern Physics* **76**, pp. 125–194, doi: 10.1103/RevModPhys.76.125.

Mac Low, M.-M. and Shull, J. M. (1986). Molecular processes and gravitational collapse in intergalactic shocks, *ApJ* **302**, pp. 585–589, doi:10.1086/164017.

Machacek, M. E., Bryan, G. L. and Abel, T. (2001). Simulations of pregalactic structure formation with radiative feedback, *ApJ* **548**, pp. 509–521, doi: 10.1086/319014.

Machida, M. N. and Doi, K. (2013). The formation of Population III stars in gas accretion stage: effects of magnetic fields, *MNRAS* **435**, pp. 3283–3305, doi:10.1093/mnras/stt1524.

Machida, M. N., Matsumoto, T. and Inutsuka, S.-i. (2008). Magnetohydrodynamics of Population III star formation, *ApJ* **685**, 690-704, doi:10.1086/591074.

Machida, M. N., Omukai, K., Matsumoto, T. and Inutsuka, S.-i. (2006). The first jets in the Universe: protostellar jets from the first stars, *ApJ* **647**, pp. L1–L4, doi:10.1086/507326.

Maciejewski, W. (2004). Nuclear spirals in galaxies: gas response to an asymmetric potential — II. Hydrodynamical models, *MNRAS* **354**, pp. 892–904, doi:10.1111/j.1365-2966.2004.08254.x.

Mack, K. J. and Wesley, D. H. (2008). Primordial black holes in the Dark Ages: observational prospects for future 21cm surveys, *ArXiv e-prints 0805.1531*.

Mackey, J., Bromm, V. and Hernquist, L. (2003). Three epochs of star formation in the high-redshift Universe, *ApJ* **586**, pp. 1–11, doi:10.1086/367613.

Madau, P. and Dickinson, M. (2014). Cosmic star-formation history, *ARA&A* **52**, pp. 415–486, doi:10.1146/annurev-astro-081811-125615.

Madau, P., Haardt, F. and Dotti, M. (2014). Super-critical growth of massive black holes from stellar-mass seeds, *ApJ* **784**, L38, doi:10.1088/2041-8205/784/2/L38.

Madau, P., Meiksin, A. and Rees, M. J. (1997). 21 centimeter tomography of the intergalactic medium at high redshift, *ApJ* **475**, pp. 429–444, doi:10.1086/303549.

Madau, P. and Rees, M. J. (2001). Massive black holes as Population III remnants, *ApJ* **551**, pp. L27–L30, doi:10.1086/319848.

Maeder, A. and Meynet, G. (2012). Rotating massive stars: from first stars to gamma ray bursts, *Reviews of Modern Physics* **84**, pp. 25–63, doi:10.1103/RevModPhys.84.25.

Magg, M., Hartwig, T., Agarwal, B., Frebel, A., Glover, S. C. O., Griffen, B. F. and Klessen, R. S. (2018). Predicting the locations of possible long-lived low-mass first stars: importance of satellite dwarf galaxies, *MNRAS* **473**, pp. 5308–5323, doi:10.1093/mnras/stx2729.

Magg, M., Hartwig, T., Glover, S. C. O., Klessen, R. S. and Whalen, D. J. (2016). A new statistical model for Population III supernova rates: discriminating between ΛCDM and WDM cosmologies, *MNRAS* **462**, pp. 3591–3601, doi:10.1093/mnras/stw1882.

Magorrian, J., Tremaine, S., Richstone, D., Bender, R., Bower, G., Dressler, A., Faber, S. M., Gebhardt, K., Green, R., Grillmair, C., Kormendy, J. and Lauer, T. (1998). The demography of massive dark objects in galaxy centers, *AJ* **115**, pp. 2285–2305, doi:10.1086/300353.

Mainetti, D., Lupi, A., Campana, S. and Colpi, M. (2016). Hydrodynamical simulations of the tidal stripping of binary stars by massive black holes, *MNRAS* **457**, pp. 2516–2529, doi:10.1093/mnras/stw197.

Maio, U., Khochfar, S., Johnson, J. L. and Ciardi, B. (2011a). The interplay between chemical and mechanical feedback from the first generation of stars, *MNRAS* **414**, pp. 1145–1157, doi:10.1111/j.1365-2966.2011.18455.x.

Maio, U., Koopmans, L. V. E. and Ciardi, B. (2011b). The impact of primordial supersonic flows on early structure formation, reionization and the lowest-mass dwarf galaxies, *MNRAS* **412**, pp. L40–L44, doi:10.1111/j.1745-3933.2010.01001.x.

Mandel, I., Brown, D. A., Gair, J. R. and Miller, M. C. (2008). Rates and characteristics of intermediate mass ratio inspirals detectable by advanced LIGO, *ApJ* **681**, 1431-1447, doi:10.1086/588246.

Maoz, E. (1998). Dynamical constraints on alternatives to supermassive black holes in galactic nuclei, *ApJ* **494**, pp. L181–L184, doi:10.1086/311194.

Mapelli, M. (2016). Massive black hole binaries from runaway collisions: the impact of metallicity, *MNRAS* **459**, pp. 3432–3446, doi:10.1093/mnras/stw869.

Mapelli, M., Ferrara, A. and Rea, N. (2006). Constraints on galactic intermediate mass black holes, *MNRAS* **368**, pp. 1340–1350, doi:10.1111/j.1365-2966.2006.10201.x.

Marconi, A., Risaliti, G., Gilli, R., Hunt, L. K., Maiolino, R. and Salvati, M. (2004). Local supermassive black holes, relics of active galactic nuclei and the X-ray background, *MNRAS* **351**, pp. 169–185, doi:10.1111/j.1365-2966.2004.07765.x.

Martin, P. G., Schwarz, D. H. and Mandy, M. E. (1996). Master equation studies of the collisional excitation and dissociation of H_2 molecules by H atoms, *ApJ* **461**, p. 265, doi:10.1086/177053.

Matsuda, T., Satō, H. and Takeda, H. (1969). Cooling of pre-galactic gas clouds by hydrogen molecule, *Progress of Theoretical Physics* **42**, pp. 219–233, doi:10.1143/PTP.42.219.

Matsumoto, T., Nakauchi, D., Ioka, K., Heger, A. and Nakamura, T. (2015). Can direct collapse black holes launch gamma-ray bursts and grow to supermassive black holes? *ApJ* **810**, 64, doi:10.1088/0004-637X/810/1/64.

Matsuoka, Y., Onoue, M., Kashikawa, N., Iwasawa, K., Strauss, M. A., Nagao, T., Imanishi, M., Niida, M., Toba, Y., Akiyama, M., Asami, N. and Bosch, J. et al. (2016). Subaru high-z exploration of low-luminosity quasars (SHELLQs). I. discovery of 15 quasars and bright galaxies at 5.7 z 6.9, *ApJ* **828**, 26, doi:10.3847/0004-637X/828/1/26.

Matthee, J., Sobral, D., Boone, F., Röttgering, H., Schaerer, D., Girard, M., Pallottini, A., Vallini, L., Ferrara, A., Darvish, B. and Mobasher, B. (2017). ALMA reveals metals yet no dust within multiple components in CR7, *ApJ* **851**, 145, doi:10.3847/1538-4357/aa9931.

Mayer, L. and Bonoli, S. (2018). The route to massive black hole formation via merger-driven direct collapse: a review, *Reports on Progress in Physics*, p. 29 doi:2018arXiv180306391M.

Mayer, L., Fiacconi, D., Bonoli, S., Quinn, T., Rovskar, R., Shen, S. and Wadsley, J. (2015). Direct formation of supermassive black holes in metal enriched gas at the heart of high-redshift galaxy mergers, *ApJ* **810**, 51, doi:10.1088/0004-637X/810/1/51.

Mayer, L., Kazantzidis, S., Escala, A. and Callegari, S. (2010). Direct formation of supermassive black holes via multi-scale gas inflows in galaxy mergers, *Nature* **466**, pp. 1082–1084, doi:10.1038/nature09294.

Mazzucchelli, C., Bañados, E., Venemans, B. P., Decarli, R., Farina, E. P., Walter, F., Eilers, A.-C., Rix, H.-W., Simcoe, R., Stern, D. and Fan, X et al. (2017). Physical properties of 15 quasars at z 6.5, *ApJ* **849**, 91, doi:10.3847/1538-4357/aa9185.

McConnell, N. J. and Ma, C.-P. (2013). Revisiting the scaling relations of black hole masses and host galaxy properties, *ApJ* **764**, 184, doi:10.1088/0004-637X/764/2/184.

McDowell, M. R. C. (1961). On the formation of H2 in H I regions, *The Observatory* **81**, pp. 240–243.

McGreer, I. D. and Bryan, G. L. (2008). The impact of HD cooling on the formation of the first stars, *ApJ* **685**, 8-20, doi:10.1086/590530.

McKee, C. F. and Ostriker, E. C. (2007). Theory of star formation, *ARA&A* **45**, pp. 565–687, doi:10.1146/annurev.astro.45.051806.110602.

McKee, C. F. and Tan, J. C. (2008). The formation of the first stars. II. radiative feedback processes and implications for the initial mass function, *ApJ* **681**, pp. 771–797, doi:10.1086/587434.

McKinney, J. C. (2006). General relativistic magnetohydrodynamic simulations of the jet formation and large-scale propagation from black hole accretion systems, *MNRAS* **368**, pp. 1561–1582, doi:10.1111/j.1365-2966.2006.10256.x.

McNamara, B. J. and Sanders, W. L. (1976). Peculiarities in some cluster color-magnitude diagrams, *A&A* **52**, p. 53.

McNamara, B. R. and Nulsen, P. E. J. (2007). Heating hot atmospheres with active galactic nuclei, *ARA&A* **45**, pp. 117–175, doi:10.1146/annurev.astro.45.051806.110625.

McNamara, B. R., Nulsen, P. E. J., Wise, M. W., Rafferty, D. A., Carilli, C., Sarazin, C. L. and Blanton, E. L. (2005). The heating of gas in a galaxy cluster by X-ray cavities and large-scale shock fronts, *Nature* **433**, pp. 45–47, doi:10.1038/nature03202.

McQuinn, M. (2012). Constraints on X-ray emissions from the reionization era, *MNRAS* **426**, pp. 1349–1360, doi:10.1111/j.1365-2966.2012.21792.x.

McQuinn, M. and O'Leary, R. M. (2012). The impact of the supersonic baryon-dark matter velocity difference on the z ˜ 20 21 cm background, *ApJ* **760**, 3, doi:10.1088/0004-637X/760/1/3.

Medvedev, M. V., Silva, L. O., Fiore, M., Fonseca, R. A. and Mori, W. B. (2004). Generation of magnetic fields in cosmological shocks, *Journal of Korean Astronomical Society* **37**, pp. 533–541, doi:10.5303/JKAS.2004.37.5.533.

Merloni, A. (2008). Cosmological evolution of supermassive black holes and AGN: a synthesis model for accretion and feedback. *Mem. Soc. Astron. Italiana* **79**, p. 1310.

Merloni, A. (2016). Observing supermassive black holes across cosmic time: from phenomenology to physics, in F. Haardt, V. Gorini, U. Moschella, A. Treves and M. Colpi (eds.), *Lecture Notes in Physics, Berlin Springer Verlag, Lecture Notes in Physics, Berlin Springer Verlag*, Vol. 905, p. 101, doi: 10.1007/978-3-319-19416-5_4.

Merritt, D. (2013). *Dynamics and Evolution of Galactic Nuclei* (Princeton: Princeton University Press).

Mesinger, A., Bryan, G. L. and Haiman, Z. (2006). Ultraviolet radiative feedback on high-redshift protogalaxies, *ApJ* **648**, pp. 835–851, doi:10.1086/506173.

Mesinger, A., Ferrara, A. and Spiegel, D. S. (2013). Signatures of X-rays in the early Universe, *MNRAS* **431**, pp. 621–637, doi:10.1093/mnras/stt198.

Mesinger, A., Greig, B. and Sobacchi, E. (2016). The Evolution Of 21 cm Structure (EOS): public, large-scale simulations of Cosmic Dawn and reionization, *MNRAS* **459**, pp. 2342–2353, doi:10.1093/mnras/stw831.

Meynet, G. and Maeder, A. (2000). Stellar evolution with rotation. V. changes in all the outputs of massive star models, *A&A* **361**, pp. 101–120.

Mezcua, M. (2017). Observational evidence for intermediate-mass black holes, *International Journal of Modern Physics D* **26**, 1730021, doi:10.1142/S021827181730021X.

Mezcua, M., Civano, F., Marchesi, S., Suh, H., Fabbiano, G. and Volonteri, M. (2018). Intermediate-mass black holes in dwarf galaxies out to redshift $\sim$ 2.4 in the Chandra COSMOS Legacy Survey, *MNRAS* **478**, pp. 2576–2591, doi:10.1093/mnras/sty1163.

Michell, J. (1784). On the means of discovering the distance, magnitude, &c. of the fixed stars, in consequence of the diminution of the velocity of their light, in case such a diminution should be found to take place in any of them, and such other data should be procured from observations, as would be farther necessary for that purpose. By the Rev. John Michell, B. D. F. R. S. In a Letter to Henry Cavendish, Esq. F. R. S. and A. S. *Philosophical Transactions of the Royal Society of London Series I* **74**, pp. 35–57.

Micic, M., Holley-Bockelmann, K., Sigurdsson, S. and Abel, T. (2007). Supermassive black hole growth and merger rates from cosmological N-body simulations, *MNRAS* **380**, pp. 1533–1540, doi:10.1111/j.1365-2966.2007.12162.x.

Miller, K. A., Bruhns, H., Eliášek, J., Čížek, M., Kreckel, H., Urbain, X. and Savin, D. W. (2011). Associative detachment of $H^- + H \rightarrow H_2 + e^-$, *Phys. Rev. A* **84**, 5, 052709, doi:10.1103/PhysRevA.84.052709.

Miller, M. C. (2009). Intermediate-mass black holes as LISA sources, *Classical and Quantum Gravity* **26**, 9, 094031, doi:10.1088/0264-9381/26/9/094031.

Milosavljević, M., Bromm, V., Couch, S. M. and Oh, S. P. (2009). Accretion onto "seed" black holes in the first galaxies, *ApJ* **698**, pp. 766–780, doi: 10.1088/0004-637X/698/1/766.

Mineo, S., Gilfanov, M. and Sunyaev, R. (2012). X-ray emission from star-forming galaxies — I. high-mass X-ray binaries, *MNRAS* **419**, pp. 2095–2115, doi: 10.1111/j.1365-2966.2011.19862.x.

Mirabel, I. F., Dijkstra, M., Laurent, P., Loeb, A. and Pritchard, J. R. (2011). Stellar black holes at the dawn of the universe, *A&A* **528**, A149, doi:10. 1051/0004-6361/201016357.

Misner, C. W., Thorne, K. S. and Wheeler, J. A. (1973). *Gravitation* (San Francisco: W.H. Freeman and Co.).

Miyaji, T., Hasinger, G., Salvato, M., Brusa, M., Cappelluti, N., Civano, F., Puccetti, S., Elvis, M., Brunner, H., Fotopoulou, S., Ueda, Y. and Griffiths, R. E. et al. (2015). Detailed shape and evolutionary behavior of the X-ray luminosity function of active galactic nuclei, *ApJ* **804**, 104, doi:10.1088/0004-637X/804/2/104.

Miyake, S., Stancil, P. C., Sadeghpour, H. R., Dalgarno, A., McLaughlin, B. M. and Forrey, R. C. (2010). Resonant H^- photodetachment: enhanced photodestruction and consequences for radiative feedback, *ApJL* **709**, pp. L168–L171, doi:10.1088/2041-8205/709/2/L168.

Miyazaki, S., Komiyama, Y., Nakaya, H., Kamata, Y., Doi, Y., Hamana, T., Karoji, H., Furusawa, H., Kawanomoto, S., Morokuma, T., Ishizuka, Y. and Nariai, K. et al. (2012). Hyper Suprime-Cam, in *Ground-based and Airborne Instrumentation for Astronomy IV, Proceedings of the SPIE*, Vol. 8446, p. 84460Z, doi:10.1117/12.926844.

Miyoshi, M., Moran, J., Herrnstein, J., Greenhill, L., Nakai, N., Diamond, P. and Inoue, M. (1995). Evidence for a black hole from high rotation velocities in a sub-parsec region of NGC4258, *Nature* **373**, pp. 127–129, doi:10.1038/373127a0.

Monaco, P., Theuns, T., Taffoni, G., Governato, F., Quinn, T. and Stadel, J. (2002). Predicting the number, spatial distribution, and merging history of dark matter halos, *ApJ* **564**, pp. 8–14, doi:10.1086/324182.

Monsalve, R. A., Rogers, A. E. E., Bowman, J. D. and Mozdzen, T. J. (2017). Calibration of the EDGES high-band receiver to observe the global 21 cm signature from the epoch of reionization, *ApJ* **835**, 49, doi:10.3847/1538-4357/835/1/49.

Montero, P. J., Janka, H.-T. and Müller, E. (2012). Relativistic collapse and explosion of rotating supermassive stars with thermonuclear effects, *ApJ* **749**, 37, doi:10.1088/0004-637X/749/1/37.

Morales, M. F. and Wyithe, J. S. B. (2010). Reionization and cosmology with 21-cm fluctuations, *ARA&A* **48**, pp. 127–171, doi:10.1146/annurev-astro-081309-130936.

Moretti, A., Campana, S., Lazzati, D. and Tagliaferri, G. (2003). The resolved fraction of the cosmic X-ray background, *ApJ* **588**, pp. 696–703, doi:10.1086/374335.

Morganson, E., De Rosa, G., Decarli, R., Walter, F., Chambers, K., McGreer, I., Fan, X., Burgett, W., Flewelling, H., Greiner, J., Hodapp, K. and Kaiser, N. et al. (2012). The first high-redshift quasar from Pan-STARRS, *AJ* **143**, 142, doi:10.1088/0004-6256/143/6/142.

Mortlock, D. J., Patel, M., Warren, S. J., Venemans, B. P., McMahon, R. G., Hewett, P. C., Simpson, C., Sharp, R. G., Burningham, B., Dye, S., Ellis, S., Gonzales-Solares, E. A. and Huélamo, N. (2009). Discovery of a redshift 6.13 quasar in the UKIRT infrared deep sky survey, *A&A* **505**, pp. 97–104, doi:10.1051/0004-6361/200811161.

Mortlock, D. J., Warren, S. J., Venemans, B. P., Patel, M., Hewett, P. C., McMahon, R. G., Simpson, C., Theuns, T., Gonzáles-Solares, E. A. and Adamson, A. et al. (2011). A luminous quasar at a redshift of z = 7.085, *Nature* **474**, pp. 616–619, doi:10.1038/nature10159.

Munari, E., Monaco, P., Sefusatti, E., Castorina, E., Mohammad, F. G., Anselmi, S. and Borgani, S. (2017). Improving fast generation of halo catalogues with higher order Lagrangian perturbation theory, *MNRAS* **465**, pp. 4658–4677, doi:10.1093/mnras/stw3085.

Muratov, A. L., Gnedin, O. Y., Gnedin, N. Y. and Zemp, M. (2013). Revisiting the first galaxies: the epoch of Population III stars, *ApJ* **773**, 19, doi:10.1088/0004-637X/773/1/19.

Nakama, T., Silk, J. and Kamionkowski, M. (2017). Stochastic gravitational waves associated with the formation of primordial black holes, *Phys. Rev. D* **95**, 4, 043511, doi:10.1103/PhysRevD.95.043511.

Nakauchi, D., Hosokawa, T., Omukai, K., Saio, H. and Nomoto, K. (2017). Do stellar winds prevent the formation of supermassive stars by accretion? *MNRAS* **465**, pp. 5016–5025, doi:10.1093/mnras/stw3114.

Nakauchi, D., Inayoshi, K. and Omukai, K. (2014). Conditions for HD cooling in the first galaxies revisited: interplay between far-ultraviolet and cosmic ray feedback in Population III star formation, *MNRAS* **442**, pp. 2667–2679, doi:10.1093/mnras/stu1042.

Nandra, K., Barret, D., Barcons, X., Fabian, A., den Herder, J.-W., Piro, L., Watson, M., Adami, C., Aird, J., Afonso, J. M. and et al. (2013). The hot and energetic Universe: a white paper presenting the science theme motivating the Athena+ mission, *ArXiv e-prints*.

Naoz, S., Yoshida, N. and Gnedin, N. Y. (2012). Simulations of early baryonic structure formation with stream velocity. I. halo abundance, *ApJ* **747**, 128, doi:10.1088/0004-637X/747/2/128.

Naoz, S., Yoshida, N. and Gnedin, N. Y. (2013). Simulations of early baryonic structure formation with stream velocity. II. The gas fraction, *ApJ* **763**, 27, doi:10.1088/0004-637X/763/1/27.

Narayan, R. and Yi, I. (1994). Advection-dominated accretion: A self-similar solution, *ApJ* **428**, pp. L13–L16, doi:10.1086/187381.

Natarajan, P., Pacucci, F., Ferrara, A., Agarwal, B., Ricarte, A., Zackrisson, E. and Cappelluti, N. (2017). Unveiling the first black holes with JWST: multi-wavelength spectral predictions, *ApJ* **838**, 117, doi:10.3847/1538-4357/aa6330.

Natarajan, P. and Treister, E. (2009). Is there an upper limit to black hole masses? *MNRAS* **393**, pp. 838–845, doi:10.1111/j.1365-2966.2008.13864.x.

Negri, A. and Volonteri, M. (2017). Black hole feeding and feedback: the physics inside the 'sub-grid', *MNRAS* **467**, pp. 3475–3492, doi:10.1093/mnras/stx362.

Neo, S., Miyaji, S., Nomoto, K. and Sugimoto, D. (1977). Effect of rapid mass accretion onto the main-sequence stars, *PASJ* **29**, pp. 249–262.

Newman, E. T., Couch, E., Chinnapared, K., Exton, A., Prakash, A. and Torrence, R. (1965). Metric of a rotating, charged mass, *Journal of Mathematical Physics* **6**, pp. 918–919, doi:10.1063/1.1704351.

Nordström, G. (1918). On the Energy of the Gravitation field in Einstein's theory, *Koninklijke Nederlandse Akademie van Wetenschappen Proceedings Series B Physical Sciences* **20**, pp. 1238–1245.

Norris, J. E., Yong, D., Bessell, M. S., Christlieb, N., Asplund, M., Gilmore, G., Wyse, R. F. G., Beers, T. C., Barklem, P. S., Frebel, A. and Ryan, S. G. (2013). The most metal-poor stars. IV. the two populations with Fe/H ¡ -3.0, *ApJ* **762**, 28, doi:10.1088/0004-637X/762/1/28.

Novak, G. S., Ostriker, J. P. and Ciotti, L. (2012). Radiative transfer and radiative driving of outflows in active galactic nuclei and starbursts, *MNRAS* **427**, pp. 2734–2756, doi:10.1111/j.1365-2966.2012.21844.x.

Novikov, I. D. and Thorne, K. S. (1973). Astrophysics of black holes. in C. Dewitt and B. S. Dewitt (eds.), *Black Holes (Les Astres Occlus)*, pp. 343–450.

Offner, S. S. R., Kratter, K. M., Matzner, C. D., Krumholz, M. R. and Klein, R. I. (2010). The formation of low-mass binary star systems via turbulent fragmentation, *ApJ* **725**, pp. 1485–1494, doi:10.1088/0004-637X/725/2/1485.

Oguri, M. and Marshall, P. J. (2010). Gravitationally lensed quasars and supernovae in future wide-field optical imaging surveys, *MNRAS* **405**, pp. 2579–2593, doi:10.1111/j.1365-2966.2010.16639.x.

Oh, S. P. and Haiman, Z. (2002). Second-generation objects in the Universe: radiative cooling and collapse of halos with virial temperatures above 10000 K, *ApJ* **569**, pp. 558–572, doi:10.1086/339393.

Ohsuga, K., Mori, M., Nakamoto, T. and Mineshige, S. (2005). Supercritical accretion flows around black holes: two-dimensional, radiation pressure-dominated disks with photon trapping, *ApJ* **628**, pp. 368–381, doi:10.1086/430728.

Oka, T., Mizuno, R., Miura, K. and Takekawa, S. (2016). Signature of an intermediate-mass black hole in the central molecular zone of our galaxy, *ApJ* **816**, L7, doi:10.3847/2041-8205/816/1/L7.

O'Leary, R. M. and McQuinn, M. (2012). The formation of the first cosmic structures and the physics of the z ˜ 20 Universe, *ApJ* **760**, 4, doi:10.1088/0004-637X/760/1/4.

Omukai, K. (2000). Protostellar collapse with various metallicities, *ApJ* **534**, pp. 809–824, doi:10.1086/308776.
Omukai, K. (2001). Primordial star formation under far-ultraviolet radiation, *ApJ* **546**, pp. 635–651, doi:10.1086/318296.
Omukai, K. and Nishi, R. (1998). Formation of Primordial Protostars, *ApJ* **508**, pp. 141–150, doi:10.1086/306395.
Omukai, K. and Palla, F. (2003). Formation of the first stars by accretion, *ApJ* **589**, pp. 677–687, doi:10.1086/374810.
Omukai, K., Schneider, R. and Haiman, Z. (2008). Can supermassive black holes form in metal-enriched high-redshift protogalaxies? *ApJ* **686**, pp. 801–814, doi:10.1086/591636.
Omukai, K., Tsuribe, T., Schneider, R. and Ferrara, A. (2005). Thermal and fragmentation properties of star-forming clouds in low-metallicity environments, *ApJ* **626**, pp. 627–643, doi:10.1086/429955.
Oppenheimer, J. R. and Snyder, H. (1939). On continued gravitational contraction, *Physical Review* **56**, pp. 455–459, doi:10.1103/PhysRev.56.455.
Oppenheimer, J. R. and Volkoff, G. M. (1939). On massive neutron cores, *Physical Review* **55**, pp. 374–381, doi:10.1103/PhysRev.55.374.
Orosz, J. A., McClintock, J. E., Aufdenberg, J. P., Remillard, R. A., Reid, M. J., Narayan, R. and Gou, L. (2011). The mass of the black hole in cygnus X-1, *ApJ* **742**, 84, doi:10.1088/0004-637X/742/2/84.
Osaki, Y. (1966). The pulsation and evolution of super-massive stars, *PASJ* **18**, p. 384.
O'Shea, B. W. and Norman, M. L. (2007). Population III star formation in a ΛCDM Universe. I. the effect of formation redshift and environment on protostellar accretion rate, *ApJ* **654**, pp. 66–92, doi:10.1086/509250.
O'Shea, B. W. and Norman, M. L. (2008). Population III star formation in a ΛCDM Universe. II. Effects of a Photodissociating Background, *ApJ*.
Özel, F., Psaltis, D., Narayan, R. and McClintock, J. E. (2010). The black hole mass distribution in the galaxy, *ApJ* **725**, pp. 1918–1927, doi:10.1088/0004-637X/725/2/1918.
Pacucci, F., Ferrara, A., Grazian, A., Fiore, F., Giallongo, E. and Puccetti, S. (2016). First identification of direct collapse black hole candidates in the early Universe in CANDELS/GOODS-S, *MNRAS* **459**, pp. 1432–1439, doi:10.1093/mnras/stw725.
Pacucci, F., Natarajan, P. and Ferrara, A. (2017a). Feedback limits to maximum seed masses of black holes, *ApJ* **835**, L36, doi:10.3847/2041-8213/835/2/L36.
Pacucci, F., Pallottini, A., Ferrara, A. and Gallerani, S. (2017b). The nature of the Lyman Alpha Emitter CR7: a persisting puzzle, *ArXiv e-prints:1702.04351*.
Pacucci, F., Pallottini, A., Ferrara, A. and Gallerani, S. (2017c). The nature of the Lyman α emitter CR7: a persisting puzzle, *MNRAS* **468**, pp. L77–L81, doi:10.1093/mnrasl/slx029.
Pacucci, F., Volonteri, M. and Ferrara, A. (2015). The growth efficiency of high-redshift black holes, *MNRAS* **452**, pp. 1922–1933, doi:10.1093/mnras/stv1465.

Paczynski, B. (1986). Gravitational microlensing by the galactic halo, *ApJ* **304**, pp. 1–5, doi:10.1086/164140.

Paczynski, B. (1991). A polytropic model of an accretion disk, a boundary layer, and a star, *ApJ* **370**, pp. 597–603, doi:10.1086/169846.

Padoan, P., Federrath, C., Chabrier, G., Evans, N. J., II, Johnstone, D., Jørgensen, J. K., McKee, C. F. and Nordlund, Å. (2014). The star formation rate of molecular clouds, *Protostars and Planets VI*, pp. 77–100doi: 10.2458/azu_uapress_9780816531240-ch004.

Padovani, P., Alexander, D. M., Assef, R. J., De Marco, B., Giommi, P., Hickox, R. C., Richards, G. T., Smolčić, V., Hatziminaoglou, E., Mainieri, V. and Salvato, M. (2017). Active galactic nuclei: what's in a name? *A&A Rev.* **25**, 2, doi:10.1007/s00159-017-0102-9.

Pakmor, R. and Springel, V. (2013). Simulations of magnetic fields in isolated disc galaxies, *MNRAS* **432**, pp. 176–193, doi:10.1093/mnras/stt428.

Palla, F., Salpeter, E. E. and Stahler, S. W. (1983). Primordial star formation — the role of molecular hydrogen, *ApJ* **271**, pp. 632–641, doi:10.1086/161231.

Palla, F. and Stahler, S. W. (1991). The evolution of intermediate-mass protostars. I — basic results, *ApJ* **375**, pp. 288–299, doi:10.1086/170188.

Pallottini, A., Ferrara, A., Gallerani, S., Salvadori, S. and D'Odorico, V. (2014). Simulating cosmic metal enrichment by the first galaxies, *MNRAS* **440**, pp. 2498–2518, doi:10.1093/mnras/stu451.

Pallottini, A., Ferrara, A., Pacucci, F., Gallerani, S., Salvadori, S., Schneider, R., Schaerer, D., Sobral, D. and Matthee, J. (2015). The brightest Ly α emitter: Pop III or black hole? *MNRAS* **453**, pp. 2465–2470, doi:10.1093/mnras/stv1795.

Pan, T., Loeb, A. and Kasen, D. (2012). Pair-instability supernovae via collision runaway in young dense star clusters, *MNRAS* **423**, pp. 2203–2208, doi: 10.1111/j.1365-2966.2012.21030.x.

Park, K. and Ricotti, M. (2011). Accretion onto intermediate-mass black holes regulated by radiative feedback. I. parametric study for spherically symmetric accretion, *ApJ*.

Park, K. and Ricotti, M. (2012). Accretion onto black holes from large scales regulated by radiative feedback. II. growth rate and duty cycle, *ApJ* **747**, 9, doi:10.1088/0004-637X/747/1/9.

Park, K. and Ricotti, M. (2013). Accretion onto black holes from large scales regulated by radiative feedback. III. enhanced luminosity of intermediate-mass black holes moving at supersonic speeds, *ApJ*.

Park, K., Ricotti, M., Natarajan, P., Bogdanović, T. and Wise, J. H. (2016). Bulge-driven fueling of seed black holes, *ApJ* **818**, 184, doi:10.3847/0004-637X/818/2/184.

Parsons, A. R., Backer, D. C., Foster, G. S., Wright, M. C. H., Bradley, R. F., Gugliucci, N. E., Parashare, C. R., Benoit, E. E. and Aguirre, J. E. et al. (2010). The precision array for probing the epoch of re-ionization: eight station results, *AJ* **139**, pp. 1468–1480, doi:10.1088/0004-6256/139/4/1468.

Pawlik, A. H., Rahmati, A., Schaye, J., Jeon, M. and Dalla Vecchia, C. (2017). The Aurora radiation-hydrodynamical simulations of reionization: calibration and first results, *MNRAS* **466**, pp. 960–973, doi:10.1093/mnras/stw2869.

Peacock, J. A. (1999). *Cosmological Physics* (Cosmological Physics, by John A. Peacock, pp. 704. ISBN 052141072X. Cambridge, UK: Cambridge University Press, January 1999.).

Peebles, P. J. E. (1980a). Nature of the matter distribution now and at Z = 1000, *Phys. Scr* **21**, pp. 720–724.

Peebles, P. J. E. (1980b). *The large-scale structure of the universe, Princeton University Press, 1980. 435 p.* (Princeton University Press).

Peebles, P. J. E. and Dicke, R. H. (1968). Origin of the globular star clusters, *ApJ* **154**, p. 891, doi:10.1086/149811.

Peng, C. Y. (2007). How mergers may affect the mass scaling relation between gravitationally bound systems, *ApJ* **671**, pp. 1098–1107, doi:10.1086/522774.

Penrose, R. (1969). Gravitational collapse: the role of general relativity, *Nuovo Cimento Rivista Serie* **1**.

Penston, M. V. (1969). Dynamics of self-gravitating gaseous spheres —III. analytical results in the free-fall of isothermal cases, *MNRAS* **144**, p. 425.

Perego, A., Dotti, M., Colpi, M. and Volonteri, M. (2009). Mass and spin co-evolution during the alignment of a black hole in a warped accretion disc, *MNRAS* **399**, pp. 2249–2263, doi:10.1111/j.1365-2966.2009.15427.x.

Peters, T., Banerjee, R., Klessen, R. S. and Mac Low, M.-M. (2011). The interplay of magnetic fields, fragmentation, and ionization feedback in high-mass star formation, *ApJ* **729**, 72, doi:10.1088/0004-637X/729/1/72.

Peters, T., Mac Low, M.-M., Banerjee, R., Klessen, R. S. and Dullemond, C. P. (2010). Understanding spatial and spectral morphologies of ultracompact H II regions, *ApJ* **719**, pp. 831–843, doi:10.1088/0004-637X/719/1/831.

Peters, T., Schleicher, D. R. G., Smith, R. J., Schmidt, W. and Klessen, R. S. (2014). Low-metallicity star formation: relative impact of metals and magnetic fields, *MNRAS* **442**, pp. 3112–3126, doi:10.1093/mnras/stu1097.

Peterson, B. M. (1993). Reverberation mapping of active galactic nuclei, *PASP* **105**, pp. 247–268, doi:10.1086/133140.

Peterson, B. M. (1997). *An Introduction to Active Galactic Nuclei* (Cambridge: New York Cambridge University Press).

Peterson, J. R. and Fabian, A. C. (2006). X-ray spectroscopy of cooling clusters, *Phys. Rep.* **427**, pp. 1–39, doi:10.1016/j.physrep.2005.12.007.

Pezzulli, E., Valiante, R. and Schneider, R. (2016). Super-Eddington growth of the first black holes, *MNRAS* **458**, pp. 3047–3059, doi:10.1093/mnras/stw505.

Pezzulli, E., Volonteri, M., Schneider, R. and Valiante, R. (2017). The sustainable growth of the first black holes, *MNRAS* **471**, pp. 589–595, doi:10.1093/mnras/stx1640.

Pfalzner, S. and Kaczmarek, T. (2013). The expansion of massive young star clusters - observation meets theory, *A&A* **559**, A38, doi:10.1051/0004-6361/201322134.

Pfister, H., Lupi, A., Capelo, P. R., Volonteri, M., Bellovary, J. M. and Dotti, M. (2017). The birth of a supermassive black hole binary, *MNRAS* **471**, pp. 3646–3656, doi:10.1093/mnras/stx1853.

Pfrommer, C., Pakmor, R., Schaal, K., Simpson, C. M. and Springel, V. (2017). Simulating cosmic ray physics on a moving mesh, *MNRAS* **465**, pp. 4500–4529, doi:10.1093/mnras/stw2941.

Planck Collaboration, Ade, P. A. R., Aghanim, N., Arnaud, M., Ashdown, M., Aumont, J., Baccigalupi, C., Banday, A. J., Barreiro, R. B., Bartlett, J. G. and et al. (2016). Planck 2015 results. XIII. cosmological parameters, *A&A* **594**, A13, doi:10.1051/0004-6361/201525830.

Pober, J. C., Ali, Z. S., Parsons, A. R., McQuinn, M., Aguirre, J. E., Bernardi, G., Bradley, R. F., Carilli, C. L., Cheng, C., DeBoer, D. R. and Dexter, M. R. et al. (2015). PAPER-64 constraints on reionization. II. the temperature of the z =8.4 intergalactic medium, *ApJ* **809**, 62, doi: 10.1088/0004-637X/809/1/62.

Popham, R. and Gammie, C. F. (1998). Advection-dominated accretion flows in the Kerr metric. II. steady state global solutions, *ApJ* **504**, pp. 419–430, doi:10.1086/306054.

Popham, R. and Narayan, R. (1991). Does accretion cease when a star approaches breakup? *ApJ* **370**, pp. 604–614, doi:10.1086/169847.

Portegies Zwart, S. F., Baumgardt, H., Hut, P., Makino, J. and McMillan, S. L. W. (2004). Formation of massive black holes through runaway collisions in dense young star clusters, *Nature* **428**, pp. 724–726, doi: 10.1038/nature02448.

Portegies Zwart, S. F., Makino, J., McMillan, S. L. W. and Hut, P. (1999). Star cluster ecology. III. runaway collisions in young compact star clusters, *A&A* **348**, pp. 117–126.

Portegies Zwart, S. F. and McMillan, S. L. W. (2002). The runaway growth of intermediate-mass black holes in dense star clusters, *ApJ* **576**, pp. 899–907, doi:10.1086/341798.

Pounds, K. A., King, A. R., Page, K. L. and O'Brien, P. T. (2003). Evidence of a high-velocity ionized outflow in a second narrow-line quasar PG 0844+349, *MNRAS* **346**, pp. 1025–1030, doi:10.1111/j.1365-2966.2003.07164.x.

Power, C., Nayakshin, S. and King, A. (2011). The accretion disc particle method for simulations of black hole feeding and feedback, *MNRAS* **412**, pp. 269–276, doi:10.1111/j.1365-2966.2010.17901.x.

Press, W. H. and Schechter, P. (1974). Formation of galaxies and clusters of galaxies by self-similar gravitational condensation, *ApJ* **187**, pp. 425–438, doi:10.1086/152650.

Prieto, J., Jimenez, R. and Verde, L. (2014). Overcooled haloes at z $\geq$ 10: a route to form low-mass first stars, *MNRAS* **437**, pp. 2320–2327, doi:10.1093/mnras/stt2049.

Pritchard, J. R. and Furlanetto, S. R. (2007). 21-cm fluctuations from inhomogeneous X-ray heating before reionization, *MNRAS* **376**, pp. 1680–1694, doi:10.1111/j.1365-2966.2007.11519.x.

Pritchard, J. R. and Loeb, A. (2010). Constraining the unexplored period between the dark ages and reionization with observations of the global 21 cm signal, *Phys. Rev. D* **82**, 2, 023006, doi:10.1103/PhysRevD.82.023006.

Pritchard, J. R. and Loeb, A. (2012). 21 cm cosmology in the 21st century, *Reports on Progress in Physics* **75**, 8, 086901, doi:10.1088/0034-4885/75/8/086901.

Pudritz, R. E. and Silk, J. (1989). The origin of magnetic fields and primordial stars in protogalaxies, *ApJ* **342**, pp. 650–659, doi:10.1086/167625.

Rafikov, R. R. (2001). The local axisymmetric instability criterion in a thin, rotating, multicomponent disc, *MNRAS* **323**, pp. 445–452, doi:10.1046/j.1365-8711.2001.04201.x.

Reed, S. L., McMahon, R. G., Banerji, M., Becker, G. D., Gonzalez-Solares, E., Martini, P., Ostrovski, F., Rauch, M., Abbott, T., Abdalla, F. B. and Allam, S. et al. (2015). DES J0454-4448: discovery of the first luminous z 6 quasar from the Dark Energy Survey, *MNRAS* **454**, pp. 3952–3961, doi:10.1093/mnras/stv2031.

Reed, S. L., McMahon, R. G., Martini, P., Banerji, M., Auger, M., Hewett, P. C., Koposov, S. E., Gibbons, S. L. J., Gonzalez-Solares, E., Ostrovski, F. and Tie, S. S. et al. (2017). Eight new luminous z 6 quasars discovered via SED model fitting of VISTA, WISE and Dark Energy Survey Year 1 observations, *MNRAS* **468**, pp. 4702–4718, doi:10.1093/mnras/stx728.

Rees, M. J. (1966). Appearance of relativistically expanding radio sources, *Nature* **211**, pp. 468–470, doi:10.1038/211468a0.

Rees, M. J. (1978a). Emission from the nuclei of nearby galaxies — Evidence for massive black holes, in E. M. Berkhuijsen and R. Wielebinski (eds.), *Structure and Properties of Nearby Galaxies, IAU Symposium*, Vol. 77, pp. 237–242.

Rees, M. J. (1978b). Quasars, *The Observatory* **98**, pp. 210–223.

Rees, M. J. (1984). Black hole models for active galactic nuclei, *ARA&A* **22**, pp. 471–506, doi:10.1146/annurev.aa.22.090184.002351.

Rees, M. J. (1988). Tidal disruption of stars by black holes of 10 to the 6th-10 to the 8th solar masses in nearby galaxies, *Nature* **333**, pp. 523–528, doi:10.1038/333523a0.

Regan, J. A. and Haehnelt, M. G. (2009). The formation of compact massive self-gravitating discs in metal-free haloes with virial temperatures of ~13000-30000K, *MNRAS* **393**, pp. 858–871, doi:10.1111/j.1365-2966.2008.14088.x.

Regan, J. A., Johansson, P. H. and Wise, J. H. (2014). The direct collapse of a massive black hole seed under the influence of an anisotropic Lyman-Werner source, *ApJ* **795**, 137, doi:10.1088/0004-637X/795/2/137.

Regan, J. A., Johansson, P. H. and Wise, J. H. (2016a). Forming supermassive black hole seeds under the influence of a nearby anisotropic multifrequency source, *MNRAS* **459**, pp. 3377–3394, doi:10.1093/mnras/stw899.

Regan, J. A., Johansson, P. H. and Wise, J. H. (2016b). Positive or negative? The impact of X-ray feedback on the formation of direct collapse black hole seeds, *MNRAS* **461**, pp. 111–125, doi:10.1093/mnras/stw1307.

Regan, J. A., Visbal, E., Wise, J. H., Haiman, Z., Johansson, P. H. and Bryan, G. L. (2017). Rapid formation of massive black holes in close proximity to embryonic protogalaxies, *Nature Astronomy* **1**, 0075, doi:10.1038/s41550-017-0075.

Reines, A. E. and Comastri, A. (2016). Observational signatures of high-redshift quasars and local relics of black hole seeds, *PASA* **33**, e054, doi:10.1017/pasa.2016.46.

Reines, A. E., Greene, J. E. and Geha, M. (2013). Dwarf galaxies with optical signatures of active massive black holes, *ApJ* **775**, 116, doi:10.1088/0004-637X/775/2/116.

Reines, A. E. and Volonteri, M. (2015). Relations between central black hole mass and total galaxy stellar mass in the local universe, *ApJ* **813**, 82, doi:10.1088/0004-637X/813/2/82.

Reinoso, B., Schleicher, D. R. G., Fellhauer, M., Klessen, R. S. and Boekholt, T. C. N. (2018). Collisions in primordial star clusters. Formation pathway for intermediate mass black holes, *A&A* **614**, A14, doi:10.1051/0004-6361/201732224.

Reissner, H. (1916). Über die Eigengravitation des elektrischen Feldes nach der Einsteinschen Theorie, *Annalen der Physik* **355**, pp. 106–120, doi:10.1002/andp.19163550905.

Reisswig, C., Ott, C. D., Abdikamalov, E., Haas, R., Mösta, P. and Schnetter, E. (2013). Formation and coalescence of cosmological supermassive-black-hole binaries in supermassive-star collapse, *Physical Review Letters* **111**, 15, 151101, doi:10.1103/PhysRevLett.111.151101.

Renzini, A., Greggio, L., di Serego Alighieri, S., Cappellari, M., Burstein, D. and Bertola, F. (1995). An ultraviolet flare at the centre of the elliptical galaxy NGC4552, *Nature* **378**, pp. 39–41, doi:10.1038/378039a0.

Richings, A. J., Schaye, J. and Oppenheimer, B. D. (2014). Non-equilibrium chemistry and cooling in the diffuse interstellar medium — II. shielded gas, *MNRAS* **442**, pp. 2780–2796, doi:10.1093/mnras/stu1046.

Ripamonti, E. and Abel, T. (2004). Fragmentation and the formation of primordial protostars: the possible role of collision-induced emission, *MNRAS* **348**, pp. 1019–1034, doi:10.1111/j.1365-2966.2004.07422.x.

Ripamonti, E., Iocco, F., Ferrara, A., Schneider, R., Bressan, A. and Marigo, P. (2010). First star formation with dark matter annihilation, *MNRAS* **406**, pp. 2605–2615, doi 10.1111/j.1365-2966.2010.16854.x.

Ripamonti, E., Mapelli, M. and Ferrara, A. (2007). The impact of dark matter decays and annihilations on the formation of the first structures, *MNRAS* **375**, pp. 1399–1408, doi:10.1111/j.1365-2966.2006.11402.x.

Ritter, J. S., Safranek-Shrader, C., Gnat, O., Milosavljević, M. and Bromm, V. (2012). Confined Population III enrichment and the prospects for prompt second-generation star formation, *ApJ* **761**, 56, doi:10.1088/0004-637X/761/1/56.

Ritter, J. S., Sluder, A., Safranek-Shrader, C., Milosavljević, M. and Bromm, V. (2015). Metal transport and chemical heterogeneity in early star forming systems, *MNRAS* **451**, pp. 1190–1198, doi:10.1093/mnras/stv982.

Roberts, W. W., Jr., Huntley, J. M. and van Albada, G. D. (1979). Gas dynamics in barred spirals — gaseous density waves and galactic shocks, *ApJ* **233**, pp. 67–84, doi:10.1086/157367.

Robertson, B., Hernquist, L., Cox, T. J., Di Matteo, T., Hopkins, P. F., Martini, P. and Springel, V. (2006). The evolution of the M_{BH}-σ relation, *ApJ* **641**, pp. 90–102, doi:10.1086/500348.

Roederer, I. U., Mateo, M., Bailey, J. I., III, Song, Y., Bell, E. F., Crane, J. D., Loebman, S., Nidever, D. L., Olszewski, E. W. and Shectman, S. A. et al. (2016). Detailed chemical abundances in the r-process-rich ultra-faint dwarf galaxy reticulum 2, *AJ* **151**, 82, doi:10.3847/0004-6256/151/3/82.

Romano-Díaz, E., Shlosman, I., Heller, C. and Hoffman, Y. (2008). Disk evolution and bar triggering driven by interactions with dark matter substructure, *ApJ* **687**, L13, doi:10.1086/593168.

Rosen, A. L., Krumholz, M. R., McKee, C. F. and Klein, R. I. (2016). An unstable truth: how massive stars get their mass, *MNRAS* **463**, pp. 2553–2573, doi: 10.1093/mnras/stw2153.

Rosswog, S. (2015). The multi-messenger picture of compact binary mergers, *International Journal of Modern Physics D* **24**, 1530012-52, doi:10.1142/ S0218271815300128.

Sadowski, A. (2009). Three-dimensional simulations of supercritical black hole accretion discs — luminosities, photon trapping and variability, *ApJS* **183**, 183.

Sadowski, A. (2011). Slim accretion disks around black holes, *ArXiv e-prints:1108.0396.*

Sadowski, A. and Narayan, R. (2016). Three-dimensional simulations of supercritical black hole accretion discs — luminosities, photon trapping and variability, *MNRAS* **456**, pp. 3929–3947, doi:10.1093/mnras/stv2941.

Sadowski, A., Narayan, R., Penna, R. and Zhu, Y. (2013). Energy, momentum and mass outflows and feedback from thick accretion discs around rotating black holes, *MNRAS* **436**, pp. 3856–3874, doi:10.1093/mnras/stt1881.

Safranek-Shrader, C., Agarwal, M., Federrath, C., Dubey, A., Milosavljević, M. and Bromm, V. (2012). Star formation in the first galaxies — I. collapse delayed by Lyman-Werner radiation, *MNRAS* **426**, pp. 1159–1177, doi: 10.1111/j.1365-2966.2012.21852.x.

Sahni, V. and Coles, P. (1995). Approximation methods for non-linear gravitational clustering, *Phys. Rep.* **262**, pp. 1–135.

Sakurai, Y., Hosokawa, T., Yoshida, N. and Yorke, H. W. (2015). Formation of primordial supermassive stars by burst accretion, *MNRAS* **452**, pp. 755–764, doi:10.1093/mnras/stv1346.

Sakurai, Y., Vorobyov, E. I., Hosokawa, T., Yoshida, N., Omukai, K. and Yorke, H. W. (2016). Supermassive star formation via episodic accretion: protostellar disc instability and radiative feedback efficiency, *MNRAS* doi: 10.1093/mnras/stw637.

Sakurai, Y., Yoshida, N., Fujii, M. S. and Hirano, S. (2017). Formation of intermediate-mass black holes through runaway collisions in the first star clusters, *ArXiv e-prints 1704.06130.*

Salpeter, E. E. (1964). Accretion of interstellar matter by massive objects. *ApJ* **140**, pp. 796–800, doi:10.1086/147973.

Salvadori, S., Ferrara, A., Schneider, R., Scannapieco, E. and Kawata, D. (2010). Mining the galactic halo for very metal-poor stars, *MNRAS* **401**, pp. L5–L9, doi:10.1111/j.1745-3933.2009.00772.x.

Salvadori, S., Schneider, R. and Ferrara, A. (2007). Cosmic stellar relics in the galactic halo, *MNRAS* **381**, pp. 647–662, doi:10.1111/j.1365-2966.2007.12133.x.

Salvadori, S., Skúladóttir, Á. and Tolstoy, E. (2015). Carbon-enhanced metal-poor stars in dwarf galaxies, *MNRAS* **454**, pp. 1320–1331, doi:10.1093/mnras/stv1969.

Salvaterra, R., Haardt, F. and Ferrara, A. (2005). Cosmic backgrounds from mini-quasars, *MNRAS* **362**, pp. L50–L54, doi:10.1111/j.1745-3933.2005.00074.x.

Sanders, R. H. (1970). The effects of stellar collisions in dense stellar systems, *ApJ* **162**, p. 791, doi:10.1086/150711.

Sanders, R. H. and Huntley, J. M. (1976). Gas response to oval distortions in disk galaxies, *ApJ* **209**, pp. 53–65, doi:10.1086/154692.

Santamaría, L., Ohme, F., Ajith, P., Brügmann, B., Dorband, N., Hannam, M., Husa, S., Mösta, P., Pollney, D., Reisswig, C., Robinson, E. L., Seiler, J. and Krishnan, B. (2010). Matching post-Newtonian and numerical relativity waveforms: Systematic errors and a new phenomenological model for nonprecessing black hole binaries, *Phys. Rev. D* **82**, 6, 064016, doi:10.1103/PhysRevD.82.064016.

Santoro, F. and Shull, J. M. (2006). Critical metallicity and fine-structure emission of primordial gas enriched by the first stars, *ApJ* **643**, pp. 26–37, doi:10.1086/501518.

Sasaki, M., Suyama, T., Tanaka, T. and Yokoyama, S. (2016). Primordial black hole scenario for the gravitational-wave event GW150914, *Physical Review Letters* **117**, 6, 061101, doi:10.1103/PhysRevLett.117.061101.

Saslaw, W. C. and Zipoy, D. (1967). Molecular Hydrogen in pre-galactic gas clouds, *Nature* **216**, pp. 976–978, doi:10.1038/216976a0.

Sathyaprakash, B., Abernathy, M., Acernese, F., Ajith, P., Allen, B., Amaro-Seoane, P., Andersson, N., Aoudia, S., Arun, K., Astone, P. and et al. (2012). Scientific objectives of Einstein Telescope, *Classical and Quantum Gravity* **29**, 12, 124013, doi:10.1088/0264-9381/29/12/124013.

Sathyaprakash, B. S. and Schutz, B. F. (2009). Physics, Astrophysics and Cosmology with gravitational waves, *Living Reviews in Relativity* **12**, doi:10.12942/lrr-2009-2.

Sazonov, S. Y., Ostriker, J. P. and Sunyaev, R. A. (2004). Quasars: the characteristic spectrum and the induced radiative heating, *MNRAS* **347**, pp. 144–156, doi:10.1111/j.1365-2966.2004.07184.x.

Schaerer, D. (2002). On the properties of massive Population III stars and metal-free stellar populations, *A&A* **382**, pp. 28–42, doi:10.1051/0004-6361:20011619.

Schauer, A. T. P., Agarwal, B., Glover, S. C. O., Klessen, R. S., Latif, M. A., Mas-Ribas, L., Rydberg, C.-E., Whalen, D. J. and Zackrisson, E. (2017a). Lyman-Werner escape fractions from the first galaxies, *MNRAS* **467**, pp. 2288–2300, doi:10.1093/mnras/stx264.

Schauer, A. T. P., Regan, J., Glover, S. C. O. and Klessen, R. S. (2017b). The formation of direct collapse black holes under the influence of streaming velocities, *MNRAS* **471**, pp. 4878–4884, doi:10.1093/mnras/stx1915.

Schauer, A. T. P., Whalen, D. J., Glover, S. C. O. and Klessen, R. S. (2015). Lyman-Werner UV escape fractions from primordial haloes, *MNRAS*.

Schaye, J., Crain, R. A., Bower, R. G., Furlong, M., Qu, Y., Rahmati, A., Sawala, T., Thomas, P. A. and Trayford, J. (2015). The EAGLE project: simulating the evolution and assembly of galaxies and their environments, *MNRAS* **446**, pp. 521–554, doi:10.1093/mnras/stu2058.

Schaye, J., Dalla Vecchia, C., Booth, C. M., Wiersma, R. P. C., Theuns, T., Haas, M. R., Bertone, S., Duffy, A. R., McCarthy, I. G. and van de Voort, F. (2010). The physics driving the cosmic star formation history, *MNRAS* **402**, pp. 1536–1560, doi:10.1111/j.1365-2966.2009.16029.x.

Schekochihin, A. A., Cowley, S. C., Hammett, G. W., Maron, J. L. and McWilliams, J. C. (2002). A model of nonlinear evolution and saturation of the turbulent MHD dynamo, *New Journal of Physics* **4**, p. 84, doi:10.1088/1367-2630/4/1/384.

Schekochihin, A. A., Cowley, S. C., Taylor, S. F., Hammett, G. W., Maron, J. L. and McWilliams, J. C. (2004). Saturated state of the nonlinear small-scale dynamo, *Physical Review Letters* **92**, 8, 084504, doi:10.1103/PhysRevLett.92.084504.

Schleicher, D. R. G., Banerjee, R., Sur, S., Arshakian, T. G., Klessen, R. S., Beck, R. and Spaans, M. (2010a). Small-scale dynamo action during the formation of the first stars and galaxies. I. The ideal MHD limit, *A&A* **522**, A115, doi:10.1051/0004-6361/201015184.

Schleicher, D. R. G., Bovino, S., Latif, M. A., Ferrara, A. and Grassi, T. (2016). The chemical evolution of self-gravitating primordial disks, *A&A* **585**, A11, doi:10.1051/0004-6361/201526356.

Schleicher, D. R. G., Galli, D., Glover, S. C. O., Banerjee, R., Palla, F., Schneider, R. and Klessen, R. S. (2009). The influence of magnetic fields on the thermodynamics of primordial star formation, *ApJ* **703**, pp. 1096–1106, doi:10.1088/0004-637X/703/1/1096.

Schleicher, D. R. G., Galli, D., Palla, F., Camenzind, M., Klessen, R. S., Bartelmann, M. and Glover, S. C. O. (2008). Effects of primordial chemistry on the cosmic microwave background, *A&A* **490**, pp. 521–535, doi: 10.1051/0004-6361:200809861.

Schleicher, D. R. G., Palla, F., Ferrara, A., Galli, D. and Latif, M. (2013). Massive black hole factories: Supermassive and quasi-star formation in primordial halos, *A&A* **558**, A59, doi:10.1051/0004-6361/201321949.

Schleicher, D. R. G., Spaans, M. and Glover, S. C. O. (2010b). Black hole formation in primordial galaxies: chemical and radiative conditions, *ApJ* **712**, pp. L69–L72, doi:10.1088/2041-8205/712/1/L69.

Schlickeiser, R. and Shukla, P. K. (2003). Cosmological magnetic field generation by the weibel instability, *ApJ* **599**, pp. L57–L60, doi:10.1086/381246.

Schmidt, M. (1963). 3C 273: A star-like object with large red-shift, *Nature* **197**, p. 1040, doi:10.1038/1971040a0.

Schneider, I. F. and Suzor-Weiner, A. (2002). Dissociative recombination of h+2, h+3 and their isotopomers: Achievements, problems and perspectives, *Contributions to Plasma Physics* **42**, 6-7, pp. 578–584, doi:10.1002/1521-3986(200211)42:6/7⟨578::AID-CTPP578⟩3.0.CO;2-Z, URL `http://dx.doi.org/10.1002/1521-3986(200211)42:6/7<578::AID-CTPP578>3.0.CO;2-Z`.

Schneider, R., Graziani, L., Marassi, S., Spera, M., Mapelli, M., Alparone, M. and Bennassuti, M. d. (2017). The formation and coalescence sites of the first gravitational wave events, *MNRAS* **471**, pp. L105–L109, doi:10.1093/mnrasl/slx118.

Schneider, R., Omukai, K., Limongi, M., Ferrara, A., Salvaterra, R., Chieffi, A. and Bianchi, S. (2012). The formation of the extremely primitive star SDSS J102915+172927 relies on dust, *MNRAS* **423**, pp. L60–L64, doi:10.1111/j.1745-3933.2012.01257.x.

Schneider, R., Salvaterra, R., Ferrara, A. and Ciardi, B. (2006). Constraints on the initial mass function of the first stars, *MNRAS* **369**, pp. 825–834, doi:10.1111/j.1365-2966.2006.10331.x.

Schnittman, J. D. (2004). Spin-orbit resonance and the evolution of compact binary systems, *Phys. Rev. D* **70**, 12, 124020, doi:10.1103/PhysRevD.70.124020.

Schober, J., Schleicher, D., Federrath, C., Glover, S., Klessen, R. S. and Banerjee, R. (2012a). The small-scale dynamo and non-ideal magnetohydrodynamics in primordial star formation, *ApJ* **754**, 99, doi:10.1088/0004-637X/754/2/99.

Schober, J., Schleicher, D., Federrath, C., Klessen, R. and Banerjee, R. (2012b). Magnetic field amplification by small-scale dynamo action: dependence on turbulence models and Reynolds and Prandtl numbers, *Phys. Rev. E* **85**, 2, 026303, doi:10.1103/PhysRevE.85.026303.

Schober, J., Schleicher, D. R. G., Federrath, C., Bovino, S. and Klessen, R. S. (2015). Saturation of the turbulent dynamo, *Phys. Rev. E* **92**, 2, 023010, doi:10.1103/PhysRevE.92.023010.

Schödel, R., Ott, T., Genzel, R., Hofmann, R., Lehnert, M., Eckart, A., Mouawad, N., Alexander, T., Reid, M. J., Lenzen, R., Hartung, M. and Lacombe, F. et al. (2002). A star in a 15.2-year orbit around the supermassive black hole at the centre of the Milky Way, *Nature* **419**, pp. 694–696, doi:10.1038/nature01121.

Schulze, A. and Wisotzki, L. (2011). Selection effects in the black hole-bulge relation and its evolution, *A&A* **535**, A87, doi:10.1051/0004-6361/201117564.

Schwarzschild, K. (1916). On the gravitational field of a mass point according to Einstein's Theory, *Abh. Konigl. Preuss. Akad. Wissenschaften Jahre 1906,92, Berlin,1907* **1916**.

Sesana, A. (2016). Prospects for multiband gravitational-wave astronomy after GW150914, *Physical Review Letters* **116**, 23, 231102, doi:10.1103/PhysRevLett.116.231102.

Sesana, A., Barausse, E., Dotti, M. and Rossi, E. M. (2014). Linking the spin evolution of massive black holes to galaxy kinematics, *ApJ* **794**, 104, doi:10.1088/0004-637X/794/2/104.

Sesana, A., Volonteri, M. and Haardt, F. (2007). The imprint of massive black hole formation models on the LISA data stream, *MNRAS* **377**, pp. 1711–1716, doi:10.1111/j.1365-2966.2007.11734.x.

Sethi, S., Haiman, Z. and Pandey, K. (2010). Supermassive black hole formation at high redshifts through a primordial magnetic field, *ApJ* **721**, pp. 615–621, doi:10.1088/0004-637X/721/1/615.

Seyfert, C. K. (1943). Nuclear emission in spiral nebulae. *ApJ* **97**, p. 28, doi:10.1086/144488.

Shakura, N. I. and Sunyaev, R. A. (1973). Black holes in binary systems. Observational appearance. *A&A* **24**, pp. 337–355.

Shang, C., Bryan, G. L. and Haiman, Z. (2010). Supermassive black hole formation by direct collapse: keeping protogalactic gas H_2 free in dark matter haloes with virial temperatures $T_{vir} > 10^4$ K, *MNRAS* **402**, pp. 1249–1262, doi:10.1111/j.1365-2966.2009.15960.x.

Shankar, F., Salucci, P., Granato, G. L., De Zotti, G. and Danese, L. (2004). Supermassive black hole demography: the match between the local and accreted mass functions, *MNRAS* **354**, pp. 1020–1030, doi:10.1111/j.1365-2966.2004.08261.x.

Shankar, F., Weinberg, D. H. and Miralda-Escudé, J. (2009). Self-consistent models of the AGN and black hole populations: duty cycles, accretion rates, and the mean radiative efficiency, *ApJ* **690**, pp. 20–41, doi:10.1088/0004-637X/690/1/20.

Shanks, T., Metcalfe, N., Chehade, B., Findlay, J. R., Irwin, M. J., Gonzalez-Solares, E., Lewis, J. R., Yoldas, A. K., Mann, R. G., Read, M. A., Sutorius, E. T. W. and Voutsinas, S. (2015). The VLT Survey Telescope ATLAS, *MNRAS* **451**, pp. 4238–4252, doi:10.1093/mnras/stv1130.

Shapiro, P. R., Giroux, M. L. and Babul, A. (1994). Reionization in a cold dark matter universe: the feedback of galaxy formation on the intergalactic medium, *ApJ* **427**, pp. 25–50, doi:10.1086/174120.

Shapiro, P. R. and Kang, H. (1987). Hydrogen molecules and the radiative cooling of pregalactic shocks, *ApJ* **318**, pp. 32–65, doi:10.1086/165350.

Shapiro, S. L. and Teukolsky, S. A. (1983). *Black holes, white dwarfs, and neutron stars: The physics of compact objects* (Wiley-Interscience).

Shchekinov, Y. A. and Vasiliev, E. O. (2004). Primordial star formation triggered by UV photons from UHECR, *A&A* **419**, pp. 19–23, doi:10.1051/0004-6361:20034579.

Shchekinov, Y. A. and Vasiliev, E. O. (2006). Formation of HD molecules in merging dark matter haloes, *MNRAS* **368**, pp. 454–460, doi:10.1111/j.1365-2966.2006.10117.x.

Shen, S., Madau, P., Conroy, C., Governato, F. and Mayer, L. (2014). The baryon cycle of dwarf galaxies: dark, bursty, gas-rich polluters, *ApJ* **792**, 99, doi:10.1088/0004-637X/792/2/99.

Sheth, R. K., Mo, H. J. and Tormen, G. (2001). Ellipsoidal collapse and an improved model for the number and spatial distribution of dark matter haloes, *MNRAS* **323**, pp. 1–12, doi:10.1046/j.1365-8711.2001.04006.x.

Shimura, T. and Takahara, F. (1995). On the spectral hardening factor of the X-ray emission from accretion disks in black hole candidates, *ApJ* **445**, pp. 780–788, doi:10.1086/175740.

Shlosman, I., Choi, J.-H., Begelman, M. C. and Nagamine, K. (2016). Supermassive black hole seed formation at high redshifts: long-term evolution of the direct collapse, *MNRAS* **456**, pp. 500–511, doi:10.1093/mnras/stv2700.

Shlosman, I., Frank, J. and Begelman, M. C. (1989). Bars within bars — A mechanism for fuelling active galactic nuclei, *Nature* **338**, pp. 45–47, doi: 10.1038/338045a0.

Shu, F. H. (1977). Self-similar collapse of isothermal spheres and star formation, *ApJ* **214**, pp. 488–497, doi:10.1086/155274.

Shu, F. H., Adams, F. C. and Lizano, S. (1987). Star formation in molecular clouds — observation and theory, *ARA&A* **25**, pp. 23–81, doi:10.1146/annurev.aa.25.090187.000323.

Shull, J. M. and van Steenberg, M. E. (1985). X-ray secondary heating and ionization in quasar emission-line clouds, *ApJ* **298**, pp. 268–274, doi: 10.1086/163605.

Sigl, G., Olinto, A. V. and Jedamzik, K. (1997). Primordial magnetic fields from cosmological first order phase transitions, *Phys. Rev. D* **55**, pp. 4582–4590, doi:10.1103/PhysRevD.55.4582.

Sijacki, D., Springel, V., Di Matteo, T. and Hernquist, L. (2007). A unified model for AGN feedback in cosmological simulations of structure formation, *MNRAS* **380**, pp. 877–900, doi:10.1111/j.1365-2966.2007.12153.x.

Sijacki, D., Vogelsberger, M., Genel, S., Springel, V., Torrey, P., Snyder, G. F., Nelson, D. and Hernquist, L. (2015). The Illustris simulation: the evolving population of black holes across cosmic time, *MNRAS* **452**, pp. 575–596, doi:10.1093/mnras/stv1340.

Silk, J. (1986). *The cosmic microwave background radiation - Implications for galaxy formation* (University of Chicago Press), pp. 143–158.

Silk, J. and Langer, M. (2006). On the first generation of stars, *MNRAS* **371**, pp. 444–450, doi:10.1111/j.1365-2966.2006.10689.x.

Silk, J. and Rees, M. J. (1998). Quasars and galaxy formation, *A&A* **331**, pp. L1–L4.

Simcoe, R. A., Sullivan, P. W., Cooksey, K. L., Kao, M. M., Matejek, M. S. and Burgasser, A. J. (2012). Extremely metal-poor gas at a redshift of 7, *Nature* **492**, pp. 79–82, doi:10.1038/nature11612.

Singh, S., Subrahmanyan, R., Udaya Shankar, N., Sathyanarayana Rao, M., Fialkov, A., Cohen, A., Barkana, R., Girish, B. S., Raghunathan, A., Somashekar, R. and Srivani, K. S. (2017). First results on the epoch of reionization from first light with SARAS 2, *ApJ* **845**, L12, doi:10.3847/2041-8213/aa831b.

Skidmore, W., TMT International Science Development Teams and Science Advisory Committee, T. (2015). Thirty meter telescope detailed science case: 2015, *Research in Astronomy and Astrophysics* **15**, 1945, doi: 10.1088/1674-4527/15/12/001.

Smidt, J., Whalen, D. J., Johnson, J. L. and Li, H. (2017). The formation of the first quasars in the Universe, *ArXiv e-prints:1703.00449*.

Smidt, J., Wiggins, B. K. and Johnson, J. L. (2016). Ab initio cosmological simulations of CR7 as an active black hole, *ApJ* **829**, L6, doi:10.3847/2041-8205/829/1/L6.

Smith, A., Bromm, V. and Loeb, A. (2016). Evidence for a direct collapse black hole in the Lyman α source CR7, *MNRAS* **460**, pp. 3143–3151, doi:10.1093/mnras/stw1129.

Smith, B. D., Bryan, G. L., Glover, S. C. O., Goldbaum, N. J., Turk, M. J., Regan, J., Wise, J. H., Schive, H.-Y., Abel, T., Emerick, A. and O'Shea, B. W. et al. (2017). GRACKLE: a chemistry and cooling library for astrophysics, *MNRAS* **466**, pp. 2217–2234, doi:10.1093/mnras/stw3291.

Smith, B. D., Turk, M. J., Sigurdsson, S., O'Shea, B. W. and Norman, M. L. (2009). Three modes of metal-enriched star formation in the early Universe, *ApJ* **691**, pp. 441–451, doi:10.1088/0004-637X/691/1/441.

Smith, R. J., Glover, S. C. O., Clark, P. C., Greif, T. and Klessen, R. S. (2011). The effects of accretion luminosity upon fragmentation in the early universe, *MNRAS* **414**, pp. 3633–3644, doi:10.1111/j.1365-2966.2011.18659.x.

Smith, R. J., Hosokawa, T., Omukai, K., Glover, S. C. O. and Klessen, R. S. (2012). Variable accretion rates and fluffy first stars, *MNRAS* **424**, pp. 457–463, doi:10.1111/j.1365-2966.2012.21211.x.

Sobral, D., Matthee, J., Brammer, G., Ferrara, A., Alegre, L., Rottgering, H., Schaerer, D., Mobasher, B. and Darvish, B. (2017). On the nature of the luminous Lya emitter CR7 and its UV components: physical conditions and JWST predictions, *ArXiv e-prints 1710.08422*.

Sobral, D., Matthee, J., Darvish, B., Schaerer, D., Mobasher, B., Röttgering, H. J. A., Santos, S. and Hemmati, S. (2015). Evidence for PopIII-like stellar populations in the most luminous Lyman-α emitters at the epoch of reionization: spectroscopic confirmation, *ApJ* **808**, 139, doi:10.1088/0004-637X/808/2/139.

Sokolowski, M., Tremblay, S. E., Wayth, R. B., Tingay, S. J., Clarke, N., Roberts, P., Waterson, M., Ekers, R. D., Hall, P., Lewis, M. and Mossammaparast, M. et al. (2015). BIGHORNS - Broadband Instrument for Global HydrOgen ReioNisation Signal, *PASA* **32**, e004, doi:10.1017/pasa.2015.3.

Sołtan, A. (1982). Masses of quasars, *MNRAS* **200**, pp. 115–122, doi:10.1093/mnras/200.1.115.

Somerville, R. S. and Davé, R. (2015). Physical models of galaxy formation in a cosmological framework, *ARA&A* **53**, pp. 51–113, doi:10.1146/annurev-astro-082812-140951.

Somerville, R. S., Hopkins, P. F., Cox, T. J., Robertson, B. E. and Hernquist, L. (2008). A semi-analytic model for the co-evolution of galaxies, black holes and active galactic nuclei, *MNRAS* **391**, pp. 481–506, doi:10.1111/j.1365-2966.2008.13805.x.

Sormani, M. C., Treß, R. G., Klessen, R. S. and Glover, S. C. O. (2017). A simple method to convert sink particles into stars, *MNRAS* **466**, pp. 407–412, doi:10.1093/mnras/stw3205.

Souza Lima, R., Mayer, L., Capelo, P. R. and Bellovary, J. M. (2017). The pairing of accreting massive black holes in multiphase circumnuclear disks: the interplay between radiative cooling, star formation, and feedback processes, *ApJ* **838**, 13, doi:10.3847/1538-4357/aa5d19.

Spaans, M. and Silk, J. (2006). Pregalactic black hole formation with an atomic hydrogen equation of state, *ApJ* **652**, pp. 902–906, doi:10.1086/508444.

Spera, M. and Mapelli, M. (2017). Very massive stars, pair-instability supernovae and intermediate-mass black holes with the SEVN code, *ArXiv e-prints*.

Spera, M., Mapelli, M. and Bressan, A. (2015). The mass spectrum of compact remnants from the PARSEC stellar evolution tracks, *MNRAS* **451**, pp. 4086–4103, doi:10.1093/mnras/stv1161.

Spergel, D., Gehrels, N., Baltay, C., Bennett, D., Breckinridge, J., Donahue, M., Dressler, A., Wallace, J. K., Whipple, A., Wollack, E. and Zhao, F. (2015). Wide-Field InfrarRed Survey Telescope-Astrophysics Focused Telescope Assets WFIRST-AFTA 2015 Report, *ArXiv e-prints:1503.03757*.

Spitzer, L. (1987). *Dynamical evolution of globular clusters, Princeton, NJ, Princeton University Press, 1987, 191 p.* (Princeton University Press).

Spitzer, L., Jr. and Greenstein, J. L. (1951). Continuous emission from planetary nebulae, *ApJ* **114**, p. 407, doi:10.1086/145480.

Spitzer, L., Jr. and Hart, M. H. (1971). Random gravitational encounters and the evolution of spherical systems. I. method, *ApJ* **164**, p. 399, doi:10.1086/150855.

Spolyar, D., Freese, K. and Gondolo, P. (2008). Dark matter and the first stars: a new phase of stellar evolution, *Physical Review Letters* **100**, 5, 051101, doi:10.1103/PhysRevLett.100.051101.

Springel, V. (2010). E pur si muove: Galilean-invariant cosmological hydrodynamical simulations on a moving mesh, *Monthly Notices of the Royal Astronomical Society* **401**, 2, pp. 791–851, doi:10.1111/j.1365-2966.2009.15715.x, URL `http://arxiv.org/abs/0901.4107`.

Stacy, A. and Bromm, V. (2013). Constraining the statistics of Population III binaries, *MNRAS* **433**, pp. 1094–1107, doi:10.1093/mnras/stt789.

Stacy, A., Bromm, V. and Lee, A. T. (2016). Building up the Population III initial mass function from cosmological initial conditions, *MNRAS* **462**, pp. 1307–1328, doi:10.1093/mnras/stw1728.

Stacy, A., Bromm, V. and Loeb, A. (2011). Effect of streaming motion of baryons relative to dark matter on the formation of the first stars, *ApJ* **730**, L1, doi:10.1088/2041-8205/730/1/L1.

Stacy, A., Greif, T. H. and Bromm, V. (2010). The first stars: formation of binaries and small multiple systems, *MNRAS* **403**, pp. 45–60, doi:10.1111/j.1365-2966.2009.16113.x.

Stacy, A., Greif, T. H and Bromm, V. (2012). The first stars: mass growth under protostellar feedback, *MNRAS* **422**, pp. 290–309, doi:10.1111/j.1365-2966.2012.20605.x.

Stacy, A., Pawlik, A. H., Bromm, V. and Loeb, A. (2014). The mutual interaction between Population III stars and self-annihilating dark matter, *MNRAS* **441**, pp. 822–836, doi:10.1093/mnras/stu621.

Stahler, S. W. (1988). Deuterium and the stellar birthline, *ApJ* **332**, pp. 804–825, doi:10.1086/166694.

Stahler, S. W., Palla, F. and Salpeter, E. E. (1986). Primordial stellar evolution — the protostar phase, *ApJ* **302**, pp. 590–605, doi:10.1086/164018.

Stahler, S. W., Shu, F. H. and Taam, R. E. (1980). The evolution of protostars. I — global formulation and results, *ApJ* **241**, pp. 637–654, doi:10.1086/158377.

Stecher, T. P. and Williams, D. A. (1967). Photodestruction of Hydrogen molecules in H I regions, *ApJ* **149**, p. L29, doi:10.1086/180047.

Stinson, G., Seth, A., Katz, N., Wadsley, J., Governato, F. and Quinn, T. (2006). Star formation and feedback in smoothed particle hydrodynamic simulations — I. isolated galaxies, *MNRAS* **373**, pp. 1074–1090, doi:10.1111/j.1365-2966.2006.11097.x.

Stone, N. C., Küpper, A. H. W. and Ostriker, J. P. (2017). Formation of massive black holes in galactic nuclei: runaway tidal encounters, *MNRAS* **467**, pp. 4180–4199, doi:10.1093/mnras/stx097.

Straub, O., Godet, O., Webb, N., Servillat, M. and Barret, D. (2014). Investigating the mass of the intermediate mass black hole candidate HLX-1 with the slimbh model, *A&A* **569**, A116, doi:10.1051/0004-6361/201423874.

Subramanian, K. (1998). Can the turbulent galactic dynamo generate large-scale magnetic fields? *MNRAS* **294**, p. 718, doi:10.1046/j.1365-8711.1998.01284.x.

Sugimura, K., Omukai, K. and Inoue, A. K. (2014). The critical radiation intensity for direct collapse black hole formation: dependence on the radiation spectral shape, *MNRAS* **445**, pp. 544–553, doi:10.1093/mnras/stu1778.

Sugiyama, N. (1995). Cosmic background anisotropies in cold dark matter cosmology, *ApJS* **100**, pp. 281–+, doi:10.1086/192220.

Sunyaev, R. A. and Zel'dovich, Y. B. (1970). Small-scale fluctuations of relic radiation, *Ap&SS* **7**, pp. 3–19, doi:10.1007/BF00653471.

Sur, S., Federrath, C., Schleicher, D. R. G., Banerjee, R. and Klessen, R. S. (2012). Magnetic field amplification during gravitational collapse — influence of turbulence, rotation and gravitational compression, *MNRAS* **423**, pp. 3148–3162, doi:10.1111/j.1365-2966.2012.21100.x.

Sur, S., Schleicher, D. R. G., Banerjee, R., Federrath, C. and Klessen, R. S. (2010). The generation of strong magnetic fields during the formation of the first stars, *ApJ* **721**, pp. L134–L138, doi:10.1088/2041-8205/721/2/L134.

Susa, H. (2013). The mass of the first stars, *ApJ* **773**, 185, doi:10.1088/0004-637X/773/2/185.

Susa, H., Hasegawa, K. and Tominaga, N. (2014). The mass spectrum of the first stars, *ApJ* **792**, 32, doi:10.1088/0004-637X/792/1/32.

Szekeres, G. (1960). On the singularities of a Riemannian manifold, *Publicationes Mathematicae Debrecen 7, 285 (1960)* **7**.

Szokoly, G. P., Bergeron, J., Hasinger, G., Lehmann, I., Kewley, L., Mainieri, V., Nonino, M., Rosati, P., Giacconi, R., Gilli, R., Gilmozzi, R. and Norman, C. et al. (2004). The Chandra Deep Field-South: optical spectroscopy. I. *ApJS* **155**, pp. 271–349, doi:10.1086/424707.

Takagi, T. (2002). Dissociative recombination and excitation of h 2 + , hd + , and d 2 + , with electrons for various vibrational states, *Physica Scripta* **2002**, T96, p. 52, URL `http://stacks.iop.org/1402-4896/2002/i=T96/a=005`.

Takahashi, K., Umeda, H. and Yoshida, T. (2014). Stellar yields of rotating first stars. I. yields of weak supernovae and abundances of carbon-enhanced hyper-metal-poor stars, *ApJ* **794**, 40, doi:10.1088/0004-637X/794/1/40.

Takahashi, S. Z. and Omukai, K. (2017). Primordial protostars accreting beyond the $\Omega\Gamma$-limit: radiation effect around the star-disc boundary, *MNRAS* **472**, pp. 532–541, doi 10.1093/mnras/stx1988.

Takeo, E., Inayoshi, K , Ohsuga, K., Takahashi, H. R. and Mineshige, S. (2018). Rapid growth of black holes accompanied with hot or warm outflows exposed to anisotropic super-Eddington radiation, *MNRAS* **476**, pp. 673–682, doi:10.1093/mnras/sty264.

Tamai, R. and Spyromilio, J. (2014). European extremely large telescope: progress report, in *Ground-based and Airborne Telescopes V*, *Proc. SPIE*, Vol. 9145, p. 91451E, doi:10.1117/12.2058467.

Tamanini, N., Caprini, C., Barausse, E., Sesana, A., Klein, A. and Petiteau, A. (2016). Science with the space-based interferometer eLISA. III: probing the expansion of the universe using gravitational wave standard sirens, *JCAP* **4**, 002, doi:10.1088/1475-7516/2016/04/002.

Tamburello, V., Capelo, P. R., Mayer, L., Bellovary, J. M. and Wadsley, J. W. (2017). Supermassive black hole pairs in clumpy galaxies at high redshift: delayed binary formation and concurrent mass growth, *MNRAS* **464**, pp. 2952–2962, doi:10.1093/mnras/stw2561.

Tamfal, T., Capelo, P. R., Kazantzidis, S., Mayer, L., Potter, D., Stadel, J. and Widrow, L. M. (2018). Formation of LISA Black Hole Binaries in Merging Dwarf Galaxies: The Imprint of Dark Matter, *ApJ* **864**, L19, doi:10.3847/2041-8213/aada4b.

Tan, J. C. and McKee, C. F. (2004). The formation of the first stars. I. mass infall rates, accretion disk structure, and protostellar evolution, *ApJ* **603**, pp. 383–400, doi:10.1086/381490.

Tanaka, T. and Haiman, Z. (2009). The asembly of supermassive black holes at high redshifts, *ApJ* **696**, pp. 1798–1822, doi:10.1088/0004-637X/696/2/1798.

Tanaka, T. L. and Li, M. (2014). The formation of massive black holes in z 30 dark matter haloes with large baryonic streaming velocities, *MNRAS* **439**, pp. 1092–1100, doi:10.1093/mnras/stu042.

Tanaka, T. L., Li, M. and Haiman, Z. (2013). The effect of baryonic streaming motions on the formation of the first supermassive black holes, *MNRAS* **435**, pp. 3559–3567, doi:10.1093/mnras/stt1553.

Taylor, P. and Kobayashi, C. (2014). Seeding black holes in cosmological simulations, *MNRAS* **442**, pp. 2751–2767, doi:10.1093/mnras/stu983.

Tegmark, M., Taylor, A. N. and Heavens, A. F. (1997). Karhunen-Loeve eigenvalue problems in cosmology: How Should We Tackle Large Data Sets? *ApJ* **480**, pp. 22–+, doi:10.1086/303939.

Tenneti, A., Di Matteo, T., Croft, R., Garcia, T. and Feng, Y. (2018). The descendants of the first quasars in the BlueTides simulation, *MNRAS* **474**, pp. 597–603, doi:10.1093/mnras/stx2788.

Teyssier, R. (2002). Cosmological hydrodynamics with adaptive mesh refinement. A new high resolution code called RAMSES, *å* **385**, 1349.

The Event Horizon Telescope Collaboration (2019). First M87 event horizon telescope results. I. The shadow of the supermassive black hole, *ApJ* **875**, L1, doi:10.3847/2041-8213/ab0ec7.

The LIGO Scientific Collaboration and the Virgo Collaboration (2018). GWTC-1: A Gravitational-Wave Transient Catalog of Compact Binary Mergers Observed by LIGO and Virgo during the First and Second Observing Runs, ArXiv e-prints 1811.12907.

Theuns, T. and Catelan, P. (1997). Angular momentum induced by tidal torques, in M. Persic and P. Salucci (eds.), *Dark and Visible Matter in Galaxies and Cosmological Implications, Astronomical Society of the Pacific Conference Series*, Vol. 117, pp. 431–+.

Thorne, K. S. (1974). Disk-accretion onto a black hole. II. evolution of the hole, *ApJ* **191**, pp. 507–520, doi:10.1086/152991.

Tielens, A. G. G. M. (2005). *The Physics and Chemistry of the Interstellar Medium, pp. ISBN 0521826349. Cambridge, UK: Cambridge University Press, 2005.* (Cambridge University Press).

Tingay, S. J., Goeke, R., Bowman, J. D., Emrich, D., Ord, S. M., Mitchell, D. A., Morales, M. F., Booler, T., Crosse, B., Wayth, R. B. and Lonsdale, C. J. et al. (2013). The Murchison Widefield Array: The square kilometre array precursor at low Rradio frequencies, *PASA* **30**, e007, doi:10.1017/pasa.2012.007.

Tollet, E., Macció, A., Dutton, A. and et al. (2016). NIHAO — IV: core creation and destruction in dark matter density profiles across cosmic time, *MNRAS* **456**, 3542.

Tolman, R. C. (1939). Static solutions of Einstein's field equations for spheres of fluid, *Physical Review* **55**, pp. 364–373, doi:10.1103/PhysRev.55.364.

Toomre, A. (1964). On the gravitational stability of a disk of stars, *ApJ* **139**, pp. 1217–1238, doi:10.1086/147861.

Toomre, A. and Toomre, J. (1972). Galactic bridges and tails, *ApJ* **178**, pp. 623–666, doi:10.1086/151823.

Trac, H., Sills, A. and Pen, U.-L. (2007). A comparison of hydrodynamic techniques for modelling collisions between main-sequence stars, *MNRAS* **377**, pp. 997–1005, doi:10.1111/j.1365-2966.2007.11709.x.

Treister, E., Schawinski, K., Volonteri, M. and Natarajan, P. (2013). New observational constraints on the growth of the first supermassive black holes, *ApJ* **778**, 130, doi:10.1088/0004-637X/778/2/130.

Treister, E., Schawinski, K., Volonteri, M., Natarajan, P. and Gawiser, E. (2011). Black hole growth in the early Universe is self-regulated and largely hidden from view, *Nature* **474**, pp. 356–358, doi:10.1038/nature10103.

Treister, E., Urry, C. M., Schawinski, K., Cardamone, C. N. and Sanders, D. B. (2010). Heavily obscured active galactic nuclei in high-redshift luminous infrared galaxies, *ApJL* **722**, pp. L238–L243, doi:10.1088/2041-8205/722/2/L238.

Treister, E., Urry, C. M. and Virani, S. (2009). The space density of compton-thick active galactic nucleus and the X-ray background, *ApJ* **696**, pp. 110–120, doi:10.1088/0004-637X/696/1/110.

Tremmel, M., Governato, F., Volonteri, M. and Quinn, T. R. (2015). Off the beaten path: a new approach to realistically model the orbital decay of supermassive black holes in galaxy formation simulations, *MNRAS* **451**, pp. 1868–1874, doi:10.1093/mnras/stv1060.

Tremmel, M., Karcher, M., Governato, F., Volonteri, M., Quinn, T. R., Pontzen, A., Anderson, L. and Bellovary, J. (2017). The Romulus cosmological simulations: a physical approach to the formation, dynamics and accretion models of SMBHs, *MNRAS* **470**, pp. 1121–1139, doi:10.1093/mnras/stx1160.

Trenti, M., Stiavelli, M. and Michael Shull, J. (2009). Metal-free gas supply at the edge of reionization: late-epoch Population III star formation, *ApJ* **700**, pp. 1672–1679, doi:10.1088/0004-637X/700/2/1672.

Tseliakhovich, D., Barkana, R. and Hirata, C. M. (2011). Suppression and spatial variation of early galaxies and minihaloes, *MNRAS* **418**, pp. 906–915, doi:10.1111/j.1365-2966.2011.19541.x.

Tseliakhovich, D. and Hirata, C. (2010). Relative velocity of dark matter and baryonic fluids and the formation of the first structures, *Phys. Rev. D* **82**, 8, 083520, doi:10.1103/PhysRevD.82.083520.

Tsuribe, T. and Omukai, K. (2006). Dust-cooling-induced fragmentation of low-metallicity clouds, *ApJ* **642**, pp. L61–L64, doi:10.1086/504290.

Tsuribe, T. and Omukai, K. (2008). Physical mechanism for the intermediate characteristic stellar mass in extremely metal poor environments, *ApJ* **676**, pp. L45–L48, doi:10.1086/587035.

Tumlinson, J. (2006). Chemical evolution in hierarchical models of cosmic structure. I. constraints on the early stellar initial mass function, *ApJ* **641**, pp. 1–20, doi:10.1086/500383.

Tumlinson, J. (2010). Chemical evolution in hierarchical models of cosmic structure. II. the formation of the Milky Way stellar halo and the distribution of the oldest stars, *ApJ* **708**, pp. 1398–1418, doi:10.1088/0004-637X/708/2/1398.

Turk, M. J., Abel, T. and O'Shea, B. (2009). The formation of Population III binaries from cosmological initial conditions, *Science* **325**, pp. 601–, doi:10.1126/science.1173540.

Turk, M. J., Clark, P., Glover, S. C. O., Greif, T. H., Abel, T., Klessen, R. and Bromm, V. (2011a). Effects of varying the three-body molecular Hydrogen formation rate in primordial star formation, *ApJ* **726**, 55, doi:10.1088/0004-637X/726/1/55.

Turk, M. J., Oishi, J. S., Abel, T. and Bryan, G. L. (2012). Magnetic fields in Population III star formation, *ApJ* **745**, 154, doi:10.1088/0004-637X/745/2/154.

Turk, M. J., Smith, B. D., Oishi, J. S., Skory, S., Skillman, S. W., Abel, T. and Norman, M. L. (2011b). yt: A multi-code analysis toolkit for astrophysical simulation data, *ApJS* **192**, 9, doi:10.1088/0067-0049/192/1/9.

Turner, M. S. and White, M. (1997). CDM models with a smooth component, *Phys. Rev. D* **56**, pp. 4439–+.

Ueda, Y., Akiyama, M., Hasinger, G., Miyaji, T. and Watson, M. G. (2014). Toward the standard population synthesis model of the X-ray background: evolution of X-ray luminosity and absorption functions of active galactic nuclei including compton-thick populations, *ApJ* **786**, 104, doi:10.1088/0004-637X/786/2/104.

Umeda, H., Hosokawa, T., Omukai, K. and Yoshida, N. (2016). The final fates of accreting supermassive stars, *ApJ* **830**, L34, doi:10.3847/2041-8205/830/2/L34.

Unno, W. (1971). Physical state and stability of super-massive objects, *PASJ* **23**, p. 123.

Valdes, M., Evoli, C. and Ferrara, A. (2010). Particle energy cascade in the intergalactic medium, *MNRAS* **404**, pp. 1569–1582, doi:10.1111/j.1365-2966.2010.16387.x.

Valiante, R., Schneider, R., Graziani, L. and Zappacosta, L. (2018). Chasing the observational signatures of seed black holes at z > 7: candidate statistics, *MNRAS* **474**, pp. 3825–3834, doi:10.1093/mnras/stx3028.

Valiante, R., Schneider, R., Volonteri, M. and Omukai, K. (2016). From the first stars to the first black holes, *MNRAS* **457**, pp. 3356–3371, doi:10.1093/mnras/stw225.

Van Borm, C., Bovino, S., Latif, M. A., Schleicher, D. R. G., Spaans, M. and Grassi, T. (2014). Effects of turbulence and rotation on protostar formation as a precursor of massive black holes, *A&A* **572**, A22, doi:10.1051/0004-6361/201424658.

Van Borm, C. and Spaans, M. (2013). The influence of magnetic fields, turbulence, and UV radiation on the formation of supermassive black holes, *A&A* **553**, L9, doi:10.1051/0004-6361/201321590.

van den Bosch, R., de Zeeuw, T., Gebhardt, K., Noyola, E. and van de Ven, G. (2006). The dynamical mass-to-light ratio profile and distance of the globular cluster M15, *ApJ* **641**, pp. 852–861, doi:10.1086/500644.

van den Bosch, R. C. E. (2016). Unification of the fundamental plane and super massive black hole masses, *ApJ* **831**, 134, doi:10.3847/0004-637X/831/2/134.

van Haarlem, M. P., Wise, M. W., Gunst, A. W., Heald, G., McKean, J. P., Hessels, J. W. T., de Bruyn, A. G., Nijboer, R., Swinbank, J., Fallows, R., Brentjens, M. and Nelles, A. e. a. (2013). LOFAR: The LOw-Frequency ARray, *A&A* **556**, A2, doi:10.1051/0004-6361/201220873.

Van Wassenhove, S., Capelo, P. R., Volonteri, M., Dotti, M., Bellovary, J. M., Mayer, L. and Governato, F. (2014). Nuclear coups: dynamics of black holes in galaxy mergers, *MNRAS* **439**, pp. 474–487, doi:10.1093/mnras/stu024.

Van Wassenhove, S., Volonteri, M., Mayer, L., Dotti, M., Bellovary, J. and Callegari, S. (2012). Observability of dual active galactic nuclei in merging galaxies, *ApJ* **748**, L7, doi:10.1088/2041-8205/748/1/L7.

Venemans, B. P., Findlay, J. R., Sutherland, W. J., De Rosa, G., McMahon, R. G., Simcoe, R., González-Solares, E. A., Kuijken, K. and Lewis, J. R.

(2013). Discovery of three z ¿ 6.5 quasars in the VISTA Kilo-Degree Infrared Galaxy (VIKING) Survey, *ApJ* **779**, 24, doi:10.1088/0004-637X/779/1/24.

Venemans, B. P., McMahon, R. G., Warren, S. J., Gonzalez-Solares, E. A., Hewett, P. C., Mortlock, D. J., Dye, S. and Sharp, R. G. (2007). The discovery of the first luminous z ~ 6 quasar in the UKIDSS Large Area Survey, *MNRAS* **376**, pp. L76–L80, doi:10.1111/j.1745-3933.2007.00290.x.

Venemans, B. P., Verdoes Kleijn, G. A., Mwebaze, J., Valentijn, E. A., Bañados, E., Decarli, R., de Jong, J. T. A., Findlay, J. R. and Kuijken, K. H. et al. (2015). First discoveries of z 6 quasars with the Kilo-Degree Survey and VISTA Kilo-Degree Infrared Galaxy survey, *MNRAS* **453**, pp. 2259–2266, doi:10.1093/mnras/stv1774.

Vink, J. S. (2008). Mass loss and the evolution of massive stars, *New Astronomy Reviews* **52**, pp. 419–422, doi:10.1016/j.newar.2008.06.008.

Vink, J. S., de Koter, A. and Lamers, H. J. G. L. M. (1999). On the nature of the bi-stability jump in the winds of early-type supergiants, *AAP* **350**, pp. 181–196.

Vink, J. S., de Koter, A. and Lamers, H. J. G. L. M. (2000). New theoretical mass-loss rates of O and B stars, *AAP* **362**, pp. 295–309.

Vink, J. S., de Koter, A. and Lamers, H. J. G. L. M. (2001). Mass-loss predictions for O and B stars as a function of metallicity, *A&A* **369**, pp. 574–588, doi:10.1051/0004-6361:20010127.

Visbal, E., Barkana, R., Fialkov, A., Tseliakhovich, D. and Hirata, C. M. (2012). The signature of the first stars in atomic hydrogen at redshift 20, *Nature* **487**, pp. 70–73, doi:10.1038/nature11177.

Visbal, E., Bryan, G. L. and Haiman, Z. (2017). What is the maximum mass of a Population III galaxy? *MNRAS* **469**, pp. 1456–1465, doi:10.1093/mnras/stx909.

Visbal, E., Haiman, Z. and Bryan, G. L. (2014). Direct collapse black hole formation from synchronized pairs of atomic cooling haloes, *MNRAS* **445**, pp. 1056–1063, doi:10.1093/mnras/stu1794.

Visbal, E., Haiman, Z. and Bryan, G. L. (2016). Formation of massive Population III galaxies through photoionization feedback: a possible explanation for CR 7, *MNRAS* **460**, pp. L59–L63, doi:10.1093/mnrasl/slw071.

Vito, F., Brandt, W. N., Stern, D., Assef, R. J., Chen, C.-T. J., Brightman, M., Comastri, A., Eisenhardt, P., Garmire, G. P., Hickox, R. and Lansbury, G. et al. (2018). Heavy X-ray obscuration in the most luminous galaxies discovered by WISE, *MNRAS* **474**, pp. 4528–4540, doi:10.1093/mnras/stx3120.

Vogelsberger, M., Genel, S., Sijacki, D., Torrey, P., Springel, V. and Hernquist, L. (2013). A model for cosmological simulations of galaxy formation physics, *MNRAS* **436**, pp. 3031–3067, doi:10.1093/mnras/stt1789.

Vogelsberger, M., Genel, S., Springel, V., Torrey, P., Sijacki, D., Xu, D., Snyder, G., Bird, S., Nelson, D. and Hernquist, L. (2014). Properties of galaxies reproduced by a hydrodynamic simulation, *Nature* **509**, pp. 177–182, doi:10.1038/nature13316.

Volonteri, M. (2010). Formation of supermassive black holes, *A&A Rev.* **18**, pp. 279–315, doi:10.1007/s00159-010-0029-x.
Volonteri, M. (2012). The formation and evolution of massive black holes, *Science* **337**, p. 544, doi:10.1126/science.1220843.
Volonteri, M. and Begelman, M. C. (2010). Quasi-stars and the cosmic evolution of massive black holes, *MNRAS* **409**, pp. 1022–1032, doi:10.1111/j.1365-2966.2010.17359.x.
Volonteri, M. and Bellovary, J. (2012). Black holes in the early Universe, *Reports on Progress in Physics* **75**, 12, 124901, doi:10.1088/0034-4885/75/12/124901.
Volonteri, M., Capelo, P. R., Netzer, H., Bellovary, J., Dotti, M. and Governato, F. (2015a). Black hole accretion versus star formation rate: theory confronts observations, *MNRAS* **452**, pp. L6–L10, doi:10.1093/mnrasl/slv078.
Volonteri, M., Capelo, P. R., Netzer, H., Bellovary, J., Dotti, M. and Governato, F. (2015b). Growing black holes and galaxies: black hole accretion versus star formation rate, *MNRAS* **449**, pp. 1470–1485, doi:10.1093/mnras/stv387.
Volonteri, M. and Natarajan, P. (2009). Journey to the M_{BH}-σ relation: the fate of low-mass black holes in the Universe, *MNRAS* **400**, pp. 1911–1918, doi:10.1111/j.1365-2966.2009.15577.x.
Volonteri, M. and Rees, M. J. (2005). Rapid growth of high-redshift black holes, *ApJ* **633**, pp. 624–629, doi:10.1086/466521.
Volonteri, M. and Reines, A. E. (2016). Inferences on the relations between central black hole mass and total galaxy stellar mass in the high-redshift Universe, *ApJL* **820**, L6, doi:10.3847/2041-8205/820/1/L6.
Volonteri, M., Reines, A. E., Atek, H., Stark, D. P. and Trebitsch, M. (2017). High-redshift galaxies and black holes detectable with the JWST: a population synthesis model from infrared to X-rays, *ApJ* **849**, 155, doi:10.3847/1538-4357/aa93f1.
von Weizsäcker, C. F. (1951). The evolution of galaxies and stars. *ApJ* **114**, p. 165, doi:10.1086/145462.
Weigel, A. K., Schawinski, K., Treister, E., Urry, C. M., Koss, M. and Trakhtenbrot, B. (2015). The systematic search for z $\gtrsim$ 5 active galactic nuclei in the Chandra Deep Field South, *MNRAS* **448**, pp. 3167–3195, doi:10.1093/mnras/stv184.
Weyl, H. (1917). Zur Gravitationstheorie, *Annalen der Physik* **359**, pp. 117–145, doi:10.1002/andp.19173591804.
Whalen, D., Abel, T. and Norman, M. L. (2004). Radiation hydrodynamic evolution of primordial H II regions, *ApJ* **610**, pp. 14–22, doi:10.1086/421548.
Whalen, D., Meiksin2, A., HARTWIG, T. and Latif, M. A. (2017). Radio power from a direct collapse black hole in CR7, *in preparation*.
Whalen, D., van Veelen, B., O'Shea, B. W. and Norman, M. L. (2008). The destruction of cosmological minihalos by primordial supernovae, *ApJ* **682**, 49-67, doi:10.1086/589643.
Whalen, D. J., Even, W., Frey, L. H., Smidt, J., Johnson, J. L., Lovekin, C. C., Fryer, C. L., Stiavelli, M., Holz, D. E., Heger, A., Woosley, S. E. and

Hungerford, A. L. (2013a). Finding the first cosmic explosions. I. pair-instability supernovae, *ApJ* **777**, 110, doi:10.1088/0004-637X/777/2/110.

Whalen, D. J. and Fryer, C. L. (2012). The formation of supermassive black holes from low-mass Pop III seeds, *ApJ* **756**, L19, doi:10.1088/2041-8205/756/1/L19.

Whalen, D. J., Johnson, J. L., Smidt, J., Meiksin, A., Heger, A., Even, W. and Fryer, C. L. (2013b). The supernova that destroyed a protogalaxy: prompt chemical enrichment and supermassive black hole growth, *ApJ* **774**, 64, doi:10.1088/0004-637X/774/1/64.

White, S. D. M. (1984). Angular momentum growth in protogalaxies, *ApJ* **286**, pp. 38–41, doi:10.1086/162573.

Whitworth, A. and Summers, D. (1985). Self-similar condensation of spherically symmetric self-gravitating isothermal gas clouds, *MNRAS* **214**, pp. 1–25, doi:10.1093/mnras/214.1.1.

Widrow, L. M., Ryu, D., Schleicher, D. R. G., Subramanian, K., Tsagas, C. G. and Treumann, R. A. (2012). The first magnetic fields, *Space Sci. Rev.* **166**, pp. 37–70, doi:10.1007/s11214-011-9833-5.

Willott, C. J., Chet, S., Bergeron, J. and Hutchings, J. B. (2011). A Lyα halo around a quasar at redshift z = 6.4, *AJ* **142**, 186, doi:10.1088/0004-6256/142/6/186.

Willott, C. J., Delfosse, X., Forveille, T., Delorme, P. and Gwyn, S. D. J. (2005). First results from the Canada-France high-z quasar survey: constraints on the z = 6 quasar luminosity function and the quasar contribution to reionization, *ApJ* **633**, pp. 630–637, doi:10.1086/462408.

Willott, C. J., Delorme P., Omont, A., Bergeron, J., Delfosse, X., Forveille, T., Albert, L., Reylé, C., Hill, G. J., Gully-Santiago, M., Vinten, P., Crampton, D., Hutchings, J. B., Schade, D., Simard, L., Sawicki, M., Beelen, A. and Cox, P. (2007). Four quasars above redshift 6 discovered by the Canada-France high-z quasar survey, *AJ* **134**, pp. 2435–2450, doi:10.1086/522962.

Willott, C. J., Delorme P., Reylé, C., Albert, L., Bergeron, J., Crampton, D., Delfosse, X., Forveille, T., Hutchings, J. B., McLure, R. J., Omont, A. and Schade, D. (2009). Six more quasars at redshift 6 discovered by the Canada-France high-z quasar survey, *AJ* **137**, pp. 3541–3547, doi:10.1088/0004-6256/137/3/3541.

Willott, C. J., Delorme, P., Reylé, C., Albert, L., Bergeron, J., Crampton, D., Delfosse, X., Forveille, T., Hutchings, J. B., McLure, R. J., Omont, A. and Schade, D. (2010). The Canada-France high-z quasar survey: nine new quasars and the luminosity function at redshift 6, *AJ* **139**, pp. 906–918, doi:10.1088/0004-6256/139/3/906.

Wise, J. H. and Abel, T. (2007a). Resolving the formation of protogalaxies. I. virialization, *ApJ* **665**, pp. 899–910, doi:10.1086/520036.

Wise, J. H. and Abel, T. (2007b). Suppression of H_2 cooling in the ultraviolet background, *ApJ* **671**, pp. 1559–1567, doi:10.1086/522876.

Wise, J. H., Abel, T., Turk, M. J., Norman, M. L. and Smith, B. D. (2012a). The birth of a galaxy — II. the role of radiation pressure, *MNRAS* **427**, pp. 311–326, doi:10.1111/j.1365-2966.2012.21809.x.

Wise, J. H., Turk, M. J. and Abel, T. (2008). Resolving the formation of protogalaxies. II. central gravitational collapse, *ApJ* **682**, pp. 745–757, doi: 10.1086/588209.

Wise, J. H., Turk, M. J., Norman, M. L. and Abel, T. (2012b). The birth of a galaxy: primordial metal enrichment and stellar populations, *ApJ* **745**, 50, doi:10.1088/0004-637X/745/1/50.

Wishart, A. W. (1979). The bound-free photo-detachment cross-section of H-, *MNRAS* **187**, pp. 59P–61P.

Wolcott-Green, J. and Haiman, Z. (2012). Feedback from the infrared background in the early Universe, *MNRAS* **425**, pp. L51–L55, doi:10.1111/j.1745-3933.2012.01298.x.

Wolcott-Green, J., Haiman, Z. and Bryan, G. L. (2011). Photodissociation of H_2 in protogalaxies: modelling self-shielding in three-dimensional simulations, *MNRAS* **418**, pp. 838–852, doi:10.1111/j.1365-2966.2011.19538.x.

Wolcott-Green, J., Haiman, Z. and Bryan, G. L. (2017). Beyond J_{crit}: a critical curve for suppression of H_2-cooling in protogalaxies, *MNRAS* doi:10.1093/mnras/stx167.

Woods, T. E., Heger, A., Whalen, D. J., Haemmerlé, L. and Klessen, R. S. (2017). On the maximum mass of accreting primordial supermassive stars, *ApJ* **842**, L6, doi:10.3847/2041-8213/aa7412.

Woosley, S. E. (2017). Pulsational pair-instability supernovae, *ApJ* **836**, 244, doi:10.3847/1538-4357/836/2/244.

Wu, X.-B., Wang, F., Fan, X., Yi, W., Zuo, W., Bian, F., Jiang, L., McGreer, I. D., Wang, R., Yang, J., Yang, Q., Thompson, D. and Beletsky, Y. (2015). An ultraluminous quasar with a twelve-billion-solar-mass black hole at redshift 6.30, *Nature* **518**, pp. 512–515, doi:10.1038/nature14241.

Wu, X.-P. and Xue, Y.-J. (2001). The X-ray background from groups and clusters of galaxies, *ApJ* **560**, pp. 544–548, doi:10.1086/322961.

Wyithe, J. S. B. and Loeb, A. (2012). Photon trapping enables super-Eddington growth of black hole seeds in galaxies at high redshift, *MNRAS* **425**, pp. 2892–2902, doi:10.1111/j.1365-2966.2012.21127.x.

Xu, H., Norman, M. L., O'Shea, B. W. and Wise, J. H. (2016a). Late Pop III star formation during the epoch of reionization: results from the renaissance simulations, *ApJ* **823**, 140, doi:10.3847/0004-637X/823/2/140.

Xu, H., Wise, J. H. and Norman, M. L. (2013). Population III stars and remnants in high-redshift galaxies, *ApJ* **773**, 83, doi:10.1088/0004-637X/773/2/83.

Xu, H., Wise, J. H., Norman, M. L., Ahn, K. and O'Shea, B. W. (2016b). Galaxy properties and UV escape fractions during the epoch of reionization: results from the renaissance simulations, *ApJ* **833**, 84, doi:10.3847/1538-4357/833/1/84.

Xue, Y. Q., Luo, B., Brandt, W. N., Bauer, F. E., Lehmer, B. D., Broos, P. S., Schneider, D. P., Alexander, D. M., Brusa, M., Comastri, A. and Fabian, A. C. et al. (2011). The chandra deep field-south survey: 4 Ms source catalogs, *ApJS* **195**, 10, doi:10.1088/0067-0049/195/1/10.

Yee, H. K. C., Gladders, M. D., Gilbank, D. G., Majumdar, S., Hoekstra, H. and Ellingson, E. (2007). The red-sequence cluster surveys, in N. Metcalfe and T. Shanks (eds.), *Cosmic Frontiers, Astronomical Society of the Pacific Conference Series*, Vol. 379, p. 103.

Yoon, S.-C., Iocco, F. and Akiyama, S. (2008). Evolution of the first stars with dark matter burning, *ApJ* **688**, L1, doi:10.1086/593976.

York, D. G., Adelman, J., Anderson, J. E., Jr., Anderson, S. F., Annis, J., Bahcall, N. A., Bakken, J. A., Barkhouser, R., Bastian, S., Berman, E. and Boroski, W. N. et al. (2000). The sloan digital Sky survey: technical summary, *AJ* **120**, pp. 1579–1587, doi:10.1086/301513.

Yorke, H. W. and Sonnhalter, C. (2002). On the formation of massive stars, *ApJ* **569**, pp. 846–862, doi:10.1086/339264.

Yoshida, N., Abel, T., Hernquist, L. and Sugiyama, N. (2003). Simulations of early structure formation: primordial gas clouds, *ApJ* **592**, pp. 645–663, doi:10.1086/375810.

Yoshida, N., Hosokawa, T. and Omukai, K. (2012). Formation of the first stars in the universe, *Progress of Theoretical and Experimental Physics* **2012**, 1, 01A305, doi:10.1093/ptep/pts022.

Yoshida, N., Oh, S. P., Kitayama, T. and Hernquist, L. (2007). Early cosmological H II/He III regions and their impact on second-generation star formation, *ApJ* **663**, pp. 687–707, doi:10.1086/518227.

Yoshida, N., Omukai, K. and Hernquist, L. (2008). Protostar formation in the early Universe, *Science* **321**, p. 669, doi:10.1126/science.1160259.

Yoshida, N., Omukai, K., Hernquist, L. and Abel, T. (2006). Formation of primordial stars in a ΛCDM Universe, *ApJ* **652**, pp. 6–25, doi:10.1086/507978.

Younger, J. D., Hopkins, P. F., Cox, T. J. and Hernquist, L. (2008). The self-regulated growth of supermassive black holes, *ApJ* **686**, 815-828, doi:10.1086/591639.

Yu, Q. and Tremaine, S. (2002). Observational constraints on growth of massive black holes, *MNRAS* **335**, pp. 965–976, doi:10.1046/j.1365-8711.2002.05532.x.

Yue, B., Ferrara, A., Salvaterra, R., Xu, Y. and Chen, X. (2013). Infrared background signatures of the first black holes, *MNRAS* **433**, pp. 1556–1566, doi:10.1093/mnras/stt826.

Yue, B., Ferrara, A., Salvaterra, R., Xu, Y. and Chen, X. (2014). The brief era of direct collapse black hole formation, *MNRAS* **440**, pp. 1263–1273, doi:10.1093/mnras/stu351.

Zana, T., Dotti, M., Capelo, P. R., Bonoli, S., Haardt, F., Mayer, L. and Spinoso, D. (2018). External versus internal triggers of bar formation in cosmological zoom-in simulations, *MNRAS* **473**, pp. 2608–2621, doi:10.1093/mnras/stx2503.

Zel'dovich, Y. B. (1964). The fate of a star and the evolution of gravitational energy upon accretion, *Soviet Physics Doklady* **9**, p. 195.

Zel'dovich, Y. B. (1970). Gravitational instability: an approximate theory for large density perturbations. *A&A* **5**, pp. 84–89.

Zel'dovich, Y. B. and Novikov, I. D. (1966). The hypothesis of cores retarded during expansion and the hot cosmological model, *AZh* **43**, pp. 758. [Transl. 1967 Soviet Ast. 10, p. 602.].

Zinnecker, H. and Yorke, H. W. (2007). Toward understanding massive star formation, *ARA&A* **45**, pp. 481–563, doi:10.1146/annurev.astro.44.051905.092549.

Zlochower, Y., Campanelli, M. and Lousto, C. O. (2011). Modeling gravitational recoil from black-hole binaries using numerical relativity, *Classical and Quantum Gravity* **28**, 11, 114015, doi:10.1088/0264-9381/28/11/114015.

Index

www.ingramcontent.com/pod-product-compliance
Ingram Content Group UK Ltd.
Pitfield, Milton Keynes, MK11 3LW, UK
UKHW020350150926
8624IPUK00004B/96

9 789813 227941